Curing the Philosopher's Disease

Reinstating Mystery in the Heart of Philosophy

Richard H. Jones

University Press of America,® Inc.
Lanham · Boulder · New York · Toronto · Plymouth, UK

Copyright © 2009 by
University Press of America,® Inc.
4501 Forbes Boulevard
Suite 200
Lanham, Maryland 20706
UPA Acquisitions Department (301) 459-3366

Estover Road
Plymouth PL6 7PY
United Kingdom

Library of Congress Control Number: 2009930801
ISBN-13: 978-0-7618-4810-3 (paperback : alk. paper)
ISBN-10: 0-7618-4810-X (paperback : alk. paper)
eISBN-13: 978-0-7618-4811-0
eISBN-10: 0-7618-4811-8

Contents

Contents

— 1 —

Exploring Mystery

Wonder is the feeling of the philosopher, and philosophy begins with wonder.
— Plato

The most beautiful experience we can have is the mysterious. It is the
fundamental emotion which stands at the cradle of true art and true science.
Whoever does not know it and can no longer wonder, no longer marvel,
is as good as dead, and his eyes are dimmed.
— Albert Einstein

A legendary college philosophy final exam consisted of only one question—
"Why?" The students scribbled away feverishly for an hour, but the only paper to
receive an "A" had simply one word—"Because!" This sums up the prevailing
attitude toward mysteries in philosophy today. Any question—at least any legitimate
question—has an answer. But is this true? Or can we ask legitimate questions of
reality that we cannot answer? Are there true mysteries to reality, and if so what
does that tell us about our knowledge?

We certainly can get along perfectly well in the world without contemplating
the mysteries of reality. For us, the universe today is not a completely terrifying
place but is, to a degree at least, controllable. Human beings have been plodding
along on earth for probably a million years. It is a struggle, but we have enough
knowledge of reality to survive and thrive. We do this mostly without being concerned with the mysteries of reality surrounding us. We do not like mystery—our
minds try to explain the unfamiliar so that we can proceed. In fact, since even the
most familiar phenomenon is encased in mysteries, we could not function very well
if we were constantly caught up in the mystery of it all or constantly reacting with
awe and wonder. How, for instance, do we do something as simple as getting our
body to move when we want it to? And why are there conscious beings around to
ask these questions? What are we, and why are we here? What is the meaning of our

1

verything come from, and where does it go? Were volved natural products? Why is there order to rder? And, the deepest mystery of all, why is there anything at all?

As we grow up, we become so accustomed to the world around us that most of us have lost any wonder we ever attached to such questions. But in our more philosophical moments, many of us do wonder about the mysteries surrounding the ordinary. The purpose of this book is to help reassert the importance of a sense of mystery encasing what we think we know.

Puzzles Versus Mysteries

The term "mystery" has become trivialized in everyday usage to refer to any unanswered questions. But the first thing to note about mystery in the philosophical sense is that, while we can experience and even fear the unknown, mystery proper only arises when we are perplexed by something we encounter. That is, it is an intellectual reaction. Mysteries are tied to our questioning, a product of our persistence in asking "why?" Thus, mystery is tied to "wondering about" something and not merely the awe resulting from "wondering at" something. Moreover, "mystery" in this sense is not merely anything obscure, vague, or baffling or any puzzle to which we do not currently happen to have an answer. It is not merely not knowing something or uncertainty or the lack of confidence in an answer for which we have some evidence. So too, human history is filled with past events of great interest to us that we may no longer be able to ascertain, but these are not mysteries in the philosophical sense. Philosophical mystery is even more than simply realizing that reality is greater than the sum of our apprehensions. Rather, the mysterious produces a question that we are not even sure how to begin to answer. Puzzles get solved or at least become smaller with study, but mysteries only get bigger the more clearly we see them. In the natural sciences, new problems are basic research scientists' stock in trade, but such new problems are not necessarily beyond what we are equipped to solve. However, whether a given problem is a puzzle or a genuine mystery is not always obvious even after some study. Indeed, the boundary in science between new research puzzles and genuine mysteries is not firmly set. History is full of examples of things once deemed philosophical or theological "mysteries" that ended up being amenable to scientific analysis. But the surviving mysteries in the philosophical sense are problems that stubbornly resist our understanding despite our repeated efforts to solve them.

In sum, philosophical mysteries are questions about reality that we are aware of, but which, at least for the present, we are not sure even how to begin to answer. These mysteries are questions at the limits of our ability to understand. They may point to aspects of reality that we either have no access to or are apparently unable to wrap our minds around—or they may only be conundrums we ourselves create by the way we conceptualize things. Genuine mysteries are questions we cannot answer either because of the very nature of reality or because of the limitations on our ability to comprehend or analyze reality. The possibilities are numerous: perhaps there is no answer to our question, perhaps we cannot even know if there is an answer or not, perhaps there is an answer but we are incapable of knowing it because of our cognitive limitations, perhaps we simply do not know the answer at present but will crack the problem eventually. Thus, this raises a problem about problems: how do we know that we are asking questions that no amount of human ingenuity will ever answer? How do we know at present what is a genuine mystery and what is a solvable puzzle? And, equally disturbingly, how do we know our questions and conceptualizations are not so totally misguided that we create problems where in fact there are none?

One possible distinction is between mysteries that are ontological in nature (i.e., part of the nature of reality with no answer in principle) and those that are epistemic (i.e., relating to our ability to know). Uncertainty is a sign of an epistemic mystery but not necessarily of an ontological one. Ontological mysteries would persist even if we were much brighter and had a different sensory apparatus. These mysteries might not be mysteries to God, if there is one, but they will remain mysteries to any beings within the universe: even if we knew all that a being could possibly know about them, we still could not fully understand them—they remain brute facts we can never explain. Today the prevailing view among philosophers, to the extent they think of mysteries at all, is that there cannot be ontological mysteries. Epistemic mysteries may remain but only because of frustrating limitations to our mental or technological capacities. Or, perhaps everything in fact is comprehensible with our rationality and technology—then, it is only a matter of time before we know everything. Nothing is really mysterious about reality in itself. How could reality have any intrinsic mysteries? Mystery is obviously only our problem. It is a matter of our understanding and our ability to know—reality just is. And since we are a product of nature, our rationality will eventually unravel all of nature. Many scientists believe there is a depth to nature that we cannot hope to reach, but others adopt this philosophical position, claiming there are no ultimate mysteries to how nature works. Many theists also believe that the world, being the creation of a perfectly rational god, is perfectly rational, and we, being created in the god's image, will be able to decipher it all. Thus, according to this position, there are no genuine ontological mysteries and all the alleged epistemic "mysteries" of how things work in the universe are really only puzzles.

On the other hand, other philosophers argue that there are problems about reality that we are intrinsically a part of—e.g., the mind/body problem—and will remain mysteries forever since we cannot get any distance from them to examine them. These have an ontological, not merely epistemic, component. In short, there will remain unfathomable aspects of even the most familiar things. In addition, there will always remain legitimate "why" mysteries related to why anything exists or why we are here. The questions surrounding these mysteries do not lose their utmost significance to us even if they cannot be answered. Thus, according to this position, our rationality cannot be celebrated as all-encompassing—genuine mysteries in fact will become only deeper and more intrenched as we study more of reality.

Only beings who can formulate questions have philosophical mysteries to solve. Only such beings are aware that there may be something more to reality that is unknown. Does this mean that we are aware of something about reality that slips through our grasp—some "known unknown" reality? Or does it mean that mysteries are no more than illusions created by our own way of thinking, mere by-products of our own conceptualizing that we mistakenly project onto reality? Some problems that seem to us at present to be philosophical mysteries may be dispelled in the future as science advances, but whether the basic philosophical ones will ever be resolved is another matter. There may well remain conundrums that we can never expect to unravel no matter how much our knowledge of reality grows.

Modern Philosophy and Mystery

Philosophers have had a love/hate relationship with the very idea of mystery. On the one hand, they have been on the forefront of emphasizing the presence and importance of mysteries in the world. On the other hand, they have also been in the vanguard of attempts to dispel all mysteries from our view of the world. Although René Descartes included wonder as one of the principal emotions, he also thought that those who understood his work would see that there is nothing to marvel at and thus wonder would cease. Since the Age of Enlightenment in the West, this latter approach has won out in philosophy, and a campaign against mystery has been waged. Understandably, philosophers, like most people, focus on what we know and want to extend our knowledge. They believe that we can maximize vision and minimize mystery. This has led most philosophers today to believe that nothing is irreducibly mysterious about the universe—all mysteries can be banished by the glare of reason. Former "mysteries" are now divided into two groups: scientific puzzles that scientists will eventually solve, and philosophical mistakes generated by conceptual systems that philosophers will unravel. Thus, there are no unanswerable questions: if there is a meaningful question to be asked of reality, we eventually will have the answer. Nothing will exceed our comprehension or remain obscure. All that is required is the proper analysis. In the end, reality will be seen as it is, free of any mystery. We may still feel awe and astonishment about some phenomenon or reality itself, but philosophy is not about feelings—all the legitimate questions about reality will be answered.

For this reason, there is very little sense of mystery today in philosophy. The same holds for Western theology: conservatives think the answers have been revealed, and liberals play down the transcendent in general and ignore any mysteries surrounding it. And in the sciences one does not advance one's career by looking at the "big questions"—many people have commented how the sciences have increasingly become more and more studies about less and less. Any puzzles circling findings in cosmology, physics, or biology are seen as only frustrating problems, not signifying a final unanalyzable ontological mystery of reality. And it must be admitted that mystery in general is certainly disquieting. The very mention of the term makes most people in philosophy uncomfortable. Many react by scoffing at the idea and just brushing it aside. Part of the problem is simply that it is hard to present a good philosophical argument concerning mysteries: mystery is not a target for direct assault—it the residue remaining after an analysis. Philosophers cannot get the type of conceptual closure that they like to get in their arguments when the topic is so murky. Accepting mystery is more like admitting a failure. Thus, the very idea of "mystery" has fallen into disfavor. Few books or articles have been written on the subject in philosophy since the rise of logical positivism. Indeed, it is seldom mentioned in most works in epistemology and metaphysics, even just to mention unraveling one. (One exception: the term "mysterians" was applied as an insult by the opponents to one position in the mind/body field, although advocates of the positions happily adopted it.) The newest philosophical encyclopedias do not include "mystery" as a topic or even as an entry in their indexes.

All in all, mystery is a very neglected subject. Nevertheless, a sense of the mystery surrounding us is important for understanding what surrounds science, philosophy, and religion, and most importantly for understanding our place in the world.

Types of Mysteries

It is human nature to ask "why?" but not all why-questions are of the same type. Philosophical mysteries about reality fall arise in three areas: metaphysics, science, and religion. The metaphysical ones are, as the cataloger of Aristotle's works saw, "beyond the sciences." They are questions about which the sciences have no direct bearing. (Science may have an indirect bearing on these questions since metaphysics has to account for scientific findings too.) Chief among these is: why is there anything at all rather than nothing? Could there have been nothing at all? No scientific finding will answer why the universe exists. For example, what finding from the subatomic level to the cosmological level could bear on that issue?

Other questions also fall in the philosophical group. For example, the nature of the "being" of reality. We all are well aware that things exist, but what is the basic nature of the "beingness" of everything? If we answer, for example, that everything is ultimately only matter, the question then becomes what exactly is "matter"? The same with "mind." Or consider values: how do our values fit with the rest of reality? Does the impact of music and beauty on us indicate anything profound concerning what is real? Or consider something as simple as looking at any ordinary material object. We only can see it from particular angles and never see it "as it really is" independent of our observations. So what is it like in itself? This leads empiricist-minded philosophers to wonder why we need to postulate "the thing in itself" at all and to stop instead merely with the appearances. Another problem applies to something very basic to us—our consciousness. Why are some material beings conscious? How do consciousness in general and self-awareness in particular arise? Scientists may end up understanding everything about the wiring of the brain, but why and how does this new layer of phenomena arise? Why is the brain organized to be at least a base for it? Indeed, this shows how something we are intimately familiar with—ourselves—can still remain very mysterious.

Or consider what scientists study. Science has many fascinating problems at the edge of research touching on the big issues. How did our universe arise? How has the present state evolved? How will the universe end (if it does end)? What are the smallest components of nature, and how do they relate to fields of energy? Why does the universe have the forces and constants it has? Is the flow of time a component to reality or merely an illusion our mind creates? Are there other "universes" beyond our possible range of observation? Why is reality organized so that higher levels of complexity emerge? How did life arise, and how has the diversity of life on earth evolved? How, in particular, did humans arise? What is the role of contingent events in the development of the universe? Or is everything in some sense necessary, so that if the universe evolved again we would end up exactly as we already have? Which of these questions will end up being answered and which will end up being mysteries?

Scientists express awe at the scales and sheer scope of the universe, and the complexity of the mechanics of nature may provoke wonder, but their response is not to remain lost in a state of wonder but to try to find how the universe works beneath the surface of things. They do not deny the wonder of it all, but they "demystify" nature in one sense: they want to know how it all works. Thus, a sense of mystery in science moves scientists forward, prompting them to discover more. However, so far every new discovery has always led to more questions. This leads to the question of what are the ultimate components and structures of nature. But even if we solve all the scientific puzzles, mysteries still lie at the foundations of science. Why is there causation in nature? Indeed, why is there an order to nature at all? And do we in fact comprehend reality at its deepest level? None of these philosophical mysteries will be answered by further scientific research on how nature works. Of course, if scientists discover all the mechanisms structuring nature, we certainly may still be struck with awe at the intricacy and vastness of it all, but we will no longer have any "how" mysteries related to how the universe works.

Nevertheless, the "why" questions would still remain. Thus, science cannot answer everything about reality, even in principle.

Religion moves from the "how" questions related to the workings of the universe to a type of "why" question related to meaning of existence. Where did we come from and why are we here? Is there a purpose to the universe? Is there meaning to my particular life? Why is there suffering? Why do we die, and what happens at death? How does the historical development of everything, and of human history in particular, fit into reality? In short, what is it all about? How do I align my life with the fundamental nature of reality? This leads to the question of whether there is something more to reality than just the everyday world and its underlying causal order. Is there a transcendental source of the natural realm? Are there many transcendental realities? Religious experiences give a sense of a power standing over against our realm, with a resulting sense of awe and even terror. This leads to questions of whether this sensed power is actually real, and, if so, what is its nature? Is it personal in nature? To believers, this also leads to new "how" questions. How did it create the world? How does it sustain the natural realm? How does it act within the world? Indeed, mysteries surround any reference to the transcendent.

The term "mystery" entered our vocabulary in connection with religion, albeit not directly connected with transcendental realities. The term derives from the Greek word "*muein*" (from which "mysticism" is also derived) meaning "to close the eyes or mouth." The initiates (*mystai*) of the Eleusinian mysteries were expected to keep silent about its secret rituals and objects. The term then entered Christianity in reference to certain doctrines—most importantly the mysteries of the incarnation and death of Jesus. To Christians, this is the revelation of the mystery that was kept secret for ages (Romans 16:25-26). In all religions, some philosophical mysteries are essential, but in the modern era mystery has been downplayed in all the Abrahamic religions. More generally, theology—understood in the general sense of the theoretical wing of any religion—can be anti-mystery, substituting reflection for a sense of mystery and the experience of awe.

Religious experiences present their own problems. Revelations end "why" mysteries for believers. With mysticism, there is a turn to the being of reality that is also of concern to philosophers but with an experiential twist. Scientists deal with the how-ness of nature and the religious in general with the "why" of it all, but those within the mystical strand of religiosity encounter through experience the mysteries of the sheer "that-ness" of the being of reality and ourselves. Since traditional mystics belong to religious traditions, they also are concerned with meaning, but their experiences lead them to a unique encounter with reality. They emphasize a type of knowing that involves completely "unknowing" the conceptual and other content of our everyday minds. The reality allegedly experienced in the resulting states of mind is characterized by negations of any properties ascribable to things in the natural realm to the point that the reality is claimed to be propertyless. This leads to claims of paradox and ineffability that point to the otherness of what is alleged experiences. All of this makes the reality allegedly experienced seem incomprehensible to those of us in our ordinary state of mind and also deeply mysterious even to the mystics themselves outside their experiences.

Mystery and Explanation

As discussed, mystery arises in the context of seeking an explanation: mystery in the philosophical sense is the product of asking "why?" or "how?"—without such questioning, there would be no mysteries. And the prevailing philosophical position toward mystery today is that, if a mystery raises a legitimate question about reality, then there is an answer to it. Thus, most philosophers do not admit much, if any, mystery to the world. When something strikes them as mysterious, for example, the mind-body problem, they push for a resolution. Each believes that at least he or she and like-minded colleagues have solved it or at least have made important strides to that end, thus dividing the field up into competing camps attacking each other and defending themselves. But competing answers, if nothing else, do help clarify the current issues for each mystery.

Thus, most philosophers begin with the working assumption that there is some explanation for any phenomenon—indeed, some would not give this assumption up at any cost. To them, understanding a phenomenon is always in principle possible. However, this can end up being the "philosopher's disease": trying to dispel the mysteries surrounding phenomena by finding a reason behind them even if there is none—i.e., the certainty that there is a "because" no matter what. (Ludwig Wittgenstein introduced the phrase "the philosopher's disease" to refer to the tendency of philosophers to nourish their diet by relying upon one kind of example [*Philosophical Investigations*, para. 593]. Since then, the phrase has been used occasionally by philosophers in different senses, e.g., to substitute theories or any abstractions for reality.) Philosophical analysis can also lead to the opposite problem: if we push philosophical analysis of anything far enough, everything begins to look fuzzy and mysterious—but are we creating a mystery where is there is none? If we look very closely at a photograph in a newspaper, all we see are an array of dots—but is there really a mystery in how our minds create the picture from the dots?

Obviously, claiming "It's a mystery" is not the explanation of anything—if something is a mystery to us, then we do not know something but have merely identified a hole in our knowledge where we think something should be. Naming a mystery may give the illusion of understanding it. Naming a problem does help us focus and organize our attention. But labeling a problem only identifies the problem and does not increase our understanding at all. In the sciences, giving something a Latin name does not mean with understand it. The method in science for an actual explanation is to "seek the causes." Explanation is often equated with the ability to predict a phenomenon's occurrence. Whether prediction is always needed for an explanation even in the sciences is open to question, but it is very hard to see prediction as even possible in the case of metaphysical mysteries. Explanation more generally is a matter of giving a reason for believing something—providing an account that "makes sense" of a phenomenon to us and puts our curiosity for a "why" or a "how" to rest. In our everyday life, we do not look for an ultimate explanation of something but are satisfied with a connection to something that we take to be unproblematic. With mysteries, however, we do not have this option. We must reach a point where we believe we have reached the ultimate justification for believing something and where no further explanation seems possible to us. Only when we are thoroughly satisfied that we have reach the bottom do we think we have finally understood something we previously found mysterious.

But this also raises a problem: are we fooling ourselves with our explanation of a mystery? Have we allayed our curiosity only by connecting a mystery in our minds to something in our ordinary experience that really is no explanation at all? How can we be sure that what satisfies us today is indeed the final answer? In the sciences and mathematics, what seems certain in one generation is sometimes overthrown in the next. Metaphysical mysteries are longer lasting, but how do we know we now have the right answer removing one? Do we have enough information at this date for any final answer? Our curiosity may be laid to rest, but how can we ever really know that our alleged explanation is correct? How do we know that we have reached a "brute fact," i.e., one that "just is" and thus is incapable of any further explanation? And aren't any brute facts mysteries in themselves? Was Karl Popper right in claiming "[t]here can be no explanation which is not in need of a further explanation"? Given any explanation, we can always ask "why that?" and so never have any final explanation. And if we finally accept something as a mystery, how do we know it is in fact a genuine mystery and not merely a puzzle that we will be capable of explaining in the future? Labeling something "a mystery" is the conclusion that we have no explanation, and how do we know we are not being premature? The "subjective" element that explanations must rely upon—our experience and our understanding—will always remain.

Affirming Mystery

The following chapters will introduce the mysteries in philosophy, science, and religion. The thrust of the discussions goes against the spirit of the age and affirms that there is more mystery in the realm of our beliefs about what we take to be fundamentally real than is usually realized: at a basic level, there are things about ourselves and the rest of reality that we may well never know in practice or that we may not be capable of knowing even in principle. There are limits to what philosophy, science, and religion can tell us about reality. In the end, we know less about the fundamental nature of reality that we like to think. Nevertheless, this has a positive side: seeing that some questions may be unanswerable, and understanding why that is so, is important even if the questions remain unanswerable—this reveals our true situation in the world.

Mystery becomes fundamentally important when we reflect on the nature of what we know and what we are. Thus, we should reinstate mystery as a basic feature of our picture of the world. Today most people at best pay lip service to the mysterious and understandably focus only on the known. We prefer some form of closure to the openness and fullness of reality. Closure does fend off mystery, but any closure in the face of a genuine mystery is only a partial, truncated view of all that is real. Moreover, we should view the reality around us with the realization that, as G. K. Chesterton put it, "nothing is so miraculous as the commonplace." Mystery frames all our claims to knowledge, and we must accept that in all likelihood it will remain a permanent part of the picture.

— 2 —

Philosophy

Reason's last step is the recognition that there are an infinite number of things that are beyond it.
— Blaise Pascal

There is no unfathomable mystery in the world.
— Moritz Schlick

The aim of philosophy, in the words of Wilfrid Sellars, is "to understand how things in the broadest possible sense of the term hang together in the broadest possible sense of the term." Philosophy has always had two components analysis of our basic ideas (beginning with Socrates's annoying queries to the Athenians) and speculation (beginning with the pre-Socratics' search for one unifying principle behind the realm of diversity and perpetual change). Each component has become more refined: analysis has evolved into the formal study of concepts and the presuppositional beliefs behind such human enterprises as art and science; and speculation has evolved into natural science and the production of elaborate metaphysical systems looking at the big picture of reality. There has also always been a tension within philosophy between the pull of our experiences and the pull of our thought: nominalism versus universals, empiricism versus rationalism—in short, our experience versus our reason. This leads to issues concerning how much we really know about reality—skepticism versus certainty—and hence to the problem of mystery.

Plato introduced the idea of wonder into philosophy. In the *Theaetetus*, in which Socrates and a student discuss the nature of knowledge, Plato has Socrates say that wondering about something (*thaumazein*) is the point where philosophy begins. It is realizing that we do not understand what usually seems self-evident to us, such as the idea of "knowledge." To Socrates, what is "self-evident" becomes a ripe candidate for mystery. If Plato introduced the idea of "wonder about," his student Aristotle introduced the idea of closure. To Plato, we should remain in the state of wondering about things since this opens up inquiry (as opposed to the awe at the "wonders" of nature); it is an unsettling state, but we should remain open to the inscrutable in the everyday world. To Aristotle, however, the goal is closure: ending wonder by finding the causes of the subject of our wonder. He valued wonder only as a preliminary step leading to the search for causes that finally ends with the First Cause. Thus, wonder is ultimately eliminated by causal knowledge. Aristotle's position came to dominate Western thought. By the time of Thomas Aquinas, wonder was seen as being only the result of our ignorance of causes, and remaining in a state of wonder and amazement was only a sign of sloth—we should keep pressing until we know all the causes except the one unknowable cause (God). By the beginning of the modern period, René Descartes listed wonder in *The Passions of the Soul* as the first of all passions, but he also insisted that this was only a means to our knowledge of things and that an excess of wonder is never anything but bad—one stuck in astonishment is not apt to investigate causes. And he felt that his method would replace all wonder with comprehension.[1]

Skepticism

Skeptics enter the fray early on by questioning whether we actually know anything at all. Skepticism is not a first-order knowledge-claim about the world (e.g., that there are quarks or that the Mets won the 1969 World Series). It is a second-order claim about the status of knowledge-claims: skeptics question whether knowledge is possible at all or at least whether we really have good grounds to accept any knowledge-claims. Skeptics ask why we think our minds could reflect anything deep about reality. (The term comes from Greek for "inquiry," not "doubt.") They go beyond arguing that all empirical knowledge is fallible to questioning the possibility of any claim at all.[2] When we make knowledge-claims, do we really see something of reality or see only products of our own mind? How do we know?

Philosophical skepticism must be distinguished from practical doubt about a particular claim—the latter method proves very valuable in our quest for reliable knowledge in science or any other matter. Doubting whether any claim to know the world can in principle be true (radical skepticism) is not the same as questioning the reliability of a particular first-order claim that leads to checkable claims in science (practical skepticism).[3] Radical skeptics deny that we have the right to assert any first-order knowledge-claims whatsoever. Some philosophical skeptics (such as Pyrrho of Elis) remain agnostic about all knowledge-claims. ("Agnosticism" here can mean either that we cannot know the truth about something or that we might know it but we are not in a position to know that we know.) Classical Greek skeptics accepted empirical claims of the observable world but withheld judgment on all claims about unobservable causes.[4] David Hume only accepted that we have knowledge of our present mental impressions and to speculate about what we cannot know is "sophistry and illusion." For example, he reduced causation to only observed regularities: we have no idea if there are any necessary connections between events in nature or what is going on in nature in general—there may well something there, but we cannot know it. Some skeptics accept solipsism (that only one's own self and its states exist, or at least that we have no good reason to believe otherwise). Today many radical philosophical skeptics question only certain categories of beliefs—such as the existence of transcendental realities or any unobservable natural entities for science—rather than all claims to empirical knowledge.

The rationalist René Descartes began modern philosophy with a quest for absolute certainty that would refute radical doubt, but he only ended up cementing doubt in modern thought. He questioned all empirical claims to knowledge with a simple argument: for all we know, a demon may be fooling us into believing mistakenly that our senses or our reasoning about our experiences exposes something of the way things really are. How do we know either our experience or our reasoning really reveals something of reality independent of us? Indeed, how do you know you are not dreaming right now? (I, like many others who have contemplated this issue, have had a dream in which I am explaining this point, certain that I was awake and that nothing could convince me otherwise, only to wake up.)

For rationalists, only what follows from undoubtable "clear and distinct" first principles that are self-evident to our reason qualifies as knowledge. In the end, that greatly limits what we can accept as true knowledge and thus only increases skepticism in general. Optical illusions raise the possibility that all our perceptions may be erroneous. But such illusions are corrected by *other perceptions*, and thus their significance in challenging *all* perception as cognitive is limited. Still, those optical illusions that remain even after we know they are illusions (e.g., a wheel spinning fast appearing to be turning backwards) do raise questions. When there are conflicting sense-perceptions, reason has to adjudicate which sense-experience is "real." Even more troubling are experiments that show that our mind "corrects" and constructs things (e.g., filling in visual blind spots). More generally, apparently our brain automatically creates a coherent, continuous narrative out of all the sensory in-put it receives. We see a reconstruction of the world, and this leads to the question of whether our visual world is only a "grand illusion." (That not all of our mental processes are under our conscious control raises another worry: all our thinking may not be under our control and may not be completely rational.)

Empiricists take the opposite tack. John Locke declared that the "real essence" of natural objects is unknowable, and all empiricists since Hume have also raised doubts about the ability of our experience or our reason to produce any certainty and thus genuine knowledge. They accept unconceptualized "sense-impressions" —Willard van Orman Quine's "surface irritations"—as the only means to knowledge of the external world and argue that no one's knowledge can go beyond their experience. Skepticism has always been prevalent among empiricists: the central difficulty they see is how to get from sense-experience to any certain knowledge of the real nature of anything beyond the appearances. Some empiricists limit knowledge-claims to our own experiences and deny we have any knowledge of the world independent of ourselves. At best, they see empirical knowledge-claims in terms of probabilities. Today radical skeptics argue that for all we know we may only be brains being kept alive and functioning in a vat—all that we take to be experiences and knowledge is merely an illusion artificially stimulated from the outside. More generally, skeptics argue that there are alternatives to any explanation of what we experience through the senses and so we can never be sure which explanation of its cuase, if any, is correct; thus, we can never be certain that the claims we choose about a reality independent of us are in fact accurate. All we know for certain are the sensations themselves, not their source.

Much of modern philosophy is an attempt to refute Cartesian doubt, but skepticism has remained irremovable part of modern philosophy, from the logical positivists to the religious (such as Søren Kierkegaard).[5] The attempt is to find irrefutable foundations—such as Descartes's *cogito ergo sum*—upon which absolutely certain knowledge could be built. (Descartes's phrase is usually translated "I *think*, therefore I am," but he meant more broadly "I am *conscious*, therefore I am.") Today such foundationalism has been generally rejected, and more broadly it is widely accepted that skepticism cannot be refuted. G. E. Moore responded by arguing that it is absurd to require a proof of the existence of the world we live in since this is precisely what we mean by "real"—we *know* an external world exists if the words "know" and "exist" and "real" mean anything at all. So too, we know with certainty that we have a body even if we cannot refute the skeptics' charge. Thus, there are some things we are certain about and in fact do not doubt, and so skepticism must be wrong, even if we cannot identify exactly where the flaw is. This common-sense response, however, has not proven very convincing—we still want to find the flaw. Other refutations have been offered, but all end up failing since skeptics can always question the premise of any argument. What is essential to skepticism is the gap between our claims and the world, and philosophers today admit that this cannot be gotten across even if we must accept our knowledge-claims for practical purposes. This does not mean that we must all become solipsists, but it does mean that we cannot refute the determined skeptic. Indeed, we cannot help but reject solipsism as a practical matter, but this does not mean that we cannot accept that it cannot be refuted in theory. The practical and theoretical issues are separate. But then again, if we reject solipsism as a theoretical matter, where do we stop? What non-arbitrary reasons can we give to accept skepticism about some claimed realities and not others?

This causes many philosophers today to limit the significance of skepticism or to dismiss it all together as a fruitless academic exercise: we obviously do know many things about the world, and the logical problem that skeptics raise is not enough to overcome this. We are as certain as we possibly can be that the earth is a slightly pear-shaped sphere even if we cannot refute the determined skeptic about the foundations of knowledge. Most philosophers agree we may legitimately hold any beliefs attained through the senses and the mind (sense-experiences, memories, and so forth) until they are shown to have been produced by an unreliable process: these beliefs are fallible but are innocent until proved guilty. The type of certainty skeptics want cannot be attained in principle and so can be ignored. However, most philosophers also accept the thrust of skeptical arguments about our claims to knowledge: we have no knock-down, logically irrefutable claims about the nature of reality. We rely upon procedures as reliable, even though we cannot prove they are reliable without relying upon those very procedures; in non-skeptics cannot prove they have knowledge without begging the question.[6] As long as realism is accepted, the possibility that we are profoundly wrong in our claims about the world cannot be gotten around. Nevertheless, the most common response among philosophers today is not so much to argue for or against any form of radical skepticism as simply to acknowledge the problem with the foundations of any purported knowledge-claim and to move on to other topics. The specter of skepticism is so firmly implanted in the modern philosophical outlook that it is not interesting enough to deal with anymore. Indeed, as discussed in the next chapter, postmodernism can be seen as what happens when a skeptical attack is launched against the claims of science and epistemology.

We can accept the point of skepticism and still go about our lives much the same as before. Certainly most people have no doubt about the reality of the world, even if coming up with a rigorous proof of its existence is not as simple as one might think. As common-sense philosophers since Thomas Reid have pointed out, nature does not allow us to live with radical doubt—we cannot be paralyzed with intellectual denial but must believe in a world out there. If we try to live with only what we can absolutely prove, we will end up with solipsism—there is no slam dunk proof of existence of an external world, that other people have minds, or that the world existed before you were born. Indeed, there is very little that can be proved beyond *all* doubt—there are disputes on what is "self-evident" even in logic. (Some ancient Greeks and Romans allegedly did try to live their lives based on the principle of acting only on evidence they could not doubt and suspending judgment on all other matters. So they ended up, for example, falling in holes since they believed that the observation of a hole in front of them was not incontrovertible proof of the actual existence of a hole.) But this does not affect the strength of the skeptics' point since it is a claim about the nature of our knowledge-claims. That is, we can still accept the second-order claim of skepticism: we must live as if the world is real, but skeptics still can argue that we cannot give a definitive reason for believing the world is so. We must live denying solipsism even if we concede that we cannot refute it logically. In short, we are in the paradoxical situation in which we can affirm the principle of skepticism while recognizing that we must actually live as if it were untrue. Nevertheless, the value of radical skepticism is that it reveals that we have to accept that we are living on faith: the faith is supported by experience, and hence is rational, but we have no certainty on the basic metaphysical matters of our everyday world.

The thrust of skepticism thus is that we are left with mystery about the fundamental nature of what is real. We are deluding ourselves when we think that we truly know with confidence anything basic of reality beyond that we exist and have mental states. If we reject radical skepticism, then there are reasons to accept some scientific claims over others as revealing something of the world's structures independent of us (as discussed in the next chapter). Even if we give up the requirement of absolute certainty for a claim to knowledge to be acceptable, still there are genuine doubts about most of our claims, not merely hypothetical ones. But philosophical skeptics do not merely question our certainty or confidence in a given first-order knowledge-claim but the entire validity of any claims to knowledge of the underlying structures of reality. This has a valuable effect in countering any dogmatic assertions or unsupported metaphysical accounts of the nature of reality, but it goes further: it undercuts knowledge-claims period. The more certain we are concerning skepticism about the entire enterprise of knowing reality, the more we are left with mystery about the real nature of things. Most broadly, skepticism raises the specter that we do not have any genuine knowledge of reality independent of us at all, and thus that all of any alleged world outside of us remains ultimately a mystery to us.

Words and Reality

Even if we reject skepticism and ask what is real, we are hit with a surprise right out of the box: there is no agreement on what exactly are the criteria for determining what is real. Reality is the sum total of everything real, but what exactly is real? What exists independently of our thoughts is a start, but any further characterization is difficult. Must it be material? Or mental? Are even those thoughts real? Or must it have the power somehow to produce an effect? Or must it be free of dependence upon anything else? Or must it be permanent or eternal? What sort of things are real: physical objects, events, ideas, sense-experiences or other mental states, space, time, God? Are numbers real? Do ideas, truths, mathematical entities, propositions, or logical relationships exist independently of conscious minds? About the only thing all people would agree on is that what is real is what we cannot get around in our explanation of the world. Beyond that, confusion reigns.

Philosophers today typically ignore the basic question of defining "reality" (hence the lack of consensus on a definition) and focus instead only on a question that seems easier to handle: whether words designate something real. The origin of the current debate can be traced to the debate in the Middle Ages over nominalism. In the modern era, nominalists have ranged from the British empiricists to Friedrich Nietzsche. They argue that all that exists are "particulars" (entities), not "universals" (categories). The latter we abstract from experience and thus only exist in our minds. For example, there are no numbers but only sets of things we number. There are red roses, but there is no other reality called "redness" that the roses "participate in" to become red. Thus, there is merely a collection of red objects in the world; "redness" is merely an abstract linguistic expression that we created and that does not apply to anything in reality. The same applies to "roseness": there are roses in the world, but the there is no one "substance," "essence," or "natural kind" that is designated by any of our terms, or any real "facts" independent of the propositions we create. The world only contains individual entities and nothing else.

This denial of universals leads to the linguistic antirealist position: words that seem to denote real objects are in fact merely our inventions and do not in fact refer to anything in the world. There are no "dogs" in the world—only terriers, German shepherds, and other breeds. And there are no generic "terriers"—only fox terriers, Airedales, and so forth. Ultimately, there are only individuals, and no terms signify anything in the world. At the extreme, antirealists believe there is no objective structure to reality; instead, we impose order on an amorphous reality by the conventions we choose. Such conventions are completely arbitrary and reflect nothing in reality. Nominalists do not deny metaphysical realism (i.e., that the world exists), but the world is populated only by a collection of individual objects.[7] To them, the puzzles about how universals exist or how universals "inhere in" particulars are fruitless dilemmas we alone have created. These are philosophical mysteries generated only by our language and have nothing to do with reality.

Thus, antirealists do not question merely the words about unobservable entities in science or transcendental realities. They deny that any noun denotes anything in reality: nouns necessarily denote categories, and all categories are merely our creations. In sum, no words apply to what is actually real. In fact, all "abstract reasoning" ends up being called into question. None of our terms reflect reality but are only inventions of our minds for our convenience, and we are left with an antirealism concerning the structure of the world and with silence concerning ultimate matters. As with skepticism, we end up surrounded by mystery with no way out.

Modernity

Since the Age of Enlightenment, philosophers have looked to reconstruct our knowledge of the world dictated solely by sense-experience and rational thinking. Rationalists since Descartes have argued that the mind somehow has direct access to the inner nature of reality's structure and so can mirror it (as evidenced in mathematics) and also that reason must be the final arbiter of what we accept as empirical knowledge since experiences conflict with each other. Rationalism's roots go back to Parmenides who argued that what is given to reason is more real than what is given in our sense-experiences since the latter conflict. (This caused Parmenides to reject the appearance of change in the world as actually an illusion; everything is actually one, permanent, and unchanging.) Thus, the causes of the phenomena we experience can be found only on levels of reality accessible only to our reason. But most rationalists, including Descartes himself, do give experience a role in establishing the first principles, even though our reasoning remains central.[8] Only reason makes experiences intelligible—the rational order of *a priori* necessity is given to our reason, not through our experience. Empiricists, on the other hand, have argued that all knowledge of the world ultimately begins and ends with sense-experience and that reason by itself is empty of content and leads only to dogmas. But only extreme empiricists deny any role for our reasoning in establishing what is seen or resolving conflicting experiences—for them, a straight straw sticking out of a glass of water that appears bent must actually be bent.

Thus, for most rationalists and empiricists both our reasoning and our experience play roles in our knowledge, although they differ on which should take priority. But either way, all knowledge according to Enlightenment thinkers was to be based on a foundation of absolutely certain premises; the same held for grounding values. Previous thinking based on faith in things unseen was to be replaced by universal reason. Humans were treated as purely natural products; the social nature of persons replaced any alleged eternal aspects as central in explaining human phenomena. Both individuals and society were seen as perfectible by reason. The world too was to be mastered by the rational mind. It was to be disenchanted of all transcendental realities, revelations, miracles, and all mystery. Mystery is excluded by Descartes's doctrine of "clear and distinct ideas," since this implies that anyone who understands a thing at all understands it fully.[9] No question about the world was seen as unanswerable. "Mysteries" became reduced to solvable problems. Thereby, all sense of mystery was to be eradicated.

Immanuel Kant's Transcendental Realism

The idea that there is a realm of objective reality behind the veil of appearances is not new. Philosophical discussions of an objective world go back to the ancient Greeks, as with Plato's realm of ideal "forms" that we cannot directly know since that realm is not accessible from our realm of shadows. Think about simply looking at a coin. From one angle, it looks circular; from another, it looks elliptical; from the edge, it looks almost rectangular. No one of these angles shows a coin "as it really is"—it "really is" some unseen reality that produces these perceptions when we look at it from various angles. Descartes used such reasoning (concerning a piece of wax) to conclude that the "objective" properties of an object are only extension in space and the capacity to exhibit a variety of forms; everything else are secondary sensory "accidents" and not part of the real "substance." Idealists countered that even the "primary" properties are of the same nature as secondary ones since we have no knowledge of them except through the senses—they all relate to the perceiver and not the perceived in itself. "Substance" as the "incomprehensible something," as the empiricist George Berkeley called it, that allegedly bears the manifest properties drops out of the picture completely, and idealism—the position that only minds and their content exist, and that the world consists only of a collection of ideas subsisting in the mind of an observer or of the eternal observer, God—is established over realism, however counterintuitive it sounds.[10] There is nothing beyond the perceptions themselves for us to intuit with our reasoning.

Immanuel Kant tried to get beyond the dispute between realists and idealists. He modernized the appearance/reality dichotomy with his distinction between things-in-itself (*noumena*) and things-as-they-appear to us (*phenomena*).[11] His basic idea is that we cannot know reality (either the world or God) as it exists in itself—we have no access to that realm through either any experience or our reasoning ("theoretical reason"). Because our reasoning leads to contradictions ("antinomies") about such matters as whether the world had a beginning or whether we have freedom of the will, we must conclude that reality-in-itself is beyond all cognition or understanding. The contradictions are not inherent in the reality-in-itself but only reveal the limits of our understanding. To avoid such contradictions, we must confine our claims to understanding to the phenomenal realm. Similarly, we cannot know the self, the subject of consciousness: it is what observes (the "transcendental self") and cannot itself be observed. The noumenal realm thus is unknowable in a different way than some distant parts of the universe may be forever unknowable to us.

This is not to deny the objective ("transcendental") reality of the object and subject but only to assert that they are forever cut off from what we can in principle know. Kant had no radical skeptical doubts about the existence of the noumenal realm: it is real and affects our senses—he simply turned his attention away from the unknowable things-in-themselves and turned to the conditions for our experience and knowledge. According to him, our minds are active in constructing the world of appearances. Appearances are created by the imposition of *a priori* categories of understanding—e.g., time, space, substance, causality—upon what is real. Such ways of organizing the world are innate to all humans and are not derived from experience or thought; we are simply born with them and cannot get around them. Even "mind" and the "external world" are already appearances and not part of the transcendental realm. Thus, even space and substance are not part of the noumenal realm—they are not "out there," existing in their own right, but are only "species of our representations."

In short, we have structured the only world we know (the phenomenal realm) with our categories of understanding. Kant did not deny that we have genuine knowledge of a real world but argued only that this knowledge is packaged in forms we ourselves supply—the *content* of knowledge derives from our experience of the real world, even though we impose the *order or form* upon that content. (Thus, "transcendental knowledge" is about how we can possibly know objects, even prior to our sense-experience, and thus supercedes both rationalism and

have knowledge of the phenomenal realm: there are certain ways things-in-themselves always appear to beings with our particular cognitive structures, and true claims reflect that. Things we experience may be a mixture of the noumenal realm and our categories of thought, but they are not subjective: we cannot arbitrarily create anything we want—the noumenal realm is objectively real and supplies one component to our knowledge. It is somewhat like subatomic realities: electrons appear like particles in some experimental set-ups and like waves in others, but we cannot make them appear any way we want, nor can we know what they are like "in themselves" apart from our experimental experience of them—all we know about the realities "in themselves" is that they are not as they appear to us; so they are neither waves or particles but have the capacity to manifest such phenomena consistently. So too, we can never circumvent our categories of perception and reason to get to the noumenal things-in-themselves directly. Even with "pure reason" we cannot reach the noumenal realm.

Thus, what common sense says is real—time, space, causation, objects—is not in fact reality as it exists in itself. We are trapped in a world our minds create. The world of appearances is made by our categories out of the things-in-themselves. The world "as it really is" is independent of our perceptions and ideas. We can posit noumena, but our reason cannot reflect the nature of noumenal reality but only the world of appearances. Since we cannot conceive the noumenal realm, language cannot apply to it. Reality-in-itself is veiled by concepts and cannot be unveiled, for any expression will always itself be another veil.[12] Our minds cannot reach the noumenal realm in any sense, and thus we have no conceptions of it—the very notion of such a realm is only a bald inference that cannot lead to any knowledge. How can we even know if the noumenal realm is divided into objects? Do numbers only apply to phenomena? If so, we cannot speak of the noumenal realm as one or many, singular or plural. So too, we see the world in terms of space and time only because we bring these categories to the world; the objective world is not structured by what are only our own mental categories.

Thus, the world of appearances is never the world-in-itself but is always constructed out of two elements: reality-in-itself and our minds. Reality is presented as formless, and our minds construct a formed world out of it. Experience too is in part given and in part constructed: the manifold of sense-experience is brought into an ordered world of objects by means of our rational mind, and we cannot neatly separate the two components. What William James called the "buzzin', blurrin', confusion" of sense-experience would remain unintelligible without our minds imposing rational order. All of this means that we see the world only through a limited perspective. Our categories of thought restrict the world we create. Thus, any attempt to know the noumenal realm will be frustrated—all we can know is the world of appearances framed by the categories that our mind brings to the experiences, not the world "behind appearances" (to use a spatial metaphor from the phenomenal realm).

The unknowability of the noumenal world solidifies a role of mystery in Kantian thought. There is simply no way of telling what the world-in-itself is like—all we can know is the ordered world we ourselves construct. We do not know if the world-in-itself is necessarily as it appears to us—indeed, if Kant is correct, it cannot be. Moreover, one can ask where the categories of our mental life (time, space, and so forth) come from. Are they "real" in the sense that all intelligent creatures would have them? And how do they interact with the noumenal realm to create appearances? Because we have no experience or intuition of the noumenal realm, there is no way even to connect it to the realm of appearances through such categories as causation—the latter remain forms of thought only for the phenomenal realm. The spaceless and timeless noumenal reality is forever inaccessible, forever hidden behind a veil of appearances we have no choice but to create. Indeed, we only can infer its existence at all. In the end, we are left with a complete mystery concerning the entire objective realm of reality.

The Twentieth Century

Logical positivists are followers of Kant in one respect here. They restrict attention to what we can know through observation, but they dismiss questions of the world apart from experience as meaningless metaphysical speculation. They also assert that scientists do not describe reality apart from our experience. However, they deemphasize any role for mental construction in observation, seeing a correspondence between statements about the world and the experiential world itself.

The non-positivist Ludwig Wittgenstein gave a slightly different spin on this point. He articulated the grandest version of the claim that our language mirrors reality. But he also affirmed that there is something more to reality that cannot be mirrored in language. Hence, the famous last line of his *Tractatus Logico-Philosophicus*: "What we cannot speak about we must consign to silence." There are no questions to ask or answer about this something more to reality since we cannot meaningfully speak about it. On the other hand, "what can be said at all can be said clearly."[13] The only proper descriptive use of language will be found in scientific propositions. (The relation of the language to the world must itself remain a mystery since this relation is not itself part of the world and therefore cannot be depicted in language.) Science will capture all that can be spoken about the world. What is left over—"the mystical"—"manifests itself" but cannot be spoken.[14] Thus, Wittgenstein's aim was to set the limit to the expression of thoughts, and to do that we have to find both sides of the limit—to see both the expressible and the unexpressible.[15] In sum, he was not denying a realm of mystery to reality but was denying it can be expressed. (Even this "expresses" the mystical and thus he would dismiss that sentence as meaningless.[16]) If anything, he preserved a domain of mystery from the intrusion of our language. It is a silence affirming a reality that shows itself, not denying any reality.

Wittgenstein affirmed another of the logical positivists' points when he said "[w]hen the answer cannot be put into words, neither can the question be put into words. *The riddle* does not exist. If a question can be framed at all, it is also *possible* to answer it."[17] But positivists draw another conclusion. In the words of Moritz Schlick, "[n]o meaningful problem can be insoluable in principle [by scientists]," and hence "[t]here is no unfathomable mystery in the world."[18] Any real question will have a logically possible solution, and scientists—at least in principle—will be able to find it. If we cannot answer a question, it is only because we have not formulated the right question with the appropriate concepts. In short, if scientists cannot possibly answer a question, there was no real question to ask in the first place. There are no limits to our knowledge except what can be legitimately asked of reality. Any mysteries about the world do not fall into that category because no observation could in principle answer them—e.g., how could we possibly tell if the world is ultimately mind or matter, and what would it matter to us if we could tell? Thus, logical positivists dismiss mysteries as illusory problems. There is nothing "unsayable" about the real world and thus no genuine mysteries. Thus, mysteries tell us nothing about the real world—"mystery" is just a word for factual or conceptual ignorance that will eventually be replaced with knowledge.

The positivists' solution to the question of mystery was neat and clean, but not surprisingly it never rang true. They never succeeded in reformulating scientific theories into sentences only about sense-experience without any theoretical commitments, and without being able to do that, the way they dealt with mysteries sounded unconvincing. Trying, in effect, to produce a new language in which metaphysical questions could not be formulated failed and did not eliminate the impression that the basic metaphysical mysteries were still legitimate. The "ordinary language" movement that followed in Anglo-American philosophy also did nothing to resolve mysteries.[19] The idea was that scientists would solve any empirical mysteries and philosophers solve the conceptual ones resulting from, to use Wittgenstein's phrase, "the bewitchment of our intelligence by means of language."[20] Thus, no mysteries would remain: all alleged mysteries would be reduced to empirical problems for scientists or conceptual ones for philosophers—all legitimate questions would be answered. That is, the alleged misuse of ordinary language was seen as the cause of philosophical problems, and the philosopher's task was to point out the confusions; the alleged mysteries would then evaporate.

According to the ordinary-language school, philosophical questions are of the form, in Wittgenstein's words, "I am in a muddle; I don't know my way" and the purpose of philosophy is "to show the fly the way out of the bottle."[21] The idea is not so much to offer a solution to a mystery as simply to show how it is to be avoided. Problems only arise when language is like an idling engine, not when it is doing actual work.[22] A simple example is the famous puzzle: "If a tree fell in the woods and there was no one around to hear it, does it make a sound?" The problem arises because of the ambiguity in the word "sound." If by "sound" we mean the objective generation of sound-waves, then obviously the tree falling makes a sound; but if by "sound" we mean the subjective sensation arising from sound-waves impinging on an eardrum, then equally obviously no sound is created. There is no dispute over the facts of the situation (the falling tree, the sound-waves, the lack of sensation) but only a confusion over a word. What is left after the analysis is a scientific description of the generation of sound-waves and their impact on the eardrum and a choice over how we want to use the word "sound"—no mystery is left. Unfortunately, substantive philosophical mysteries such as the problem of free will and determinism have proven totally resistant to all such analytical efforts. Analytical philosophy has helped to clarify the concepts and issues but not to resolve the problems. Such mysteries, in short, do not appear to be pseudo-problems after all.

Metaphysics

In sum, mysteries have survived modern philosophical analysis unscathed into the contemporary philosophical scene—in fact, they only have become clearer. But this has not lessened the negative attitude toward philosophical mysteries in philosophy today. The problem is that these mysteries are metaphysical in nature, i.e., they go beyond empirical findings to a degree that scientific theories do not and thus are resistant to empirical testing. Metaphysics goes "beyond the sciences," but not all metaphysics goes "beyond experience" in the sense of being about unexperienced transcendental realities alone, nor are metaphysical ideas the product of "pure thought" independent of experience—such ideas must account for (or at least be consistent with) our experiences and scientific findings. But since metaphysical explanations apply to all relevant phenomena, they are beyond testing by empirical predictions as with science. More generally, metaphysics relates to our broadest picture of what is real—our fundamental ontology and a picture of how the elements of reality fit together. What is reality? Why does the world exist? Why does the world have the general features it has? What things within the world are ultimately real? How do we fit into the scheme of things? Is there a transcendental source to what we experience, or is the natural realm studied in science all there is?

Such systems have been in general disfavor with Anglo-American philosophers for well over a hundred years (with some notable exceptions such as Alfred North Whitehead). Philosophers led a skepticism-inspired attack on any ideas not closely tied tightly to experience—in particular, those not tied to the possibility of empirical checking. Logical positivists dismissed metaphysical propositions as literally meaningless since no set of observations that we could in principle make would determine the truth or falsity of the propositions. Thus, they did not argue over which metaphysical system was best but dismissed the entire undertaking as a delusion. But this suspicion was more wide-ranging than just among positivists. The only metaphysical principles that many philosophers would accept are those

ntific research and thus have some empirical relevance, not "first principles" that are presented as *a priori* or otherwise conceptually independent of other claims. Metaphysics in general earned the reputation of, in F. H. Bradley's phrase, "the finding of bad reasons for what we believe upon instinct."[23] Fundamental beliefs were no longer justified but explained away by psychological or sociological causes. Today any attempt to create any system, grand or small, that mirrors the deep structure of the world is greeted with doubt.

In sum, the atmosphere of skepticism has led to an unprecedented denial of the possibility of metaphysics at all, even though the schools produced in the twentieth century that raised the doubts turned out not to be very persuasive. Nevertheless, even the most skeptical of philosophers operate with specific beliefs about the nature of reality—a worldview of what is deemed real and of how the world is constructed that provides an orientation for how they live their lives. One valuable point that postmodernists make is that no one operates free of such frameworks of beliefs—no one sees the world free of any perspective. Our thinking and speaking embody basic ontological commitments. In particular, philosophy is always attached to some worldview. Most philosophers today may see materialism (that everything real is made only of matter) or naturalism (that only what natural scientists can in principle study is real) as obviously true. But these implicit belief-commitments about the nature of reality are not metaphysically neutral—e.g., the universe is accepted as uncaused and life after death is rejected. And, as discussed in Chapter 4, science does not prove any metaphysical position. They are no less metaphysical than their predecessors, even if discussing their status is of no interest to their subscribers. Getting away from "speculative metaphysics" is simply not possible. In fact, the attempts to deny metaphysics only made their presence clearer at the foundations surrounding the knowledge that scientists supply.

Metaphysics also has a problem that science and religion do not: it is based on our reason alone. All experiences are placed in a framework of understanding and do not provide any testing of those ideas. Our speculations remain our speculations. Perhaps there is in fact no deep order to reality for our mind to grasp and express. Perhaps reality is not put together in such a way that any of our grand systems of thought could capture the true nature of things. We are not in a position to examine the totality of reality as an object from the outside, and thus perhaps we cannot tell which system is correct. Or perhaps there is such an order and our minds simply cannot grasp it.

Moreover, when we take an aspect of reality and use it as a "root metaphor" for understanding the basic nature of reality (e.g., the brain as a computer or the universe as an organism),[24] does the image end up being a Procrustean bed controlling what we can see and thus end up inhibiting rather than helping us to understand all that we could understand? If we start to view persons as machines to learn about the body, do we end up reducing persons to that image and not seeing all there is to being human? Explanations and speculation may satisfy our mind, but how, the skeptic in us should ask, do we know that our metaphysical systems—the elements of our beliefs most removed from empirical testing—are not merely illusions of knowledge shielding us from the unknown rather than genuine insights? At a minimum, how do we prevent ourselves from getting caught up in nothing but our own ideas? Is metaphysics nothing but an attempt to deny any mystery to reality? But can we accept that we do not know something fundamental about reality? Or, if we were left without the shield of some metaphysical beliefs, is the mystery too immense for us to function in the world?

To see some of the problems even without grand schemes, consider two metaphysical mysteries of fundamental importance: the beingness of the universe and consciousness.

Beingness

The most basic mystery of them all is why there is something rather than nothing—the mystery of existence itself. It does not concern the structures at work in reality but merely the bald fact that anything exists at all. Why is there something and not nothing? As Wittgenstein put it, the mystical is not *how* the world is—this is the subject for science—but *that* it exists.[25] This will be discussed in Chapter 5. Second to this is the mystery of what is the basic nature of the very "stuff" the world is made of. What constitutes "beingness," i.e., the sheer "is-ness" of things, the "power of things to be"?

Obviously, we are all aware that something exists. We may infer the existence of noumena or unobservable entities, but we do not infer that something exists because of its effects on us—we are immediately aware that something exists (including ourselves). The mystery is not the product of any special experience but is a matter of reflection on the most basic awareness of all: we are immediately aware that something existence. But what is the nature, in Buddhist term, of the "such-ness" of the world apart from our conceptual division of it into objects? It is a prime instance of the mysteriousness of the ordinary. Beingness is not undifferentiated matter but is what gives matter its existence — along with mind and whatever else is part of the universe. The philosophical mystery of "matter" is not to identify the lowest level subatomic particles or how higher levels of phenomena emerge from lower level structures or any other issue related to how the world is — those are subjects for science — but simply to identify what "matter" is. What is the beingness of matter? What gives matter its power to be? Does "matter" mean anything more than the absence of biological and conscious powers in some thing that exists? Are there different types of being? Is the beingness of living or conscious beings something more or different than the being of atoms or of, for example, numbers? Even if beingness has some inherent properties (e.g., mass), still what is it? What is a "superstring," or whatever is the smallest bit of matter, apart from the structures it embodies? How do the structures of nature (e.g., gravity and electromagnetism) relate to the beingness upon which they operate? What is more fundamental to reality—the beingness or those structures? Is beingness here just to give a platform to the structures that really run the show? Are being or structures somehow the product of the other?

What the world is made of in a metaphysical sense (as opposed to a scientific issue of the most fundamental entities and fields) is no longer a major topic of interest in philosophy. Is the stuff of the world matter, fields of energy, conscious-ness, "spirit," love, ideas in "the mind of God," or God itself? But then again what is, for example, "matter"? Or "energy"? In philosophy, this mystery became reduced to another issue: whether "being" is a property things have simply because they exist. And to most philosophers today, "being" is simply an empty term. To positivists, the immediate givenness of experience is not amenable to further analysis and any speculation on the ultimate nature of reality is meaningless. To Quine, "to be is to be the value of a variable."[26] Beyond that, philosophers are no longer interested in the notion of being. The focus is on the objects of the world and our conceptualizations, not what gives them being. Being simply is, and attention is directed to other issues.

Can we explain or analyze anything about being? How can we analyze the substance the world is made of, whatever it is? How can we get distance from it to analyze it? What can we say beyond that it is whatever it is that instantiates things (i.e., gives them being)? It is not testable by pushing and pulling, as are casual structures, since it is uniform to everything. Are all attempts to express the nature of beingness—to identify what is beyond the instantiation of objects—attempts to transform beingness to a distinct object? However, will all attempts to grasp beingness fail since beingness is not a particular object among objects? Do we have here something like the problem that the "sense of the world" lies outside the world, as Wittgenstein said,[27] and so we cannot explain it or even describe it from within? We would have to step outside the universe, which we obviously cannot. To philosophers in general, if beingness is beyond all possible rational analysis, the answer has to be that we cannot say anything about its nature, and thus the question loses any frame of reference and thus is devoid of sense.[28] There is only a perplexing and ultimately nonsensical question with which we plague ourselves.

Milton Munitz asks whether we can even *speak* of beingness—"boundless existence" in his terminology—since it is not an object or set of objects.[29] Beingness never presents itself as a phenomenon. Beingness, Munitz notes, "shines through" the known universe but is not identical with it or an entity of any type. It is not a thing or combination of things or the totality of things. Unlike an object, it is not "conceptually bound." It has no properties, qualities, or structures to discover—it has nothing to describe. It is utterly unique in that it is not an instance of any category whatsoever. Thus, "beingness in itself" is unintelligible, since intelligibility requires the applicability of descriptive or explanatory concepts.[30] That is, intelligibility relates to *what* something is or *how* it is, not to the underlying *that-ness* of reality. We live in a world of differentiated objects and see and speak only of those objects. Beingness itself remains beneath any conceptual map we could apply to it to create order.[31] Once we start speaking of beingness—or even just naming it—we make it one object among objects, which "it" is not. That is, we see trees and cars, not "being" and we cannot formulate propositions about "it." If Munitz's position is correct, any explanations or understanding of the beingness of our world would be foreclosed. As with Wittgenstein, beingness manifests itself and hence we are aware of it, but it is unutterable and incapable of being conveyed in language. Thus, the proper response to our awareness of beingness is silence.[32] If this position on language is correct, we are left with only mystery about the very beingness of the world. That is, beingness cannot be expressed or "captured" by any language, and thus is a mystery.

So is the beingness of the world itself indeed beyond our understanding? One issue is how exactly are we aware of beingness. We experience our bodies, floors, wind, and so forth, but do we experience *being*? When we walk into a wall, do we experience the beingness of a wall or just its objectness? If we are aware only of objectness, is our sense of beingness a product of reflection, i.e., a posited reality beneath what we sense? Do mystical experiences put us in contact with beingness "as it really is" independent of our conceptualizations and points of view? But we could not avoid the presence of beingness even if we successfully confine our lives just to a world of the objects we conceptually differentiate. Indeed, beingness is self-evident. We have no idea what it is except as we are continually aware of it, strange as that might sound. Being a constant in all experience, we cannot step away and examine it. Any inquiry will be confined to some topic within nature that is more limited in scope, and upon reflection we are aware that it is limited. This means we are aware that there is something more, even if we cannot comprehend it. Thus, our reflection upon our awareness creates the mystery of beingness. We end up asking a question—what is the nature of being?—that is not answerable, but that we cannot eliminate either.

A related mystery also must be mentioned: our sense of both permanence and constant change—of timeless "being" and temporal "becoming." Each is a mystery in itself, and their relation also remains a mystery. We think being is changeless; yet the world is constantly dynamic, a realm of perpetual change. Does only the dynamic becoming exist, and is the idea of changeless being simply an illusion our mind has created? Is it like our sense of personal identity that persists through constant changes as we age?

Becoming also introduces the mystery of time. To Aristotle, time is simply the measure of change. No rate of this change can be even imagined—what would the units be? Hence, the question "How fast does time flow?" is a classic example of questions that positivists dismissed as meaningless. The sense of the reality of the flow of change may seem undeniable, but philosophers from Parmenides to John McTaggart have introduced alleged paradoxes connected to the process of change and have concluded that change is only apparent and that reason tells us it is only an optical illusion—"tensed facts" in terms of past, present, and future are not objective features of reality and thus do not actually make up part of what is real.

....ded also have concluded that there is no reality—"time"—that "flows." In fact, as discussed in Chapter 4, today some physicists also argue that the "passing" of time is an illusion caused by our mind, even though the before/after sequence of events is not. Add to this the mystery of "the present"—that elusive moment "in" time that is not part of the measurement of time but is the space on the temporal continuum within which we are aware of reality. As with all mysteries of the ordinary, time is, as Augustine said, something we understand until we try to explain it.

Consciousness

Nothing is more obvious to us than the fact that we can make our body move. But nothing is more mysterious than trying to grasp how the mind actually accomplishes this. We know no more than did the ancient Greeks about how we get our body to walk or think. This leads to the topics of consciousness and what it is to be human. What in the structure of reality permits there even to be beings here to ask these questions? How has what exists become able to reflect upon itself? Why is subjective experience "attached" to some physical events? Forget about the scientific problems of how consciousness arose, why is there consciousness in the universe at all? Why was evolution capable of producing it? Why are we not zombies (i.e., beings who are physiologically identical to us and do the same acts as we do but have no consciousness)?[33] Is there one center to consciousness, or just lots of different areas of the brain hard at work? Why do we experience it as unified and not disjoined? Is there a unified entity called the "self," or, as Hume and Buddhists believe, no center to the constantly changing "bundle of different perceptions"? Or, to cite the Upanishads (with echoes in Kant and Wittgenstein), is the "seer of seeing" nevertheless real even if we cannot see it? How do we link such a Kantian "transcendental self" with the flow of our changing experiences and memories? Or is the "self" merely a fiction we posit to become an entity in a world of entities? Do we think it is real just because we have words for "it"? That is, are we simply misled by grammar into creating the notion of a "self" when there is in fact no such entity?[34] Or do we need a center—a unified agent—for "free will" to work, or is "self-determination" another example of the grammatical mistake? Is mind a separate reality from matter? Is, as William James and Henri Bergson believed, our mind a "reducing valve" that normally lets into our consciousness only enough of reality to enable us to survive? And what exactly happens to the conscious subject when we die? The brain dies, but does the mind? What exactly ceases? Does anything real cease at death or ever?

That the nature of consciousness is itself a mystery is revealed by the fact that the scientists and philosophers studying it cannot even agree on what exactly they are studying or come up with a common definition. Consciousness includes decidedly different activities: awareness of our environment through our senses, awareness of having a body, thinking, a sense of being here and now, emotions, memory, an awareness of seeing from a particular point of view (subjectivity), and an awareness of being aware and a sense of being the subject of awareness (self-consciousness). In the West, some philosophers have taken consciousness to be as equally fundamental as matter in the makeup of reality. Some philosophers see personhood as a fundamental category of reality. Panpsychists see all matter, even on the smallest scale, as having mental properties inherent in it. However, Descartes set up the fundamental dichotomy of mind and matter that has shaped the modern era.

With self-consciousness, we are aware of the content of our consciousness and can examine our thoughts and emotions, but it is not a sensory awareness. Studies suggest that young babies and perhaps even some animals have a sense of an individual self—this suggests that self-consciousness, like the rest of conscious-ness, is pre-linguistic even if the content in adults becomes shaped by language. The fundamental nature of self-awareness and its place in reality is a matter of metaphysical speculation. In general, classical Indians did not see a mind/body problem as modern Westerners do. Advaita Vedantins take the same experience of thinking that Descartes thought proved the existence of an individual thinking self as an experience of the nonindividual consciousness underlying everything (Brahman). Thus, Advaitins disagree with what Descartes took to be the one certain empirical claim. On the other side, Buddhists and most material reductionists deny that there is any self at all—i.e., no distinct executive in charge of all the mental functions happening in the brain. This raises the issue of what exactly do we experience in "self-consciousness"? And what is its relation to the rest of reality?

Consciousness does not seem to us to be material; it does not seem to occupy space, nor is it something we can push or pull. On the other hand, it also does not feel unattached to the body. The mind seems to participate in the world by being somehow embodied in a body, rather than being a transcendental disembodied "soul" or "self." Thoughts may seem disembodied, but they also seem to be coming from our head; emotions such as joy, excitement, or anger seem even more attached, permeating more of our body. We also know that affecting the brain can affect consciousness. Illness and drugs can alter states of mind, and operations that disconnect parts of the brain can affect everything from perception to a sense of morality. Today many scientists and philosophers argue that our bodies greatly influence how we think. All of this has led most philosophers to see the mind as somehow dependent upon the body. There still are a few prominent mind/body dualists (e.g., Karl Popper and John Eccles) who believe the mind is something independent of the brain: we walk with our legs—our legs do not walk us—and we think with our brain, but the brain does not do the thinking. However, the vast majority of philosophers today endorse a form of monism in which our conscious-ness, like everything else, is the product of matter—materialism.[35]

For materialists, matter is all that exists, and thus consciousness is a form of matter or material activity.[36] They are more impressed by rocks and other inanimate objects than living entities or conscious experiences in deeming what is ultimately real. Thus, they look for the smallest bits of matter to be the foundation of their ontology. Not that materialists analyze what exact "matter" is—indeed, they show remarkably little interest in that metaphysical question, despite their title. To them, whatever physicists study is the end of the issue. For materialists, all alleged mental properties are in fact physical properties.[37] It is a matter of *causation*. How could a nonmaterial "mind" interact with "matter" to cause it to move? Conversely, how could a material event like suffering a burn cause pain if mental states were not material? All thought, feeling, sensation, the unconscious, and any paranormal abilities (if genuine) are nothing but electrochemical reactions in the brain—no more than the firing of neurons. The brain generates mental events merely through the complex interaction of matter operating under physical forces. Our mental life, including the self-monitoring by being aware that we are aware, is no different in nature than the turning of a plant toward sunlight—the interactions of matter occurring when our brain senses things are simply much more complex.

To reductive materialists, mental events are either identical to physical ones or are at most the useless epiphenomena of physical events with no causal power of their own, like a steam whistle apparently causing the train to move.[38] The material brain does the actual work, and those neurological actions actually happen a fraction of a second prior to our "conscious" decisions.[39] Everything about our personal identity—our intelligence, emotions, moral character, memories—is firmly rooted in our brain physiology and would not exist without it, and so there is no separate "self." The brain simply monitors its own activity, but it also creates the image of a controlling agent—a "self"—that does it. There is no "I" at work—no central agent controlling our mental life—but at most a small horde of material agents, each controlling part of the activity. Thus, the sense of a unified nonmaterial "self" is an illusion. Consciousness simply records our prior material actions and gives us new material events to react to. Some reductionists go so far as to deny that there really is any consciousness at all—we are no more *conscious* than is a television set. Memories and the other content of consciousness are just the mechanical recordings of inanimate processes like a VCR recording and playing back a television program. When we finally engineer robots to do everything we can do in our mental life, the illusion that consciousness is real will go the way of a vitalistic substance in biology.

So too is the whole idea of a "free will," despite the valiant efforts of "compatibilists," who, following Thomas Hobbes, see free will in terms of acts that are simply physically unrestrained rather than uncaused by physical events. There are only physical events and no mental power exists to cause physical events. Thus, we are left with a purely deterministic chain of physical events. How could a material universe operating under determinism produce beings with an nonmaterial consciousness, let alone free will? Because we do not know now what will happen next, our choices and actions seem free. But what we think are acts we freely willed are in fact merely a string of inanimate events. We are not the author of our

why does this connection exist?

actions—nonconscious matter is in control. Our convictions that when we act "we could have done otherwise" and that we act for reasons are groundless. Events in the brain that we think we initiate by free will are only caused by neurological events. Actions happen to us, and the feeling of a conscious "will" is only an illusion created by the brain to create a sense of ownership.[40]

Even if materialists accept that consciousness is real, they see the body as dominating. Our bodies and brains have evolved to survive in a specific environment, and our consciousness is a natural product of the same process and is shaped by our bodies and actions.[41] There is nothing particularly special about human consciousness, or being human in general, and nothing mysterious or miraculous: to become aware of oneself and one's actions has an obvious advantage for survival, and so it evolved. We may not know yet how consciousness arises, but our consciousness is no different in type from that of other animals, even though language is an evolved tool that permits us to expand ours greatly. The ability to reflect—to create and manipulate inner pictures of things—simply arose through evolution as a response to stimulations that help us survive, and the sense of self simply arose when consciousness evolved enough that it could reflect on itself and recognize that there is a store of information in the mind about previous experiences and mental states.[42] Dualists merely have an oversimplified concept of matter and do not know what matter can do. At present, it is a waste of time to try to answer the big question of how consciousness arose—we do not know enough about that yet—and so it is better to concentrate on just finding neural correlates of different types of conscious events. Eventually we will know enough, and the "mystery" of consciousness will be reduced to a solvable "problem."

The other branch of the materialists—the antireductionists—give mental phenomena greater reality.[43] They emphasize that a fundamentally new level to reality emerges with the appearance of consciousness. But consciousness is still a purely natural emergent feature of a complex biological being. No other realities are involved. It is not that the mind is a different "substance" than matter but that a new level of organization is structuring matter in conscious beings in addition to the physical and biological ones. The brain/mind is one indissolvable unity within a material being. Consciousness is caused by neurobiological processes and is as much a part of the natural biological order as any other biological feature such as the digestive acids secreted by the stomach—in effect, consciousness is just "secreted" by the brain.[44] Consciousness is no less real for being so: it is still real and has causal powers and is part of the natural causal order.

But for antireductionists felt experiences, subjectivity, and the other features of consciousness reveal *something genuinely new to reality*, something that inanimate matter can never have and that cannot be eliminated from our picture of the world. True, the greenness of grass is not there without us: it is not independent of our neural system but a matter of the interaction between us, light, and the grass. And it is true that different mixes of wavelengths of light can produce the same colors and that people have different subjective reactions (as color-blindness indicates). Perhaps if I could plug your optic nerve into my brain, I would see what you call "green" as what I call "yellow." Nevertheless, the experience itself is no less real than the grass and light-waves involved in perception. Reality is not colorless or otherwise devoid of sense-qualities. Nature is not, in Alfred North Whitehead's characterization, "a dull affair, soundless, senseless, colourless"—seeing color requires beings capable of a sense of color, but the world is not colorless any more than objects in a drawer are invisible just because they are not being sensed right now. Nature always generated all that was necessary for color and simply had to wait for part of its creations (color-sensitive subjects) to develop to sense it. Such sensations are still non-accidental parts of reality because we are a non-accidental part of reality. Our "subjective" experience is as much a part of reality as what is experienced. The sense-qualities involved in seeing the Mona Lisa are as real and irreducible as the canvas and paint. Sense-experience is not the "appearance" of some other reality but is itself a feature of reality in its own right.

Thus, a blind scientist may have complete knowledge of the physics of vision but is still missing something real. To antireductionists, mental properties are not the same as other properties within a material being—pain is something more than just the actions of nerve cells—and we cannot describe the world fully only in terms of the latter. Mental properties have an "inside" that objects do not. Thus, science, limited as it is to a third-person point of view, will necessarily omit this first-person element to reality and therefore can never give a complete account of the world. Subjectivity is outside the scope of science, although some mental powers are part of the natural causal order. Even some reductionists accept the reality of consciousness (as a material product) and admit consciousness, in the words of the reductionist Stephen Hawking, is not a quality that can be measured from the outside.[45] Nevertheless, sense-experiences and the rest of consciousness are purely natural products of matter, even if how exactly they fit into the material world remains a problem.

In sum, consciousness according to the antireductive materialists is not reducible to nonconscious levels of organization operating within matter. The brain may only produce consciousness the way charged particles produce electromagnetic fields, but consciousness nevertheless remains one of the irreducible features of reality. Its reality is nonnegotiable, and there remains an unbridgeable gap between its material base and the conscious level of reality. Reductionists may agree so far, but antireductionists part company here in arguing that consciousness has *causal power* and is not merely an epiphenomenon: it is a new level of organization to matter and can interact with other levels in one universal order of causation—physical, biological, and mental causes are all integral parts of one causal order, and there is no reason to give the nonconscious causes any priority. The mind adds no new type of energy to the natural system. Consciousness evolved because it had some causal role to play in our survival, not just to record other material events. This view also permits free will—genuine choice and agency—as a component of the overall causal system determining our actions, and our choices must be taken into account for any complete explanation of our actions. Free will could evolve in a world that is deterministic before the appearance of conscious life only if consciousness is a separate level of organization. We have at least some degree of choice in our actions, and any explanation of our actions would then require reference to our intentions as causes, not just to the physiological conditions of the event.

But antireductionists concede that mental causation remains a mystery and will remain so as long as emergence and the different levels of organization shaping matter remain mysteries. For antireductionists, how consciousness emerges is a mystery, at least for the present (as discussed in Chapter 4). But the new level of reality also adds a "why" mystery and not just a "how" mystery. What empirical finding could ever suggest that subjective experience would emerge from physical facts? How could we predict our sense of the greenness of grass from the physics of light? Many problems of consciousness may be comparatively easy to study (e.g., sensory stimulation or recording memories), but subjective experiences —explaining consciousness itself—is the "hard problem," as David Chalmers puts it. Why is subjective consciousness part of the natural world at all? How does matter generate any subjective state in the first place (if that is what happens)? If the brain is just a soggy, meat-based computer, how does it generate sense qualia and other subjective phenomena? More generally, trying to give a unified account of such conscious phenomena as subjectivity, agency, and intentionality, and the persistency of a sense of self is not yet even on the horizon. (Chalmers is attracted to panpsychism to bridge the gap between matter and conscious experience, thereby making consciousness as basic a property as mass or an electrical charge. This introduces a dualism of properties but not of substances.) For reductive materialists, consciousness is not a philosophical mystery at all but simply a puzzle for empirical research: subjective feelings are only epiphenomenal material states with no causal power, and we will someday understand how the brain produces them. Such states only seem mysterious because conscious states do not feel like material objects. What we really need is philosophical therapy to overcome our deep intuition of dualism, not any new metaphysics.[47]

Some materialists, however, see possible limitations on our ability to understand. Colin McGinn and the "mysterians" argue that our brain has evolved to navigate us through our physical environment and not to reflect upon itself and thus we will never know very much about consciousness.[48] For them, how the aggregation of insentient neurons generates subjective awareness will remain insolvable in principle: we need a new type of category that links the mental and physical, but we cannot think in categories other than the mental and the physical. How could we step back from our situation and come up with the missing way of thinking? It is a limitation on our cognitive abilities, like not being able to visualize four dimensions in mathematics. Evolution simply did not design our brains to understand that process. Thus, there are physiological causes—a sort of "innate ignorance"—for the limitations in our ability to understand all things mental, including the determinism/free will problem and more importantly how consciousness arises. No part of the world is inherently any more mysterious than any other part—there simply are questions that fall outside the "cognitive space" of beings like us and thus will remain mysteries to us. It is a matter of psychology, not metaphysics. Most philosophical issues will remain mysteries, McGinn believes, because of our physiological limitations. Beings from other planets with a different physiology may have entirely different sets of mysteries.[49]

All in all, for reductionists and antireductionists alike, how the body produces consciousness—how a collection of neurons gets a new property, how electrochemical activity becomes mental "information"—is a mystery at present and perhaps forever. Still, for materialists, consciousness is a perfectly ordinary product of the universe even if we cannot get enough distance from it to understand its nature or how it arose. Our minds are no different in kind from the minds produced by evolution in other animals. Just because we do not understand this natural process is no reason to start invoking souls or spirits or a creator god. Consciousness, in sum, may be something we were not designed to understand, but nothing about us involves dual substances or transcendental realities or is otherwise mysterious. The incomprehensibility of how it arises simply lies solely with us.

Discussions of the mind/body problem quickly degenerate into a stark conflict of metaphysics. Some philosophers cannot believe their opponents truly believe what they are saying. But it is amazing how confident most philosophers are in the positions they hold on what constitutes a human being and on what will happen at death. In candid moments, some reductionists have stated that they cannot imagine anything that would convince them that their version of materialism is wrong, thereby admitting the metaphysical nature of their position. But as of yet all that reductionists have presented are unfulfilled promises dictated by their metaphysics of how they believe reality is—and all current reductive theories have major philosophical problems. Nevertheless, reductionists brush aside what they cannot explain, and, in the words of Francis Crick, simply "hope for the best."[50] They are so enthralled by the success of science that they deny that anything allegedly nonmaterial (and thus not part of a scientific explanation) could possibly be real—in fact, they think Descartes seriously warped our point of view by seeing things in terms of two distinct substances.

But dualists and antireductive materialists see any attempt to explain consciousness in terms of nonconscious elements as a fundamental mistake. And it must be admitted that science has not proved that consciousness is not distinct from matter: all scientists produce is a correlation of mental events with matter—it is a *metaphysical* assumption that matter somehow causes the mental events instead of simply supplying a base necessary for mental events to appear. Of course, with their success in the study of brain activity, it is easy to see why neuroscientists may simply ignore (or be unaware of) the philosophical issues and claim to be producing a "theory of consciousness." But as things stand, neuroscientists only study something closely associated with consciousness—its neural and other biological bases—and not consciousness itself. We are left having trouble determining even how to frame the issues of mind and matter—a sure sign that we are surrounded by a mystery (or by nonsense). In the spirit of science, we should leave the metaphysical issue open.

However, as a practical matter, we must commit to belief on some of these issues. We cannot in practice remain agnostic on all matters and set them aside as merely metaphysical. In particular, we cannot but presume to have free will: even determinists admit we must act 'as if' free will is real, and fatalists still must seem to themselves to choose actions because they do not know in advance which actions are predetermined or the result of fate—in sum, if you fell off a boat into the ocean, you would still try to save yourself, no matter what you think about fate and determinism. So too, accepting the idea of personhood seems essential to how we understand and interact with others. And even Hume admitted that nature has left us with no choice but to the believe in the reality of at least our body. We may be able to remain agnostic about some issues but not on issues touching practical matters such as those.

Nevertheless, we must accept that mysteries surround all the issues related to just being human. Consciousness may seem obvious, something we can be aware of whenever we like. But grasping it is hard to do. We do not know its nature, or whether it is a fundamental feature of reality or instead is reducible to some other feature. We do not know how consciousness arises or whether it can come to an end. We do not know how mental causation—if it is genuine—actually works. Is there an agent causation between indeterminism and determinism permitting our control of some bodily events, or is no "self" needed at all?[51] How does the mental "software" interact with the bodily "hardware"? We may eventually learn everything there is to know about the wiring of the brain and be able to correlate every conscious event with some specific brain activity, but we will still be left with the vital mysteries of how and why this new level of reality was produced. Neither dualists, idealists, reductive materialists, nor antireductive materialists deny that in the final analysis there are mysteries here. As with beingness, the mind is something we know directly and intimately, but it remains a mystery.

The Philosopher's Disease

Philosophers from the beginning have always recognized limitations to our ability to understand reality. For Plato, the fact that we inhabit an imperfect material world limits our ability to comprehend the real world of the realm of perfect forms—we are prisoners in a cave who can see only moving shadows on the cave walls and cannot see their real causes. For Kant, our mental abilities cannot comprehend the world-in-itself. But as a rule, philosophers also suffer from a compulsion: trying to dispel any mystery surrounding phenomena by finding reasons for them—even if there are none. At a minimum, this creates conceptual barriers between us and reality by focusing on concepts, providing reasons, giving a general account of all of reality, and in general directing our attention away from direct contact with reality and toward our conceptions.

This condition is certainly true with regard to the mysteries of reality. One would think that Kant's idea of a noumenal realm would firmly plant mystery at the very heart of all modern philosophical thought: reality as it is "in itself" is forever inaccessible to us and hence unknowable even in principle. Wittgenstein's idea of "the mystical" should have had the same effect. However, just the opposite is true: modern philosophers have tried to find ways around the mysterious. But attempts to eradicate philosophical mysteries by translating them into manageable puzzles have not proven successful. A more common response has been simply to ignore what must be mysterious and to focus instead on what seems potentially knowable—the structures of the world as described in science. It is, of course, very important to develop the latter, but the point is that the surrounding mysteries do not disappear just because it is convenient to look the other way. We get caught up in questions of how the world works, and the related issues of power and control over our environment. We miss the fullness of reality beyond our comprehension or deny the possibility that there may be things about reality we cannot in principle know. We forget even the obvious mystery of the simple fact that the world exists. The effect is the demise of a sense of mystery—the wrenching of all mystery from our view of reality that leaves a truncated view of all that is real.

Metaphysical systems can be seen as beginning with a helpful insight into an aspect of reality and then blowing that insight out of proportion when it is taken to be the final picture of all reality. Isaiah Berlin described the general history of all thought as a series of liberating ideas that became suffocating straight jackets. Moreover, there is no cumulative progress in metaphysical thought but only a succession of different points of view (usually based on scientific changes). Each point of view highlights problems with other positions, but it itself becomes a problem when it is taken to be the final insight into the nature of reality. This does not mean that all metaphysical thought must be rejected out of hand as unjustifiable speculation, but we do have to remember that with enough imagination we can always produce reasons for any position we like. The insight of empiricism is that we have to keep thought as close to experience as possible and that the further removed we are from the check of experience the greater the risk that we are fooling ourselves. That nature is constantly surprising scientists should give metaphysicians pause about their projects when they think that they can know the fundamental nature of reality by basically *a priori* thought alone. What "seems plausible to us" has changed for scientists over the years, and this will always indirectly affect metaphysics. Reason and experiences will always be in tension, and the latter is a corrective in scientific thought but only more indirectly in metaphysics. Ideas will always be important to us for maneuvering in our environment, but we must remember that they are always our ideas and that they may become a barrier between us and what is actually real. It may also be that reality is not put together in such a way that any of our systems of thought could represent the final nature of things. Thus, we may be left with the possibility that our thought may not be able to reflect the basic nature of reality even in principle.

The insight of practical skepticism is related: we can never be certain about our claims to knowledge, especially on fundamental metaphysical matters. The more radical philosophical skeptics take their arguments as negating all claims to theoretical knowledge. Philosophers usually concede this insight about the foundation of knowledge but plow ahead anyway. However, as Quine has said, it is doubtful that we are any further along than where Hume left us—"the Humean predicament is the human predicament."[52] Yet it is important to distinguish practical and philosophical skepticism here: accepting the latter means only that we do not have absolute foundations and the resulting certainty, not that all our first-order knowledge-claims are wrong or that we do not know anything. The same applies to analyzing concepts: if we look hard enough at any concept through Socratic questioning we are left not with "clear and distinct ideas" but blurs—but this does not mean our concepts may not reflect something in reality. We obviously do know many things about our experienced world, even though we may be wrong about them in principle. As discussed in the next chapter, just because we can think of possible alternatives does not mean we are wrong in our beliefs.

How much philosophical skepticism should we embrace? Just about transcendental objects? Theoretical objects in science? Should we become solipsists at least in principle? In the end, is the decision to accept or reject any philosophical skepticism merely arbitrary? Do we side with Hume and his skepticism, or with the common-sense philosophers such as Reid and Moore? Still, the problem is twofold. First, why should we expect that beings like us would be able to know ultimate matters? We are, after all, just one very limited product, either of God or of nature. Second, it might be the case that our reason can in fact reflect the ultimate level of reality since we are its product—i.e., our consciousness has not merely evolved so that we can interact with our environment more effectively in matters of survival, but it also participates in the ground of reality. If so, we may be able to come up with correct ideas about ultimate matters at least related to testable how-mysteries. But considering our situation and how we are constructed, how can we ever know that these ideas are correct? And how can we ever know answers to the ultimate why-mysteries of reality (e.g., the nature and source of beingness or why consciousness arose) when we cannot turn them into objects for examination? Total silence in these matters may not be warranted, but, combined with the insight of skepticism, a little more modesty and less hubris may be in order when our reason alone is involved.

Thus, we are left with the two competing drives in philosophical reflection: toward an awareness of mystery and toward quashing it. The impulse to understand mysteries—to comprehend the uncomprehended—will always be there. But upon reflection we are aware that there is more to reality than just what we abstract out—we are aware that there is something beyond our conceptual understanding.[53]

Reality is more than our conceptual packaging would suggest. But reality is not just more than the sum total of our attempts to understand: something there resists our conceptual grasp. We know we do not live in a world shorn of mystery, even if we in the modern world most often ignore it. In the end, we may be able to formulate legitimate questions to which we may never be able to supply answers.

But perhaps philosophy is better seen, not as attempting to uncover new truths as other disciplines do, but as freeing us to think outside the box. That is, its value is in the questions asked rather than any purported answers. Indeed, maybe the real role of philosophy today is not to try to penetrate mysteries but to make us aware of the "something more" to reality framing all our attempts to understand—to help us realize that there indeed are more things in heaven and earth than are dreamt of in our philosophies. However, as will become apparent in the next chapter, philosophers today of different stripes continue to ignore mystery completely.

Notes

1.   Rubenstein 2008, p. 15. See ibid., pp. 1-24 for a survey of the history of wonder in philosophy. She disagrees with the usual conclusion that modern philosophy calls for the end of wonder and concludes instead that Western philosophy has not expunged wonder so much as it internalized it and installed a transcendental subject as its source (ibid., p. 187).

2.   Skepticism should not be confused with *epistemological relativism*, i.e., the claim that all beliefs about reality have an equal claim to being true. Skeptics question the possibility that any claims could be true. If skeptics are correct, there would be not a cacophony of competing voices, as postmodern relativists would like, but silence.

3.   Since most practical skeptics are materialists or naturalists, practical skepticism in practice becomes a defense of materialism or naturalism or science against all attacks, not a critical examination of all claims.

4.   Hankinson 1995, chapter 17.

5.   For a rigorous defense of skepticism, see Unger 1975. For more recent discussions, see Landesman 2002 and Grayling 2008.

6.   Landesman 2002, p. 202. Following Moore, Robert Landesman does say that we do possess genuine knowledge even if we cannot prove it; thus, we must recognize our fallible condition and be wary of speculative claims. "Our claims to knowledge should be constrained by an understanding of the limits of our ability to prove the reliability of the procedures we use to get in contact with the external world" (ibid.).

7.   "Metaphysical realism" is the claim that the world does not depend for its existence upon the minds of an individual conscious being—something exists independent of our thoughts, and so if we remove all conscious beings from the universe something still remains. In short, it is *anti-solipsism*: the world is not a subjective illusion. There is a mind-independent world and also a mind that is real, if perhaps dependent on the body. That is, metaphysical realism is not just the claim of an "objective" reality—the "something out there" when no one is experiencing it—but can also include consciousness as a part of what is ultimately real, whether or not the content of subjective consciousness corresponds to anything real. Such common-sense realism does not have any built-in epistemology of a

correspondence theory of truth or a metaphysics of materialism, reductionism, or naturalism.

8. For Descartes, knowledge involved certainty, and the only certainty we can have is from the innate ideas God implanted in our minds that we can see clearly and distinctly. These ideas necessarily conform absolutely to the fundamental laws of nature (since God would not deceive us). Thus, the fundamental laws of nature can be *known* (i.e., known with certainty) only by pure reason. But Descartes conceded that which mechanisms God freely chose to use within nature could only be discovered by experiment. Once the fundamental laws are discovered, our minds can discover *a priori* all the different forms the laws take from the innate ideas God has implanted in us. Thus, empirical observation and experiment still played a central role in his thought.

9. Foster 1957, p. 22 n. 3.

10. The basis for the rejection of idealism is some basic questions: if the universe consists only of ideas, why can't we make it anyway we want? And if the universe consisted only of ideas, do things exist when we are not observing them? And why should what we can in principle perceive be the criterion for is real anyway? The idealists' reply is that the ideas are in the mind of God or are otherwise somehow fixed. All that is denied is the lack of *matter* "behind" appearances. For a contemporary treatment of idealism and panpsychism, see Leslie 2001.

11. See Kant 1997, 1998.

12. However, silence is hard to maintain since we can fix our mind on the idea of that realm, and even Kant has to apply the category of "existence" to it, although he said the categories of phenomenal reality, including substance and cause, lose all meaning when applied. As Graham Priest puts it, it is a contradiction inherent in the limits of thought. Priest 1995, pp. 90-91. In the end, Kant tried to stick to a purely negative notion of noumena. (The problem with the *via negativa* will be discussed in Chapter 7.)

13. Wittgenstein 1961, Preface, p. 3. More fully, he said in the Preface that the whole import of his book could be summarized as follows: "what can be said at all can be said clearly; and what one cannot talk about one must pass over in silence." On Wittgenstein on silence and mystery, see B. F. McGuinness, "The Mysticism of the 'Tractatus,'" *Philosophical Review* 75 (July 1966): 305-28; Russell Nieli, *Wittgenstein: From Mysticism to Ordinary Language: A Study of Viennese Positivism and the Thought of Ludwig Wittgenstein* (Albany: State University of New York, 1987); Frederick Sontag, *Wittgenstein and the Mystical: Philosophy as an Ascetic Practice* (Atlanta: Scholars Press, 1995).

14. Ibid., Proposition 6.552.

15. Ibid., Preface, p. 3.

16. The idea behind his claim is that anyone who understands the *Tractatus* will eventually recognize that the propositions are nonsensical (Prop. 6.54)—i.e., once we have climbed the ladder, we will throw it away. Language is a picture of the logical structure of reality; each element of a statement corresponds to an element of reality (Prop. 2.15). Thus, the limits of our language are the limits of our world. However, this "picture theory" of language cannot itself be pictured since it is not part of the world. But, as discussed in Chapters 3 and 7, if we reject a "picture theory" as the proper relation between language and reality, we can reject this claim. Of course, a statement may at first seem meaningful but upon reflection be seen to be nonsensical, but the picture theory of meaning is not the cause. The ladder, after all, is *real* and throwing it away does not make it any less real. Similarly, the words explaining the nature of language are still meaningful even if they do not "picture" something in the structure of nonlinguistic reality. The problem is only with the picture

theory itself—the use of language to explain language is still legitimate.

17. Wittgenstein 1961, Prop. 6.5. He adds that skepticism is not irrefutable; rather, it is *senseless* (6.51). Doubt can only exist where there is a question, and there is no legitimate question here.

18. Quoted in Foster 1957, p. 21.

19. One school of continental philosophy—phenomenology—does not so much deny mystery as simply screens it out. Phenomenologists focus only on our experienced world and bracket all issues of causes, so the issue of mystery does not arise at all.

20. Wittgenstein 1967, no. 109.

21. Ibid., no. 309.

22. Ibid., no. 132.

23. Bradley 1897, p. x.

24. See Pepper 1942.

25. Wittgenstein 1961, Prop. 6.44. He adds "the feeling of the world as a limited whole is the mystical feeling" (Prop. 6.45). Wittgenstein later claimed that the sentence "I wonder at the existence of the world" is a misuse of language (because we cannot imagine it *not* existing). He limited the sense of wonder to only what is *in* the world and saw the experience of wondering at the existence of the world as seeing the world as a miracle. Wittgenstein 1965, pp. 8-10. But it is not clear why wonder should be restricted to questions of the how-ness of things or other matters within the world rather than to their basic that-ness of the world itself, even if we cannot imagine it not existing.

26. Quine 1960, p. 42. Note that this ties what we deem real to conceptual schemes. To Quine, one is ontologically committed to the variables that a theory entails. Quine 1969.

27. Wittgenstein 1961, Prop. 6.41.

28. Edwards 1967.

29. Munitz 1986, 1990. Also see Priest 1995 on "closure" versus "transcendence" and Lawson 2001 on "closure" versus the "openness" of reality. Also see Rubenstein 2008.

30. Munitz 1986, p. 274. Also see Martin Heidegger on "Being" in Cooper 2002, pp. 292-95; Rubenstein 2008: 25-60.

31. The postmodernist Hilary Lawson (2001) sees reality as totally amorphous, permitting us to impose any conceptual scheme onto it, thereby creating different sets of discrete objects. But two dimensions of reality must be distinguished. While *beingness* may be amorphous, there is another dimension to reality—the *structures* that are instantiated in being—and whether this latter dimension is totally amorphous (as antirealists must argue) or has a form we can at least partially discern (as realists argue) will be discussed in the next two chapters.

32. Munitz 1986, p. 278; see Wittgenstein 1961, Props. 6.522, 7.

33. Zombies have become a point of contention in philosophy. Reductionists are forced to argue that zombies are not only physically impossible but that it is impossible even to *conceive* a perfect zombie: if a being is physiologically exactly like us, *it cannot be nonconscious* but must be conscious too. But even if this is so, this does not mean that there is not "something more" to consciousness than material interactions: consciousness simply would accompany the physiology of zombies as it does with us.

34. Some philosophers believe that our belief in a self may be merely a trick of language. We are not fooled by the statement "It is raining" into thinking that there is an entity—an "it"—that does the raining: there is just the rain and its various causes. But we

posit a reality to correspond to the words "I," "me," and "mine" when in reality there is no such thing—no permanent substratum to experience that "has" thoughts and feelings. That babies may have a sense of self prior to the acquisition of language is a problem for this claim.

35. For a neuroscientist who endorses a mind/body dualism, see Wilder Penfield 1975.

36. It is hard to see how one can consistently argue for reductive materialism since a rational argument presupposes choices while reductive materialism reduces everything to a deterministic chain of mindless material events. All ideas are caused by electrical processes in the brain or some other material event, and thus no *reason* is involved. Hence, there can be no rational argument. There is the appearance of an argument, but only physical events are actually involved. Matter determines everything, and thus no real argument is involved, and rationality cannot enter the picture. We have no choice in our beliefs but are compelled by events beyond our control (if that phrase is even meaningful) to believe whatever we happen to believe. There can be no rationally held positions or any freely chosen ones. This does not mean that materialism is wrong, but only that we in fact have no choice in accepting or rejecting it. (Of course, if reductive materialists are correct, accepting or rejecting this point is also determined.) Naturalists who reject a material determinism do not have this problem.

37. See Dennett 1991.

38. What seems intuitive to reductive materialists is "identity theory": mental states and brain states are not related but are in fact *identical*—i.e., when we talk about the mind, we are really talking about physical events in the brain. Neurological and mental accounts are simply two descriptions of the same thing. This approach has not been successful yet for even sensations, let alone other mental events. Moreover, there might not always be the same type of physical cause for a given type of mental event. However, all materialists insist that there is some physical cause for each mental state and each change in a mental state—the mind is simply a physical process.

39. This is based on experiments by Benjamin Libet. Some have suggested these results mean that the unconscious mind informs the conscious mind after it has started the action. Others have suggested that it simply takes time for the conscious mind to register its actions in the brain. To some, these explanations suggest a mind/body dualism.

40. See Wegner 2002. He does not deny any role of the mind in events, only free will.

41. Those who accept life after death may agree that the mind is a natural product tied to the body and thus is not what survives death. For example, the official Catholic theology is that the body evolves and the soul is implanted at conception by God. This transcendental soul is not a natural product, and it survives death; everything natural is external to it and does not affect it. So, if the soul and the mind are distinct, with the cessation of the body and brain functions at death, the mind may cease but something transcendental survives.

42. If parapsychological powers are in fact genuine, then the universe is more complex than is currently supposed, but there may be a materialistic explanation of those powers. However, *near-death experiences* do present a challenge to materialism. If these experiences are what the experiencers claim them to be, then conscious experiences occur when the brain is not functioning. If so, there would be no correlation of brain states and all experiences—indeed, it would be hard to argue that must be a physiological basis for all conscious experiences. Thus, if genuine, they would suggest that the mind is independent of the brain.

43. See Sperry 1985; Searle 1992; Jones 2000, chapter 3.

44. Searle 1992, p. 90.

45.   Stephen Hawking, "The Objections of an Unashamed Reductionist," in Roger Penrose, et al., *The Large, the Small and the Human Mind* (New York: Cambridge University Press, 1997), p. 171.

46.   To use a recent analogy, the mind/body relationship is like that of the software and hardware of a computer: they work together, but one is not reducible to the other. Our first-person reasons for our actions is one explanation (the software); a third-person account of bodily movements is another (the hardware); and both accounts are compatible. But the analogy makes a dualism of consciousness and matter, and thus materialists would reject it. Another new metaphor for the mind/body relation is that of a "consciousness field" arising from neural activity like a magnetic field arising from an electrical circuit run through a wire coil. But electricity does not react back on matter, as antireductive materialists assert for consciousness (since it is part of one causal order of nature). This illustrates the problem of metaphors: they give us new ways to think about things, but if they control our thought they can lead to problems. Metaphors will be discussed in Chapters 6 and 7. In a hundred years, some other technology will probably be the basis for the metaphors for consciousness.

47.   David Papineau, *Thinking About Consciousness* (Oxford: Clarendon Press, 2002), pp. 1-4.

48.   See McGinn 1999.

49.   McGinn 1991.

50.   Francis Crick, *The Astonishing Hypothesis: The Scientific Search for the Soul* (New York: Charles Scribner's Sons, 1994), p. 256.

51.   For the argument that free will remains a mystery even if there is agent causation, see van Inwagen 2002.

52.   Quine 1969, p. 72.

53.   As Wittgenstein put it, to be able to draw a limit to thought we have to find both sides of the limit to thought. Wittgenstein 1961, Preface, p. 3. We are aware of the mystery beyond—it somehow *shows* itself—but, if his view of language is correct, we still cannot say anything about it. Of course, this still makes mystery a "something" labeled "a mystery." The problem with Wittgenstein's theory of language was discussed above in note 14.

Postmodernism and Critical Realism

The natural world has a small or non-existent role in the construction of scientific knowledge.

— Harry Collins

The displacement of the idea that facts and evidence matter by the idea that everything boils down to subjective interests and perspective is—second only to American political campaigns—the most prominent and pernicious manifestation of anti-intellectualism in our time.

— Larry Laudan

A given for contemporary philosophy is that the quest for certainty in our knowledge and values that began with Descartes has failed. There are no rock solid foundations in reason or experience upon which to build claims to knowledge or ethics. There are no incorrigible sense-experiences. There are no *a priori*, necessary, or self-evident truths—all claims about the world are open to revision in light of new experiences. No claims to truth in science or philosophy are beyond doubt; all are fallible. There is no such thing as absolute proof even in mathematics: the proofs of one generation are the fallacies of the next.[1] As Kant's ideas have evolved, all experience is now seen as shot through and through with conceptual structuring supplied by one's particular language and so does not give us the world in a bare form. We cannot distinguish the contribution our language makes and the contribution reality makes to our claims. In Richard Rorty's phrase, there is nothing given prior to our conversations. So too there are no universal standards of reasoning that transcend different cultures and eras. There is no "disembodied reason" or "transcultural rationality" but only different standards of reasoning firmly based in different cultures and eras.[2] Values too depend totally on particular cultures—there is no ultimate or eternal standards of the "Good" or "Beauty" that grounds any values. In ethics, we fall back on convictions for which we can give no final justification. Deciding to be moral is a fundamental value-choice we either do or do not make and no further justification is possible—no complete proof any value-system is possible.

Postmodernism

This situation has spawned two general responses. The more radical reaction is that of the postmodernists.[3] (Because "postmodernist" has taken on negative connotations in philosophical circles, few people admit to the title—they prefer "antifoundationalists" or "antirealists."[4]) They adopt radical skepticism with a vengeance, denying all theoretical knowledge and indeed denying any knowledge at all is possible.[5] Science has no better claim to knowing the world than astrology. Since we cannot access the world free of an interpretation, postmodernists proclaim in a dramatic fashion "the end of philosophy" (in the sense of the end of the search for unassailable foundations and certainty and the demise of *a priori* truths).

Postmodernists utilize Willard van Orman Quine's image of the structure of knowledge as a "web of beliefs"—the totality of what we call knowledge is a man-made fabric that impinges experience only along the edges; a conflict with experience occasions re-adjustments, but the total web is so underdetermined by the boundary conditions of experience that we have great latitude in choosing which beliefs we adjust; no particular experiences are linked with any particular statements in the web except indirectly through considerations of equilibrium that affect the web as a whole. Our statements about the external world face the tribunal of experience only as a whole, not individually. This means that any statement can be held true, come what may, if we make drastic enough adjustments elsewhere in our conceptual system. Physical objects are imported into a situation as convenient intermediaries; they are posits, comparable epistemologically to the gods of Homer; the myth of physical objects has proven more efficacious than other myths as a device for working a manageable structure into the flux of experience, but it still differs only in degree and not in kind from other myths.[6] Experiences are likely to change beliefs along the edge of the web, while the more theoretical and logical beliefs lie deeper within the web and thus are less open to change. But, Quine argued, all beliefs are still open to change, even the laws of logic; there are no sharp divisions between different types of beliefs; in fact, adjustments anywhere to accommodate experiences affect the web as a whole.[7] The unit of empirical inquiry thus is the whole of science (Quine) or indeed even the whole of a culture (Rorty).[8] But in the end, it is we who decide how to accommodate the new experiences—experiences themselves do not determine a belief.

For postmodernists, society is the key to everything. Nothing is free of the context of a particular cultural and historical setting. There is no "human nature"—no fixed, transcultural structures to humans. Who we are—our entire consciousness—is thoroughly shaped by our social environment and especially by our language. Even concepts we use to describe our inner, subjective life are public categories we learn. In Rorty's words, socialization "goes all the way down" —there is nothing "beneath" socialization that defines a human." This applies equally to knowledge. There is no "objective" perspective—no privileged 'God's point of view" or "view from nowhere"—available to us but only particular, limited, social perspectives. The mind, in sum, cannot mirror the world. In fact, there is no "reality" to mirror but only "social constructions" we mistakenly accept as reality. As Friedrich Nietzsche said, "there are no fact but only interpretations." For Hilary Putnam, "objects" do not exist independently of our conceptual schemes—*we* cut up the world into objects when we introduce one or another scheme of description.[10] Different languages are not merely translations of the same labels from different cultures. Rather, they encode different inventories of the objects making up the world. So there is no one answer to the question "what objects does the world consist of?" but many answers, all equally invalid in the end. If we ever reach a transcultural consensus on an inventory of what types of things make up the world, it will because of social factors related to political power, not because of the nature of the world. We think our own ideas reflect reality "as it really is" when in fact they are simply the product of our historical contingencies and nothing else. Indeed, we can believe anything with impunity—reality in no way constrains our concepts and hence our beliefs. We simply adjust our web of beliefs when we meet any unexpected or inconvenient experience. In this way, we decide what is real.

Some postmodernists are also deconstructionists. They discount the alleged cognitive content of beliefs altogether and instead look for social explanations for why someone holds those beliefs. That is, justifications in terms of reasons are rejected as rationalizations of the real causes and replaced by sociological explanations in terms of, for example, which social class interests are really being defended under the guise of "disinterested knowledge." Beliefs, in short, are a matter of power, not reason. The whole modernist "quest for certainty" was only a product of the social and theological chaos of early modern times, and modernity's real inner "will to power" became fully apparent in the imperialism of the modern era and especially in the horrendous violence, commercialism, and ecological damage in the West in the twentieth century. Any claim the modern West had to moral or epistemic superiority was consumed by these events.

This is part of the postmodernists' broader political program. They see the negative aspects of modern life as a consequence of a misplaced optimistic faith in reason as a cure-all. This they see as sufficient justification to reject the modern outlook as a basis for a global reformation and to accept instead a pluralism of cultures and a pluralism of ways of encountering the world. What we need is Paul Feyerabend's "free society" in which all traditions have equal rights and equal access to the centers of power." In particular, science is not the standard for all —B societies to follow: even if it does reveal something of reality, it is only one way of knowing. We should not philosophize about the relationships between such points of view or try to produce a grand unifying scheme. To know the most of reality, we

ctual points of views, not reduce them. (In light of
s, this reduces to a relativism of equally invalid

Postmodernists use a literary metaphor to explain human understanding: "all
is text." (One can deconstruct postmodernists by noting that, since they are acade-
mics, books figure prominently in how they see things—to them, of course, "all is
text.") Reality, history, experience, humans—all are human texts that must be
interpreted. We have no access to a "world" independent of our conceptual
systems, and so, even if it exists, it cannot figure in our cognitive claims. There is
no uninterpreted reality for us to reach—it is interpretation "all the way down." We
may think that we could "read" reality without adding any element of our
own—that our thoughts are a simple reflection of what is there. But we do not see
reality as it is in itself but only as we structure it.

One can see how Kant led to postmodernism: the mind cannot mirror any order
to reality since our minds create that order. There is no "way the world is"
independent of theories, and thus it is meaningless to ask whether our concepts hit
something totally free of conceptualization. Postmodernists, however, go further:
they deny that noumenal reality plays any role at all in our thought. According to
Rorty, talk of it is "purely vacuous" because it does not work. To pragmatists, such
as Putnam, the noumenal realm is "quite empty" since it plays no role in any
scientific inquiry. For all postmodernists, our view of phenomenal reality is totally
amorphous—it is indeterminate and infinitely malleable. There are no "facts" about
reality; we only create such groundless artifices with our own categories.

In sum, think of three elements: reality, the subject, and the act of perception
or thinking. The untutored view is that reality completely determines our perception
and thought; modern philosophers give weight to both reality and our conceptual
structuring in our mental life; but postmodernists move to the opposite extreme and
deny that reality in the end figures in our perception or thought at all. In fact,
postmodernists would say there is no difference between the three elements
because society totally shapes the subject and also determines the view of reality
and hence the perception or thought; any alleged reality drops out of the picture
altogether. In the end, we are all theorists, but there is no reality to theorize about.
Everything is interpretation with nothing left to interpret. We have gone from the
idea that our concepts objectively mirror what is real to the other extreme: what we
accept as true mirrors only ourselves.

This makes everyone an "author." Each reader creates his or her own text—no
"facts" stand in our way. The same with values: there is no objective value-order
to reality. There is no one set of "right" values, no way to compare different values,
and everyone is left feeling justified with their own values. Nothing grounds
values: all values are just arbitrary commitments we choose. Thus, we are left with
nihilism. In science, reality falls away as any type of control on scientists'
theorizing: theories are products of particular social settings and of negotiations
within the scientific community. This is an attack on the actual content of alleged

scientific knowledge. "Facts" are not discovered but are simply our manufactured fabrications. Any Kantian thing-in-itself is beyond language and hence has no referent. Thus, there is no reason to bother with "the world as it really is"—if there is such a thing. We are left with only socially-constructed "texts."

Each conceptual system sets its own standard of truth, knowledge, evidence, and reasoning. Thus, all arguments are ultimately circular within a particular web of belief. All our justification will be within a web of belief: any belief will be justified by the entire web of beliefs, but the web as a whole cannot be justified. Concepts and beliefs cannot relate to any ineffable reality but only to other concepts and beliefs. That is, whenever we try to justify a belief we always end up invoking another belief or an experience already conceptualized within our web of beliefs. Truth is a matter only of statements and thus depends for justification on other statements. As Rorty said, only a sentence can be relevant to the truth of another sentence.[12] The justification of beliefs thus becomes a matter exclusively of other beliefs, and the coherence of our web of beliefs is all that matters. There is no "fit with the world" or "coherence with experiences"—justification is not grounded in experiences or any feature of reality or any logical foundation but only in other beliefs within the same web. Justifications arise only after we adopt a point of view. Thus, there can be no final justification—no reasons or evidence—for holding the point of view itself. In other words, all basic belief- and value-commitments are also ultimately groundless.

Similarly, postmodernists do not think much of the idea of "truth." How could any historically-conditioned language state timeless, universal truths? Instead, there is a relativism of different beliefs, all equally useful in dealing with the world and none a timeless truth. Beliefs are not about "truth" but about our actions in the world and our practices in society. Social conditions determine what survives and thus determine what we accept as "truth." Thus, "truth" becomes merely a reflection of the power of whatever group is in charge of a society at the time. At best, a social consensus, not the world, finally determines what is "true" and what is "false." Reality cannot enter the picture. "Empirical fit" would seem to be part of any web of beliefs, but the postmodernists' claim is that we can sustain any claim we deem important—we can adjust our web to absorb any new empirical findings. No nonconceptual reality, if there is any, can exercise any control on our theorizing, for we cannot compare theories to an unconceptualized reality.

Postmodernists also reject all metaphysics, both in the sense of the assertion of such alleged transcendental realities as souls or gods or fixed "essences" and also in the sense of any broad, encompassing systems that attempt to describe all of reality or "reality as it is in itself" or paint a big picture of how all things hang together. Indeed, there is no "objective world"—no way the world is independent of our conceptual schemes. To postmodernists, all "grand narrative" accounts of the world (e.g., those legitimating science) are suspect. At worst, all such narratives add nothing to our understanding and should be rejected out of hand. At best, none should be treated as the final word on the subject of reality but should compete in

except their own

conversation with others.   1 8.

But despite their rhetoric about metaphysics, postmodernists do not really deny that there is a world or claim that we literally "create" worlds. Some postmodernists claim that they are only speaking metaphorically when they speak of "constructing worlds" or "living in different worlds." But what is it a metaphor for? And without the startling claim, they are not saying anything very remarkable about the mundane fact that different people see things differently and that metaphysical systems conflict. Postmodernists do not deny metaphysical realism—i.e., that if we remove all human perceptions and ideas, some reality would still exists. At least when they are confronted with the issue, they admit that something exists prior to theorizing. As Putnam says: "*of course* there are tables and chairs, and any philosophy that tells us that there really aren't—that there are really only sense data, or only 'texts', or whatever—is more than slightly crazy."[13] The important thing that postmodernists point out is that we simply cannot get around our conceptual schemes to get to that reality in itself. What exists independently of our conceptual frameworks is the Kantian reality-in-itself about which we can know nothing.

Postmodernism and Mystery

One would expect that the idea of fundamental mysteries to reality would play a central role in postmodernism since postmodernists deny any knowability to reality. For them, the whole quest for certainty or truth or knowing reality-in-itself "behind" our conceptual schemes is misguided. The whole noumenal realm is removed from the picture. The how-ness, what-ness, and that-ness to reality are all beyond what we can know. Anything not open to conceptualization is, according to Rorty "gets us nowhere" and according to is Putnam "quite empty." We have a diversity of conceptual schemes that help us maneuver around the world, and that is all. Indeed, postmodernists add a new angle: since everything in the phenomenal realm is infinitely malleable to our interests, even it too must be a total mystery. We can merely adjust the "facts" anyway we like—what we are structuring is beyond knowing.

Yet postmodernists do not exalt mystery. There is no exhilaration at the openness entailed by the alleged malleability of phenomenal reality. Nor do we have the capacity even to speak of a noumenal mystery. Thus, mystery does not figure in postmodern thought at all. Postmodernists could argue that we really know nothing basic about reality and that scientific accounts are, in the words of Erwin Goodenough, only curtains we drop between us and the mystery of reality because we cannot face the truth, or, along with Abraham Maslow, that we have a psychological need to believe we know. But such an attack on realism would involve invoking the mystery of reality and so postmodernists do not employ it. At best, postmodernists embrace mystery only in a very superficial way: it is merely the lack of certainty concerning our claims about the occurrences in the phenomenal world.

Postmodernism must be superficial both epistemically and ontologically since post-modernists see no depth to reality. There may be a sense of learning to live with unexplained events as an intractable part of the human predicament, but there is no sense of deep mysteries underlying the nature of things.

In fact, postmodernists cannot see any mystery to reality at all. We are trapped by our conceptual schemes—we never get to reality itself but can only deal with our own creations. It is pointless to try to deal with any mysteries to reality, i.e., something beyond the control of our conceptual schemes. To postmodernists, all we have are our constructs and we cannot infer what the world is really like from any of them. Indeed, there is no deep structure to reality but only our interpreta-tions, which we cannot get around. There is no openness to reality—all we have are the closures of our various conceptual schemes. We are trapped by our languages: we can have no sense of any reality beyond our conceptualizations, i.e., of some-thing to reality that any language could not capture.

Thus, there is nothing mysterious in reality because all we have are our con-ceptual schemes. How could we even be aware of anything else if we are totally confined by the conceptual? We cannot even express the fact that we are confined within "a form of life" or "language-game" unless we are aware of something more outside it, and that by definition is ruled out. If any mystery did somehow magically try to show itself, we would just absorb it by adjusting our web of belief and thus it would never be a mystery. Our conceptualizations are the only reality there is. In sum, mystery represents the unconceptualized, but to postmodernists everything is conceptual all the way down and nothing is left over. Hence, in the end, there can be no mystery since there is no question of something behind our conceptions.

Critical Realism

The other reaction to our epistemic situation—critical realism—is less radical. "Critical realism" is a loose label, like "postmodernism." It refers to those thinkers who adopt a form of realism about knowledge, science, language, and thought in general: things are what they are independent of our thought, and our statements are true or false independent of what we believe. But critical realists reject any "naive realism," i.e., that perception and beliefs are determined by reality alone and thus all languages encode a straight forward representation of the world-in-itself. Rather, all our knowledge requires our conceptualizing, and this is dependent on a framework of concepts. Science's built-in self-correcting method of checking by others renders the origin of the ideas irrelevant. Social factors influence how science as an institution advances or falters, but they do not influence the component to theories supplied by reality—its empirical content. And unlike postmodernists, realists contend that, despite the presence of social factors in the production of knowledge, our constructs can still give genuine knowledge of reality beyond the manifestations to our senses—in particular, of the unobservable struc-

...order that scientists note in appearances.[14] All our claims to knowledge are fallible and revisable, but the role of doubt in the advance of science does not lead to the radical philosophical skepticism of postmodernism: through empirical testing we can see if our ideas pan out. Also, the presence of possible alternative claims and the pull of skepticism in general does not mean that no claims are better than others. Realists can adopt a metaphor from Karl Popper: the facts of science are like pilings driven into a swamp rather than a solid foundation. This enables science to give us some incomplete and revisable but nevertheless genuine knowledge of reality.

Thus, like postmodernists, critical realists deny what Wilfrid Sellars called "the myth of the given"—i.e., that reality is simply given to us free of our conceptual efforts. Nothing enters our consciousness without being processed. Our concepts structure what is experienced into something manageable. Nor is there any empirical knowledge given by unconceptualized sense-experience. We cannot divide knowledge neatly into separate experiential and conceptual components: there is no way to free the part that the world contributes from the part that we contribute. The two components remain inexorably mixed, but one component to our experiences is fixed: it is supplied by reality independent of us. The conceptual element structures the content, but it does not supply or control the content. Realists can accept the antirealists' insight that we cannot step outside all frameworks. We cannot capture the world free of some element contributed by us—obviously, we cannot *describe* the world independently of using some language to describe it. But to realists, reality constrains what we think. Reality is not infinitely malleable—we cannot "make reality" any way we please. For example, gravity pulls by an inverse square law for animals with different sensory powers or for people with different conceptual systems, not just for scientists with certain beliefs, and there is nothing we can do about it.

In addition, all awareness is not, as Sellars says, a "linguistic affair." Animals and babies obviously have awareness and move about in the world even if they have no linguistic concepts. They also have some type of knowledge and memories. So too, they can feel pain even if they have no concept for it. Thus, none of these states are completely dependent on linguistic abilities. Moreover, language does not totally control perception. Concepts do not always affect perceptions themselves. It is one thing to claim that language *describing* perceptions is theory-laden (even if the theory is low-level) but another thing to claim that *perceptions themselves* are also theory-laden. For example, traditional Hopi Indians have only one term for all "flying things." But does this mean that they *see* insects, airplanes, and pilots as the same thing or see them in any way differently than English speakers do? Or consider conflicting scientific theories: Copernicus did not *see* something different in a sunset than did Ptolemy. It is not comparable to *seeing* a duck or a rabbit in the Gestalt figure. Adopting heliocentrism does not change our perception: the sun still seems to move, and the earth still seems to stand still—our perceptions are still structured, but the new cosmology does not penetrate them. Postmodernists must do more than simply note a role for concepts in our knowledge to conclude that we actually *experience* the world in different ways, let alone live in "different worlds."

Similarly, our concepts are involved in knowledge-claims, but reality is not an amorphous lump we can crave up anyway we want. For realists, conceptual systems are not arbitrary: some systems make sense and some do not because of the nature of reality. Postmodernists think any concepts will do—any cookie cutter works. All terms are like a term we could invent that groups together only bread, mailboxes, and quasars. But such an artificial category does not illuminate anything about reality precisely because of the way reality is structured. Our thought is not a "perfectly arbitrary imposition" upon a virtually unknown reality, as the nominalist John Locke thought. In addition, once we accept any conceptual scheme, reality determines results. That is, our concepts are socially-created tools, but they can state objective facts about reality. The category "trees" may be something we have made up and not reflect a natural division to reality and thus not be part of the ultimate furniture of the universe, but once we adopt that category reality determines, or at least vastly constrains, our options in conceptualizing "facts." Thus, for example, there are no trees growing in the room in which I am typing at the moment—that fact is not a social product even if the categories are. Our language thus says something true, even though it is our language and not a system dictated by reality. The same applies to the term "self." Our sense of self may be socially-constructed. We may be, in Daniel Dennett's phrase, "a web of words." But, as Dennett realizes, there are limits. I cannot spin a "story" in which I am a pitcher for the New York Yankees and still claim to be sane—the facts simply stand in the way.

Truths can only be stated in cultural forms, but that does not mean that the content is contingent upon the forms. If we had eight fingers rather than ten, our numbering system would probably be in base eight, but that does not in any way affect the number of things actually counted. We devise the measuring system, as it were, but reality provides the measurements. Postmodernists may say that light does not really travel at around 186,000 miles per second because "miles" and "seconds" are only our social creations—there are no "real" units—and thus the speed of light is only a matter of social convention. But that is wrong: what categories we choose for distance and time will, of course, be social creations, but the speed of light is still what it is and it can be measured in any categories we like. We cannot make light go faster or slower by the categories we choose—the categories in no way affect the issue. We know the real speed of light even if we have to use our categories to express it. Not all differences between systems of description are simply like the difference in measuring systems, but the point remains that, while our concepts are necessary to knowledge-claims, they in no way determine the answers reality supplies. As Hilary Putnam puts it, we cannot talk of facts without specifying the language being used, but the facts are there to be discovered, not legislated by us.[15]

Obviously, we cannot describe something without adopting a framework of understanding. Moreover, reality does not dictate a point of view. There is no one transcultural, timeless point of view but only more limited points of view reflecting particular interests and values, one of which we have to be operating with whenever we think or see. Scientists too must adopt a point of view: reality is examined only through the filter of a particular theory that scientists construct. But we do not need "God's point of view" or to be detached from our context to gain real knowledge of the world. There may be no one neutral account of reality, but there are still true (and hence also false) claims to be made. Our points of view may be limited and conditioned by social and physiological factors, but they can still reveal features of reality. We need not transcend our historical situation to make genuine knowledge-claims about reality. Our points of view may also be contingent on historical facts and may reflect certain values, but for postmodernists to speak of them as "biased," "subjective," or "prejudiced" distorts this situation badly: although we must adopt a point of view, a perspective can still lead to genuine knowledge. Our all-too-human, conditioned, finite points of view can still enable us to know aspects of reality. In sum, realists contend that the culturally-conditioned questions can lead to answers that are not determined by cultural beliefs.

Realists, unlike postmodernists, speak of truth, of getting something right about the world. There can be conflicting claims about the same thing, and some claims are wrong. For example, either the earth really is flat or it is not; the truth or falsity of the claim is not up to what we believe or our categories. In addition, just because our concepts are involved does not mean that every conceptual scheme is equal—reality determines otherwise. There is the element of coherence of beliefs in the justification of what we believe, but we cannot eliminate the element of correspondence with what is real; coherence in the end is relevant to how we justify claims but not to the question of what is true. Reality has a causal influence on our beliefs even though naive realists are wrong in believing that this is all there is to the issue. Thus, the claim "the earth moves" is true because the earth moves, regardless of what we believe or our point of view. It is not a matter of consensus—we were wrong in believing the earth does not move despite thousands of years of consensus. Nor is it merely a matter of adopting a particular web of beliefs. It is true for everyone even though we must adopt a certain way of speaking to express it. True statements obviously do involve our formulations—they are *statements*, after all—but the world determines their truth, and our conceptual nets hit some genuine features of reality. Thus, that the earth moves can only be expressed in a statement with certain concepts, but this does not mean that we are not stating something that is true. To make a true statement, we must adopt some point of view, but the statement will be made true by the nature of reality and will then be true for all people.

However, we cannot measure "verisimilitude" or the "degree of similarity" to reality because we cannot see reality from the outside. We may be "converging on truth" in the sense that we are learning more and more about reality. This means that many current statements in science may only be "limited and approximate

truths" and never the "final truth." Scientists do not sp
but as "well-tested" or "well-supported" or "getting at the truth." Nevertheless, our
true statements tell us something about reality, even if there is more to reality to be
learned or things that we will never learn. The problem is how can we be sure of
anything, since statements involve our conceptualizations and they are always open
to further revision in light of new experiences. But the idea that we cannot claim
to know something unless it is transparently true or otherwise absolutely certain
and final is simply false. So too we may be shown later to be wrong about some
claims to knowledge, but that does not mean we do not know other things. And
unless we know everything there is to know of reality, we humans will remain in
the awkward position of never knowing what are the things we actually know and
what is wrong in our claims to truth.

Realists thus, unlike postmodernists, can provide a reason to study reality even
though our knowledge is always subject to revision or even rejection. They espouse
not only a metaphysical realism (that the world exists independently of our ideas)
but a scientific realism (that there are real structures to reality and scientists are
discovering facts about them). Scientists are not simply relegated to correlating
observations. Scientific theoretical claims remain our guesses at the causal workings of reality that are responsible for what we observe—what entities, processes,
properties, and states exist independently from their empirical manifestations to us.
To use Einstein's analogy, nature is a closed watch: we can see the face and the
hands moving and can make guesses as to how it works, but we can never open the
casing to observe the workings and so we cannot compare our picture to the real
mechanisms. Our explanations—our theories—may remain incomplete and
tentative, but realists believe that there are definite structures inherent to reality that
shape our knowledge-claims. At least some of these structures are intelligible to us
and await our discovery. Thus, something like the structures modeled in theories
actually exist, and scientists are, on the whole, learning more about them. This
means that some of our posits are better than others in our attempts to know a
reality independent of our linguistic creations. Realism means that the world-in-
itself somehow interacts with us and constrains our ideas of what we deem real.
Even if our knowledge is only of a Kantian phenomenal realm arising from the
uniquely human way of experiencing the world, we still have genuine knowledge
of the world independent of our creations.

According to realists, we formulate questions for reality—questions which
must be the product of our concerns and thus will be social creations—but reality
supplies the answers or at least constrains how we can conceptualize things. There
is no way we could conceptualize all the data gathered on the nature of the earth
and still produce a flat earth theory—there simply are no edges to sail off. Reality
determines that, and, to use Quine's image, no amount of "adjusting the content of
the web of our beliefs" will change that fact. (Quine's overall thought, incidently,
is much closer to the positivists' than the postmodernists'. He thought that we
could reform our web of beliefs in light of new discoveries, not that we could

change it on a whim or make up anything we wanted because reality is infinitely malleable.) When encountering new data, scientists do have to make decisions on how to understand them (and thus how to conceptualize them), but scientists do not have the free rein that postmodernists seem to think. Scientists from another planet may well have a very different science, since the questions its scientists ask will reflect both their own concerns and any differences in physiology (e.g., they may see a different range of the electromagnetic spectrum).[16] But if they end up asking the same questions we do, they will end up with the same answers our scientists do—assuming that our scientists have been getting the right answers. Thus, if they are interested in gravity, they will end up with Newton's inverse square law for the pull of objects on each other and eventually with Einstein's theory.

Similarly with history: just because there is no study of history without interpretations does not mean historians can defend any position. We cannot "adopt a form of life" or "reconfigure our worldview" in a way that will change the fact that Julius Caesar lived before Abraham Lincoln. Historians in each generation may come up with different theories of the causes of historical events and thus "history as explanation" changes in every generation— in that sense, all history may be contemporary history, as Bendetto Croce said—but the events themselves remain fixed signposts (assuming historians have identified them correctly). Doing history is not like Stalin doctoring photographs to fit his current political needs. Similarly with literature: how we read Shakespeare today will depend on our frame of reference, but that does not mean that he did not mean something specific—we may have to study Elizabethan England to understand him or we may unfortunately no longer be able to understand him fully, but that does not change this fact. Indeed, it is not at all clear why anyone would bother to read Shakespeare if we simply make him conform to our ideas and can make anything we want out of his work. There must be something "objective" to it—and if the same applies to reality, as postmodernists claim, it is not clear why anyone would practice science.

Realists can argue that we can transcend our cultural limitations enough to see other points of view: languages and cultures may present points of view for our thinking, but we can transcend the limitations that they seem to impose. In fact, how else would we see problems and devise alternatives? How would scientific revolutions ever occur if old theories totally controlled scientists' thinking? How could Johannes Kepler get past the prevailing idea that only the "perfect" form of motion (circular) could exist in the heavens and come up with the idea of elliptical planetary orbits? Or consider something as basic as Euclidean geometry. It may be "common sense" in the West, but mathematicians were able to devise non-Euclidean geometries. More generally, we can hold as problematic what we had previously accepted as obvious. Indeed, we can recognize that we are in fact holding a specific point of view only if we can transcend it. As Ernest Gellner said, "[t]ranscendence of custom is the most important fact in the intellectual history of mankind."[17] We are not prisoners of old points of view; we can at least partially abstract ourselves from our cultural situation; it is not a matter of destroying our prior ideas but a matter of not being held rigidly by them. Thus, if naive realists operate with "the myth of the given," then postmodernists operate with, in Karl Popper's phrase, "the myth of the framework." The world thereby can still exercise some control over our changes in conceptualization and provide grounds for asserting one belief over another. There may be no non-circular way to justify *all* beliefs in our "web of beliefs," but circles get less vicious as they get larger and incorporate matters we have no reason to doubt. We can then legitimately speak of justifying particular beliefs and not limit rationality to different standards operating within different webs.

Language for realists, as for postmodernists, plays a central role in our thought. But realists believe concepts do not totally control how we see the world—we can transcend the limitations of language. That Plato and Aristotle, Hume and Hegel, and Newton and Einstein all used Indo-European languages and yet came up with very different views of what is real certainly suggests that language does not control thought. We can also revise a language in light of the way the world really is: we can see both the language and the world, step back, and revise the former as needed. Some statements in some languages may not be translatable into other languages. For example, the ancient Egyptians' only word for "to go north" was "to go upstream" (the direction in which the Nile flows). So when their armies first encountered the Euphrates (which flows south), they had to describe it paradoxically as "that circling water that goes downstream in going upstream."[18] Note two things. First, the Egyptians' perception was not controlled by their language: they could *see* the Euphrates flowed south ("upstream")—they could experience the dissonance that reality did not conform to their concepts even though they could not express it in their conceptual system without paradox. Second, the statement they used was not true in one language and *false* in the other—the truth is simply not expressible without paradox in one of the languages.

Realists can also accept that reality is continuous and constantly changing while our terms are discrete and static. Thus, language cannot "mirror" what is real. But the continuity in the flow of reality is great enough and the changes slow enough that terms are useful and can be learned. The impermanence of reality certainly does not mean that words have no meaning or can be used interchangeably—"car" and "person" still designate different parts of the world even if all of reality is impermanent and interconnected. Critical realism does not require a one-to-one mapping of distinct concepts onto discrete aspects of the world but only that sentences tell us something true about the world. Unlike the early Wittgenstein, realists need not believe that words simply label what is real and that the structure of reality is essentially linguistic. Languages make distinctions we can use to state true facts even if there are no permanent isolated referents in reality. The words "Hudson River" designate something in reality even if the "shoreline" and the molecules in the "river" are constantly changing. Even if there are no discrete entities in reality to designate, nevertheless the sentence "The Hudson River is in the eastern United States" tells us something true about the world while the sentence "The Nile River is in the eastern United States" tells us something false.

In addition, claims at the edge of scientific research may change as scientists explore more, but ideas deeper within our webs of belief are not likely to change, and noting that in principle they *could* change is not grounds to reject them. Too much evidence seems to point most easily to the idea that the earth is revolving, orbiting the sun, and also moving in other ways with the solar system and galaxy to believe that the earth does not move. Some enterprising young astrophysicist might be able to come up with a new absurdly complex Ptolemaic version of the universe in which the whole universe swings around a stationary earth at the center, but that is not a reason to question the belief that the earth is moving.[19] Just because convoluted alternatives are logically possible does not mean all are equally likely. There is no certainty here—we cannot get outside the universe to see from a fixed framework—but this still does not mean all theories are equally plausible since other considerations are involved in theory-choice (as discussed below).

To realists, the entire idea of "the social construction of reality" is wrong. At best, it is the social construction of what a culture accepts as real. We create the concepts we use to navigate through reality—hence, our "conceptual worlds" are socially-created—but we do not in any sense create the reality we encounter. Postmodernists distort the situation by improperly conflating language and reality to conclude that "the real world" is constituted by social and specifically linguistic realities. Nor do we in any sense "create realities" by coming up with new concepts.[20] Scientists did not create "dinosaurs" when Richard Owen coined the term in 1842 as a new way to classify some extinct animals—if dinosaurs existed, they lived millions of years before and did not depend in any way upon our concepts for their existence. When modern astronomers revised their view of the heavens, they did not destroy "stars" and create "planets." Postmodernists may see something profound about our conceptual systems and what we consider "knowledge" in astronomers reclassifying Pluto today as less than a full "planet," but realists point out that nothing changes in reality or in our knowledge of the nature of Pluto. The question "how many planets are there in our solar system?" is affected, but classifying Pluto as, in effect, a "stone" or a "pebble" does not affect science. At best, postmodern claims about concepts are only a very convoluted way of saying merely that modern scientists made up new concepts for phenomena: obviously no concepts exist independently of us, and so no reality would have been classified as a "dinosaur" without us. Realists do not deny the obvious—that the concepts are human inventions—but for them this does not foreclose the issue of whether the new concepts in some way capture a feature of reality we previously had not noticed. Concepts are social and mental in nature, but their referents in the universe are not (except, of course, for social and mental realities). That our knowledge remains *our* knowledge is not denied, but its content is not constituted by social factors.

Realists are more receptive to the idea of mystery than are postmodernists. Postmodernists, with their emphasis on the lack of knowledge, should make mystery a central feature of their program, but, as discussed above, they do not even consider any possibility of a reality apart from our social constructions. Realists, on the other hand, make claims about a reality apart from our conceptions and thus can affirm a mystery to a noumenal reality-in-itself apart from our explorations through different perspectives. Realists can assert either that reality-in-itself is somehow mirrored in the phenomena of reality-as-it-appears-to-us as their cause or that we have no idea if the noumenal world is at all like phenomenal reality. They may also accept that there may be truths about the workings under-lying the phenomena of nature that will forever lie beyond our ability to know and thus that how-mysteries within our phenomenal realm may also remain forever.

In any case, that reality has fixed, real structures and is not totally amorphous is affirmed. And as long as the possibility of truth is a consideration and we have a sense that there is "something more" to reality than we have captured by our con-ceptions, then mystery is part of the picture. Thus, realism leaves an opening for mystery that postmodernists cannot accept, since according to them our cultural conceptions completely control our perceptions and thoughts. Nevertheless, con-temporary realists also do not make mystery a major topic—the scientific quest to remove mysteries is. Thus, they too do not emphasize mysteries.

Problems for Postmodernists

Until postmodernists are willing to admit that reality plays a role in determining our scientific and everyday knowledge, that not all conceptual divisions are arbitrary, and that some beliefs about the world are true and some false, there is a substantive dispute between realists and these skeptics and relativists.[21] But postmodernists have two major problems. First, they can make no claim to truth in any universal sense. They cannot argue that their skepticism or relativism is "true for everyone." To do so would be to admit that there are in fact truth-claims that are dependent for their formulation on particular cultural contexts but are nevertheless universally true, independent of specific cultural or historical contexts or what beliefs people happen to hold. The claim "all is socially-constructed" becomes only their own second-order claim about first-order knowledge-claims; it amounts to no more than the belief of their own group and does not hold "true" for all.

Of course, postmodernists are perfectly happy with this result: we are left with an epistemic relativism of beliefs, with truth no longer in the picture. But how post-modernists could recognize even this assertion as *true* is problematic. Post-modernists seem to believe their position is *correct* and others' positions are *wrong for everyone*. Thus, they have the dilemma that anyone who denies universal truths has: their claims—"all is interpretation," "there are no transcultural truths"—all sound like truths being asserted for everyone, not just "truths" for those who adopt a postmodern stance. How postmodernists cannot contradict themselves is not clear. Moreover, postmodernists are in the self-defeating position of not being able to defend their claims: all beliefs are determined by social factors, and thus reason and evidence have no place to operate. Postmodernists cannot advance reasons for adopting their position to outsiders since reasons only operate within conceptual schemes, not between them. One must already be inside their web of beliefs for their arguments to make sense. Nor is it clear how anyone could have adopted the postmodern point of view in the first place since by their own admission we all would be trapped in our prior point of view—our previous perspective should have completely controlled our beliefs, and so we could not see any alternatives. In the end, postmodernists must deny that freely adopting such a position is even possible: we are caused to adopt postmodernism by social or other factors and do not freely choose to accept or reject it "rational" considerations.

The problem with truth may not dent the postmodernists' self-assuredness, but there is a second problem. If all our concepts are simply social constructions and all justifications of belief can appeal in the end only to other beliefs within a cultural system, how do we anchor a web of beliefs anywhere in reality? How do we get out of a circle of our own constructions to the real world? If a term's reference is only "a sociological matter," as Rorty claimed, then there is a problem. As John McDowell says, this is a "frictionless holism" cut loose from any external constraints. Realists can admit that the *meaning* of terms may be inextricably tied to the meaning of at least some other terms within a language or a scientific theory, but postmodernists go to the extreme of claiming that the real unit of meaning is a language as a whole. And if meaning determines *reference* then we are hopelessly trapped within a circle of concepts with no way to connect to the world. Semantic relationships are only among concepts and not between a concept and the world—any relation to reality is impossible.[22]

But obviously our language does help us move around reality. How can postmodernists explain the simple fact that language works at all? They remain trapped within their webs of beliefs with no way to connect language to reality. Pragmatists believe language does not involve making representations of the world but functions only as a tool for getting around the world—the interaction between our linguistically-driven behavior and our environment is the end of the story. Still, they have no explanation for how language "hooks onto" the world to work in the first place. Why do some vocabularies help and some do not if all are interlocking webs? Postmodernists end up with an air-tight, interlocking holism that makes the entire idea that language works in the real world appear miraculous. Realists can invoke the idea mentioned above that language refers to impermanent but fairly stable configurations. In any case, if the world were very different from what it is, our conceptual frameworks would also differ, but postmodernists cannot explain why this would be since to them conceptual systems make references only to themselves. In short, there is something wrong with the postmodern theory of language if it entails that we cannot do what we obviously can do.

Postmodernists cannot present a viable alternative to realism until they can somehow anchor the webs of belief in reality—but any attempt to "mesh" such free-floating webs with reality will end up as a form of realism. In the end, they have to deny that reality plays any role in our web of beliefs. Thus, there is a fundamental dilemma that sinks the postmodernists' position, even if they cannot express this dilemma from within their webs of belief.

The Problem for Realists

Realists have one big problem that postmodernists think is insurmountable: scientific revolutions. History is full of examples of people being certain about something (e.g., that the earth is flat) only to be shown that they were drastically wrong. When Henri Poincaré surveyed the history of science, he saw only "ruins piled upon ruins" and thus the "bankruptcy of science." Certainly, a consensus or our sense of certainty at any stage of history is not a sure sign of truth. Scientific revolutions do not preserve all of the insights and findings from older theories but radically reconceptualize our view of things. Newton might be a "good approximation" of Einsteinian physics for practical purposes, but this cannot get around the fact that Einstein provided a radically new view of the nature of reality, any more than a road map treating the earth as flat is an "approximation" of the theory that the earth is a globe—on the structures of reality, Newton got it wrong. If we have worldview-shattering changes like the switch from the Ptolemaic to the Copernican cosmology, aren't all our scientific beliefs in danger of being overthrown? Do we have any reason to believe that "electrons" and "genes" will not go the way of "phlogiston" and "the ether"? In light of such revolutions, how can we be sure that we have any definite knowledge of the nature of reality? Also remember that the best theory at any time is only the best among current competitors—how do we know that some better theory will not be coming along later? Certainly that quantum physics and relativity are currently incompatible is good reason to conclude that we do not know all that is going on on these levels and thus is a solid ground to reject realism here. Don't we have every reason to expect that *all* our best current theories will eventually be replaced by more comprehensive theories with fundamentally different conceptual frameworks? Why should we think the current conceptual tools are final? Isn't it just hubris to think we genuinely *know* anything today in any of the sciences?

The moral that postmodernists draw from this is that science is not progressive: there is no accumulation of facts or advances in our understanding with each revolution but merely changes in the questions scientists ask. With each revolution, new data are not added to an established wall of knowledge—rather, the wall is redesigned, down to remolding the bricks (the "facts"). Theories determine what is relevant and thus what gets described and how it is described. As Thomas Kuhn said, there is no theory-independent way to reconstruct phrases like "really there."[23] Theories thus determine what is accepted as the "facts." As Nelson Goodman puts it, a theory is a large fact, and a fact is a small theory. Scientists literally construct the world by determining what the "facts" are by negotiating among themselves what is collectively accepted as the prevailing theory of a subject for the moment. In the end, science is circular: the web of beliefs determine the "facts," and the resulting "facts" justify the web of beliefs. The world drops out of the picture, and only the community of scientists remains. Science changes like Darwinian evolution: there is no purpose or goal—a "final truth" about reality—but only random changes adapting to more complex findings. There is no "true adaption" in nature or "true theory" in science. The bottom line is we are not gaining more knowledge of reality but just randomly hopping to new questions. Scientists are good at coming up with new ways to solve puzzles they themselves devise, but there is no sense in which we are closing in on a more complete picture of a reality independent of each new linguistic system. Science is best seen as providing us with useful knowledge for getting around in the world—it only gets into trouble when scientists try to delve into the workings of reality underlying appearances.

Empiricists in the postmodernist camp are impressed with the predictive accuracy and mathematical precision of scientific laws, but they reject the idea that the theorizing about unobservable structures and entities to explain why the predictions are accurate is providing us with any "true" picture of the workings of reality. Rather, theories "work" by being empirically adequate. Their admonition is that scientists stick to correlating phenomena and stay out of metaphysics. We cannot go beyond what our senses tell us.[24] In the words of Bas van Fraassen, we have no valid clues into the hidden structure behind phenomena, and realism is an unwarranted and unhelpful intrusion of metaphysics that adds nothing to science.[25] Empiricists remain agnostic about the causes of phenomena—something in reality causes them all right, but we do not know what. We cannot know what we cannot experience. In science, it does not matter if "electrons" are real as long as the electron model works. Science is ultimately just a matter of correlating pointer readings, not insights into underlying causal mechanisms. After all, every eclipse for a thousand years before Copernicus "confirmed" the Ptolemaic picture. Such explanations ultimately go beyond what can be observed or tested into the metaphysics of what is behind appearances and thus should be rejected as an unnecessary appendage to science. Such theories are just our endless attempts to reinvent the world. On the other hand, a pluralism of empirical approaches should be encouraged to expose more correlations of phenomena.

But scientific revolutions pose a problem for postmodernists too. There should be no way empirical anomalies can get through the web controlling perceptions, and thus there should never be a reason for changing theories. According to Kuhn, since each theory defines a different "world," scientists with different theories are practicing in different worlds.[26] Scientists disagree on what is a problem and what is a solution. Thus, no reasons can be offered to one's opponents for why they should change their position, and a scientist's change in theory is thus a Gestalt-like switch—a conversion.[27] But the switch is ultimately without reason since there is no common rationality between theories to use. Indeed, there can be no reason for any scientific change at all since each conceptual scheme controls our understanding and scientists will simply adjust their existing webs of belief to accommodate every new finding. Scientists may change their theories if they like, but there could be no "scientific progress" in any sense. Scientists would just be jumping from one "world" to another totally arbitrarily and literally for no reason.

If empirical consequences were all there is to science, different theories would all be equal.[28] Data do underdetermine theories, but realists insist that by using the theory-criteria scientists have developed they can in fact identify some of the workings in nature that produce the lawful, observable events. This is where the strength of rationalism comes in. Scientists must invoke criteria other than "compatibility with experience" to select the best available theory. This involves adopting conventions such as the belief that a simpler theory probably reflects reality better than one slapped together. This legitimately imports metaphysics—something "beyond" the empirical considerations— science. Kuhn himself presented a set of such internal values: empirical and predictive accuracy, self-consistency, scope, simplicity, and fruitfulness for new research topics.[29] That is, empirical findings remain central, but when it comes to choosing among competing theories having equivalent predictions, factors such as the simplicity of what is postulated as the workings of reality rightly matter in determining which theory to accept. According to realists, these criteria are not simply our conventions; they help us reflect reality in our theories. According to postmodernists, these criteria are totally subjective. For example, why should we think that beauty—what is aesthetically pleasing to beings like us—should reflect the nature of reality? But that all such criteria are *metaphysical* assumptions within science does not make them arbitrary or subjective. The extra-empirical criteria are always utilized with empirical data, and in that context they lead to new predictions that are sometimes confirmed and are rejected when they are repeatedly are not; thus, testing supports the criteria as leading to new information about reality. But postmodernists contend that it is only a *coincidence* that the subjective criteria worked and that the confirmation had nothing to do with them. But according to realists, reality always has a check upon our theorizing.

If postmodernists are correct that external, social causes are responsible for theory-change, then indeed such change is irrational and Paul Feyerabend was correct in speaking of "brain-washing" when some scientists convince others to switch theories.[30] But for realists, scientists can rationally decide that one theory is superior to another in depicting reality and thus that there is genuine progress in science as scientists learn more about a phenomenon. If we use the metaphor of evolution, then science is Lamarckian, not Darwinian: science advances, not by random "mutation" alone (i.e., guesses), but by incorporating advances in fitness to reality that scientists devise in each generation. However, as Kuhn pointed out, these criteria do not form any sort of algorithm for theory-choice.[31] Choices still must be made. But that some such reasoning is required does not mean that it is necessarily arbitrary and cannot lead to finding the workings behind appearances.

The Case for Realism

As a rule, working scientists believe they are hitting the causal processes of reality even if they do not have complete knowledge of them.[32] Realists can also counter postmodernists' basic points. They can agree that all evidence and reasoning is evaluated from within points of view. We cannot test a theory against the world "as it is in itself" since we cannot get in between a theory and the world to see if they correspond. Nor is there a neutral conceptual system in which to compare and evaluate all claims. With a change in theory, all theoretical terms change part of their meaning. But there is a fundamental problem here for postmodernists: if, for example, Ptolemy and Copernicus meant different things by the word "sun" and were in fact referring to different realities, then why couldn't we hold both the Ptolemaic and Copernican theories at the same time? After all, under this view, they were talking about apples and oranges, and thus their views on one subject can have no bearing on their views on the other. But the fact remains that theories like the Ptolemaic and Copernican obviously do compete in some way or we could hold them both at once. In some sense they are talking about the same subject. Even Thomas Kuhn in later clarifications said that he did not mean to suggest that incommensurability between theories meant that they could not be compared.

In addition, even if perception is "theory-laden," it is not necessarily laden with a theory in dispute. Instead, it may be structured with a more general and common level of structuring. For example, different viewers of the Necker cube can still discuss how long the lines are even as they dispute the total figure. So too, Ptolemy and Copernicus can both agree the sun is a "celestial orb" even if they radically disagree on its nature—their speech about the "celestial orb" would then be *theory-neutral* for their dispute, even though still not *theory-free*. How else can we explain the fact that Johannes Kepler, a Copernican, could use data from Tycho Brahe, who held a modified Ptolemaic theory, to devise his theory of elliptical orbits? The theories are genuine competitors about the same subject, and any "incommensurability" of conceptual systems, if it exists, is irrelevant.

That scientists can make novel predictions that sometimes come true—indeed, that experiments can produce consistent results at all—suggests that scientists have hit something in reality causally responsible for our observations even if theories of the nature of what is hit change as scientists conduct more studies. Postmodernists, however, remain unimpressed. Scientists may have hit something, but what? What continuity of reference remains through theory-change at least in the unobservable subatomic realm where theory rather than direct experience predominates in our understanding? Theories dictate, not merely the meaning of terms, but their referents; so with a change in theory, there is necessarily a change in what is referred to. However, realists can still argue that the fact that we can manipulate the subatomic realm means our terms like "electron" are designating something real, especially if the same theoretical term figures in different types of experiments, even if our understanding of the reality changes radically as we learn more. We do

how can POMO's know?

learn that sometimes theoretical terms do not in fact refer to any reality (such as "phlogiston"), but in other cases as our understanding expands the entities referred to by old terms are still believed to exist and the terms become more established. Earlier data become incorporated and explained.

Realists also point to another phenomenon: the world constantly surprises scientists. We cannot force reality to produce what we want. Not only can scientists not force reality to do whatever they want, reality sometimes hits them with the unexpected when making new observations or experiments. If our web of beliefs controlled our expectations, as postmodernists argue, how could that be? If our theories determined the "facts," no theory would ever fail. A prevailing web of beliefs would absorb all blows. The most that postmodern scientists would need to do is to revise existing theories with endless epicycles to save their web of belief by readjusting the "facts." But realists take realities' recalcitrant anomalies as unexpected new discoveries that may be a reason to replace entire theories. If surprising empirical findings stubbornly resist existing theories, they force us to rethink our ideas—we are aware that our ideas have failed. In many cases, reality stubbornly resists our efforts to make it conform to our theories.

Anomalies also bring up another point about science: reality does not dictate how we conceptualize it. For example, when the Michelson-Morley experiments on the "ether" did not comport with expectations, scientists had to decide what to do. Henri Poincaré, for one, believed we could not give up Euclidean geometry under any circumstances, but scientists found a form of non-Euclidean (Riemannian) geometry to be helpful. To postmodernists, this means we decide if space is curved or not. To realists, it means that reality only gives scientists clues—we have to figure out how it works. But when reality is consistent in the answers it gives to our questions, and when scientists can approach the same phenomenon from different angles and get consistent answers, it is reasonable by any standard to conclude that scientists have hit some structure underlying appearances. Thus, even with leeway in theory-construction (since we do not know everything), reality still constrains theory-construction in a way postmodernists reject. The literary scholar Stanley Fish likens scientific theories to the rules of baseball.[33] But while we can easily imagine umpires changing the strike zone each season, it is harder to imagine physicists changing Newton's inverse square law for gravity. No one could make a plausible case for such a revision, even if imaginative physicists could come up with convoluted equations for it. Pragmatists would say that no one would bother to try to do that because the puzzle of gravity has already been solved. But realists would reply that the reason the puzzle has been solved is precisely because gravity really obeys the inverse square law—physicists have discovered something real in the workings of nature. Theories that are "pragmatically progressive" by increasing in their problem-solving capacity are so because they reflect more of reality. Reality is not infinitely adaptable to our ideas, and our theories are not a matter of this season's fashion.

An accurate prediction does seem to indicate that scientists are on to a structure of reality. Realists see realism as the best explanation of the accuracy of predictions and the correlation of data—indeed, even Hilary Putnam says that realism is the only philosophy that does not make the success of science a *miracle*. Without structural realism, why would methods reflecting human interests and values lead to successful predictions? Of course, postmodernists will remind us how long the Ptolemaic theory prevailed: the predictions of eclipses may have shown that scientists were on to something, but the cosmology they used to explain them was dramatically wrong. They note that in the final analysis most scientists have been mistaken most of the time. What could make us think that scientists at this particular moment have finally produced "the true picture" of any aspect reality? Postmodernists believe that, if history is any indication, all current theories are likely to be false at least in some respects and will eventually be refuted.

Realists can readily admit that scientists have made glaring mistakes in the past. But that does not mean that there are no real structures to be found and that later scientists have not gotten onto the right track eventually. Our conceptual frameworks initially shape our scientific quest, but nature can readjust our frameworks and affect the quest: its answers to our questions can affect what questions we ask next. With this feedback from reality, we can identify the mechanisms at work in reality with ever-increasing accuracy. Copernicus got some major things wrong—the sun moves and is not near the center of a small universe, and the planets do not have circular orbits—but he made a major step in the right direction. His advances were not simply a matter of mathematical simplification but about the structure of the cosmos. Scientific advances are always like that: they are "closer to the truth" or "more accurate" but seldom are the last word. There may always be more to learn about anything, but to say that "all theories are wrong" or "there is no scientific knowledge" distorts this situation.

However, theories do remain our guesses at the workings of reality, and reality responds to our inquiries without telling us what is true or what questions we should be asking or what we should ignore. Postmodernists can add a related problem: theory-choice is always a matter between available theories, and even if we can tell which of the rivals is better we cannot tell if a better one may not eventually come along. Rarely scientists reject all competing theories in light of empirical evidence; usually they take sides. If it is usually a matter of "I'll believe this until I hear something better," how can we know if any of our conceptions are ever *true*? How can we conclude anything is true or false if all the evidence is not currently available? Even the most confirmed theoretical accounts may turn out to be incomplete or even false. In short, even if we have great grounds for asserting something now, how can we be certain that we *know* anything? Our limited perspectives may mean that we will never have anything more than approximations of the true structures of reality. Nevertheless, realists insist that we are getting at those structures, if not the full story. We have genuine, if incomplete, knowledge of the real structures of reality and are gaining more.

Postmodernists believe that the history of science compels us to accept some form of antirealism, however counterintuitive it may seem. Some form of agnosticism about the existence of theoretical entities is at best all we are entitled to accept. Postmodernists are correct in saying that we can never be certain we are on the right track, but they cannot take solace in the fact that scientists typically do not say that an accepted theory at the edge of research is "true"—they usually say "it's interesting" or "it's good" or "it's the best we have." Scientists may still remain realists who believe the theory has hit upon some of the structures of reality, but they know that our knowledge is imperfect and so more is at work then we know. They can be more tentative, open, and less firmly committed to the conceptual side of their theories—even skeptical of the theories they use—than postmodernists seem to realize, and yet they are still realists. (Some philosophers purpose adopting realism toward theoretical posits on a case-by-case basis, since the frontiers of quantum physics may not be representative of all science.) All knowledge may be fallible, but this means only that we should hold the claims tentatively, not that theories are necessarily off target.

It is certainly reasonable to claim that theories that have been battle-tested have earned the right to a critical realistic claim. The entities in such well-confirmed theories are theoretical in the sense that they are not observable, but if there is evidence of them from many independent tests then we should accept them—i.e., something like the posits of these theories should be tentatively accepted as actually existing. Quarks may go the way of phlogiston if physicists try other approaches to that level of reality that make better predictions, but the well-established causal interactions on the atomic level (e.g., to explain Brownian motion) give us a solid reason to believe atoms will never be replaced at their own level of interaction—they are causal entities in their own right and a permanent fixture in science, even though they have lower level constituents. If we focus on realms other than the frontiers of research, most would agree that there is a solid core of knowledge, For example, the earth moves. So too, the theory of evolution is clearly on to something in the development of life on this planet, even if neo-Darwinians may be missing some structures at work.

Realists can certainly concede that scientists must have a perspective.[34] Our science will remain *our* science. We cannot escape seeing the world through our eyes. Scientists must begin with some idea of what they are looking for: experiments and scientific observations are designed to look for something specific. Thus both empiricism and rationalism have in-put in science: experiments are not blind but are guided by ideas, even if theories are also devised after observations to explain those observations. Experiments do not dictate in advance what will found or not found. The data do "speak for themselves," but they only answer questions we ask. And the guiding principles within a particular science's course of theorizing—its "metaphysics"—are open to revision in light of scientific findings. That is, empirical pressures can cause revisions of these principles, even if they do not dictate the exact changes scientists must make. We learn which questions are fruitful and which are not. Reality determines if the course we have chosen is a dead end or not.

Where Do We Stand?

So how much mystery is there in any scientific theory? Well, the answer depends on the philosophy of science one adopts, and critical realists make the stronger case. By pushing and pulling at things in experiments, scientists certainly appear to be hitting those causes, even if they must make guesses about their nature. We must give up claims to absolute certainty, but we need not leap to the opposite extreme of the radical skeptic's despair over all claims. In the end, the underlying causal workings of the world are not a totally amorphous mystery: scientists can make a compelling claim to know something of them, and therefore some form of scientific realism must be accepted. Thus, we need not remain totally agnostic about the mechanisms underlying appearances.

Realists have the basic intuitions that reliable predictions would not be possible unless there were real causal structures revealed in reproducible patterns and that we humans must be able to know something of reality since we are part of it. As Ernan McMullin says, "the long term success of a scientific theory gives reason to believe that something like the entities and structure postulated by the theory actually exist."[35] Scientists may only produce models of aspects of reality, but they are hitting some genuine aspects of it. Postmodernists are not being intellectually more honest with their agnosticism about the possible workings of nature—a "something we know not what." But we must accept the possibility that the mechanisms to reality that scientists postulate may be wrong either in the details or in their basic nature. Nevertheless, there is an established body of evidence that suggests that scientists have made contact with the causal workings underlying accurate predictions in certain areas. And when scientists are hit with the unexpected, genuine progress in our knowledge is possible. The new experience can alter our web of beliefs or even shred it. Of course, complete knowledge of how nature works may be forever beyond our grasp. Science of necessary can only develop within the confines of our limited cognitive abilities and our conceptual capabilities. We have no reason to believe that beings with the particular mental capacities that we have today will shortly solve all the mysteries surrounding nature's workings. Science is fallible and incomplete, and we are left with proximate knowledge of the workings but genuine knowledge nonetheless.

But this discussion of critical realism and postmodernism does demonstrate how hard it is to defend the very idea that we have some genuine knowledge of reality and that how nature works is not a total mystery. It also demonstrates how philosophers today can deal with issues impinging on mystery and yet not deal with it at all. Each scientific statement is accurate only "as far as we know," and this leaves a mystery surrounding scientific claims. But there is less how-ness mystery for realists than for postmodernists: to the latter, we cannot know anything of the world apart from appearances; to the former, we can know at least something of the world's workings. For realists, we can still talk of whether a problem is a mystery or only a puzzle, while for postmodernists all such talk is dismissed. (Of course, we cannot eliminate the possibility that the noumenal "world as it is in itself" is totally unlike the causal order revealed in science, but then we would have the mystery of why such an order would appear to us at all.) Empiricists are right to be wary of claims to know unobservable realities, and realists have to be cautious about the ontological commitments scientific theories entail. But neither group shows much interest in the mysteries surrounding the how-ness of the workings of reality, let alone the mystery of the beingness and basic orderliness of reality: realists focus on what is allegedly known and on solving the next research riddle, and postmodernists deny any such knowledge is possible at all. The need for closure spawned by the philosopher's disease strikes again, and that there is more to reality than what is allegedly known is not a topic of interest today.

Notes

1.   Kline 1980, p. 318. Mathematics is a matter of our "refined and organized intuitions" and can be wrong (pp. 307-27). No truths are "self-evident." What we take to be a "clear and distinct idea" or "obvious" changes with changes in our knowledge and metaphysical beliefs. Logic itself is not as certain as most believe. What we take as a consistent deduction may later be seen simply as an error. Even in a simple inference—if a, then b—we decide in the end what is a valid deduction. As our knowledge and beliefs change, so may what we accept as a valid inference.

2.   One problem not usually addressed in philosophy is that cultures are not hermetically-sealed, self-contained entities, especially today. The cross-cultural interactions may make the entire idea of distinct "cultural points of view" inoperative. Some anthropologists also see universals to cultures and less variation in beliefs and values than do postmodernists. See, e.g., Donald E. Brown, *Human Universals* (Philadelphia: Temple University Press, 1991). The philosopher Peter Strawson also argues that there is a massive core of human "common sense"—a common conceptual scheme that has no history (e.g., the idea of personhood) and is indispensable to even the most sophisticated persons. Many linguists also argue that all languages are syntactically similar. In sum, there may in fact be more things innate to being human and to cultures than postmodernists believe.

3.   See Natoli and Hutcheon 1993; on science, see Bloor 1991; Latour and Woolgar 1992; Collins and Pinch 1995.

4.   There are a great variety of antirealists in philosophy today—positivists, empiricists, pragmatists, instrumentalists, operationalists—questioning the reality of different things (theoretical entities or all material objects). Philosophers take antirealism seriously but postmodernism is greeted with derision. According to Brian Leiter, "Postmodernism is non-existent in all the leading philosophy departments throughout the English-speaking world, where it is regarded, with justice, as sophomoric skeptical posturing." Brian Leiter, ed., *The Future of Philosophy* (New York: Oxford University Press, 2004), p. 19.

5.   Most postmodern philosophers reject the label "skeptic" since they see this label as based on *a reaction* to a claim of absolute knowledge or foundationalism—i.e., once the claim to certainty is abandoned, they think that the whole problem of radical skepticism

dissolves. But if skepticism is tied to *claims about the world*, the issue remains for realists.

6.   Quine calls something a "posit" only to identify its epistemic status. Posits are "real from the standpoint of the theory that is being built." Quine 1960, p. 22. Quine later gave the posits of physics a higher status than other posits—he ended up with a "robust realism" toward physics. That is, the posits are real if the theory is true—we are ontologically committed to whatever entities (including mathematical ones) are indispensable to a theory we accept as justified. *Epistemically* all posits within different theories have the same status, but *ontologically* some represent real features and some not.

7.   Quine 1961, pp. 42-44. He notes that even the law of the excluded middle might be expendable in order to simplify quantum physics (although that idea has not caught on). He later backtracked on whether such basic laws of logic as the law of noncontradiction could be revisable even in principle; he saw that these laws remain general laws governing all revisions of belief. See Quine and Ullian 1978. Such laws are just a matter of the consistent use of language and we are not free to change that. Lewis Carroll also argued that we must treat the laws of inference as being a different order than the inferences they generate or no actual inferences (and hence no justifications) can ever occur. "What the Tortoise Said to Achilles," *Mind* 4 (1895): 278-80. To Gottlob Frege, logical laws were prescriptive "laws of truth," not descriptions of how we think.

8.   Quine later regretted his "needlessly strong statement of holism"—it is not the whole of science that confronts experience at one time but "chunks of it" sufficient to imply an observable effect of an experimental condition. But he still endorses a "moderate holism." Quine 2004, pp. 57-58.

9.   Rorty 1989, p. xiii.

10.   Putnam 1981, p. 52.

11.   See Feyerabend 1978.

12.   Even logical positivists can say the same. Otto Neurath said "statements are compared with other statements, not with 'experiences,' not with a 'world' nor with anything else." "A statement is called correct if it can be incorporated [into 'the totality of existing statements']." So too sense-experience is always conceptualized into "protocol sentences" in this process. But the positivists did not end up with a holism as do postmodernists.

13.   Putnam 1989, pp. 3-4. As the joke goes, if postmodernists truly believe that we "create" reality, then they should try jumping off a tall building without a parachute and see if they can "adopt a worldview" that keeps them from crashing into the ground.

14.   According to Kant, the whole noumenal realm is unobservable. But the distinction here is between tables and chairs, which are observable, and atoms and quarks, which are posited as causes but are unobservable at present and thus only theoretical. It also brings up the issue of what constitutes "observation" (e.g., is an electron microscope a type of observation?) and the shifting boundary between observable and unobservable in science. Whether the two types of posits (noumenal for tables and chairs versus theoretical for quarks) are really different or whether unobservables really change the old "appearance versus reality" problem is an issue.

15.   Putnam 1989, p. 114.

16.   See Nicholas Rescher, "Extraterrestrial Science," in Rescher 1999, pp. 197-220. If beings experienced a different range of the electromagnetic spectrum, they might, for example, not see the sun as a ball but as the center of a vast range of radiation.

17.   Ernest Gellner, "Wittgenstein on Rules and Private Language," *American Scholar* 53 (1984): 243-63.

18. John A. Wilson, "Egypt," in Henri Frankfort . _
Intellectual Adventure of Ancient Man (Baltimore: Penguin Books, 1946), pp. 45-46.

19. This is not as farfetched as it sounds. Fundamentalist Christians can cite respectable scientific papers raising doubts about whether the universe is really expanding and interpreting data to present an earth-centered cosmos. See John Byl, "The Role of Belief in Modern Cosmology," in Jitse M. van der Meer, ed., *Facets of Faith and Science*, vol. 3 (Lanham, Md.: University Press of America, 1996), pp. 47-62. They might cite relativity for the idea that an earth-centered framework has as much a claim to be real as any other framework, but they must also contend with mechanics and explain how the entire universe could revolve around a stationary earth. To argue that the earth is not rotating, they would have to contend with the Foucault Pendulum and the Coriolis Effect. In addition, relativity would not in fact accomplish what the fundamentalists want: it would not establish the earth to be unmoving at all—rather, all frameworks would be relative to each other and none could be privileged as the "true" one. In that sense, relativity says nothing about what is "really" moving or unmoving.

20. Postmodernists disagree. Andrew Pickering in *Constructing Quarks* (Chicago: University of Chicago, 1984) means that scientists *literally* construct quarks—his is an *ontological* claim about what constitutes the world, not a claim about scientists devising the *concept* of quarks. Physics could have been constituted in such a way that denies the reality of quarks—the simplification and explanatory power of the concept and its basis for interesting new research findings does not convince Pickering that any reality is being designated by the term "quark." Some postmodernists have backtracked in light of the Sokal hoax, but many in "science studies" (such Harry Collins and Bruno Latour) have not.

21. Epistemic relativism, like skepticism, is a second-order claim about the nature of first-order empirical claims—namely, that no first-order claim is any better than any other since standards of rationality and truth vary from culture to culture and are determined totally by each particular culture. One cannot be a consistent, thorough-going relativist who includes all second-order claims in the same batch (since this would have to include relativism itself), but one can be a consistent relativist about first-order claims about the world. Thus, relativists would have to argue that relativism is true for everyone. This means that relativists must accept that some second-order claims are in fact true and others false regardless of one's framework, even though they are products of one particular contingent cultural point of view.

22. Epistemological holism (concerning how the change of one belief within the web of belief affects the whole) and semantic holism (concerning how the meaning of one term in a language depends on the meaning of all others) are hard to distinguish in postmodernism.

23. Kuhn 1970, p. 206.

24. Empiricists disagree among themselves as to what exactly is given in the senses—are causes, space, and time given, or are they constructed ideas? They must reject all historical sciences such as evolution and areas of cosmology since these are not open to observable checking today. Empiricists also have trouble coming to grips with optical illusions since on their terms illusions must be as "real" as any other perception. There is no way to adjudicate between conflicting experiences without bringing in reason, and so they must accept them all. In general, they have a problem with the idea of "appearances versus reality" since all they will accept as cognitive are appearances: we have the color, feel, and taste of an apple but no apple—or at least we cannot say anything about it.

25. Van Fraassen 1980, 1989. Van Fraassen is prepared to accept as real only what can be seen using the unaided senses.

26. Kuhn 1970, p.150. As the linguist Edward Sapir put it, "[t]he worlds in which different societies live are distinct worlds, not merely the same world with different labels attached."

27. Ibid., pp. 94, 150-51.

28. To give an example of the situation at the frontiers of science, consider the current dispute in cosmology over the number of dimensions to our universe. Under one theory, the universe has three spatial dimensions, while under another theory the universe is a holographic projection of a two-dimensional universe. See Juan Maldacena, "The Illusion of Gravity," *Scientific American* 293 (November 2005): 56-63. The competing theories are equivalent—each entry in one theory has a counterpart in the other, and each theory makes exactly the same predictions. So, at present, is the choice between them only a matter of perspective? Is the answer to how many dimensions there are to our universe not a matter of science but of metaphysics?

29. Kuhn 1977, pp. 320-39; Kuhn 1970, pp. 199-200. The role of inconsistencies in the development of theories and of science more generally has recently begun to receive attention. See Meheus 2002.

30. By "farewell to reason," Feyerabend (1987) meant that there was no abstract norm everything must conform to; he was not actually inviting an irrationality in which "anything goes."

31. Kuhn 1970, p. 200.

32. Some quantum physicists have adopted the instrumentalist or empiricist position and simply do not worry about whether the theoretical entities postulated in a theory "really" exist. They forego any of the different interpretations of quantum physics and "shut up and calculate." But most quantum physicists today believe they are hitting something real, even if we do not know everything about the subatomic realm and must remain agnostic about what is hit. This combines a metaphysical realism with a structural antirealism..

33. Stanley Fish, "Professor Sokal's Bad Joke," in *The Sokal Hoax: The Sham That Shook the Academy*, edited by the editors of *Lingua Franca* (Lincoln: University of Nebraska Press, 2000), pp. 81-84. Postmodernists distort science when they call it a "*work of fiction.*" Science is a work of "fiction" in the trivial sense of being a human product but not in the more common and stronger sense of being "*fictional,*" i.e., made up. Science is a human activity and does involve our concepts, but the ambiguity in the word "fiction" does not warrant lumping science in the same class with novels and poems—scientists' interaction with nature *qua* scientists is radically different is radically different than novelists'.

34. Realists can also accept that science is a social product. They obviously could agree that social conditions are responsible for the rise and flourishing of science and can agree that ideology often distorts knowledge. But postmodern sociologists of knowledge try to explain beliefs in terms of the social causes for what is accepted as "knowledge." Those in the "strong program" do not differentiate true and false beliefs but explain both the same way; their explanations are not about whether a claim is true or not but only about the social conditions for why a particular claim arose and was held. Those in the "weak program" claim only to explain errors interfering with the truth, but postmodernists cannot differentiate "truth" from "error" since they do not admit "right" answers in science.

35. In Leplin 1984, p. 26.

— 4 —

Science

The eternal mystery of the world is its comprehensibility.
— Albert Einstein

Science's job is to reduce all mysteries to trivialities.
— Niels Bohr

Many great scientists see a sense of wonder and mystery as driving their research. Albert Einstein said "The most beautiful and profound emotion we can experience is the sensation of the mysterious. It is the strongest and noblest of source of all true science." But there are different types of "wonder."[1] There is the metaphysical type of wonder generated by the simple fact that anything exists—the that-ness of reality—and there is also the wonder of the amazing variety and intricate details of reality and the beautiful order of it all. Looking up at a starry night sky or examining insects can provoke either. But it is intellectual curiosity for the "how-ness" of things—wondering about how something works—that drives science. That is, what makes a scientist a scientist is not wonder but his or her response to wonder: trying to figure out how nature works. Thus, "wonder" in the sense of being astonished by nature must be distinguished from "wondering why" in the sense of looking for the reasons and causes of things. Scientists need not be enthralled by the philosophical sense of wonder at the mere existence of anything or be awe-struck by nature to be scientists, although they may well have moments of these when they reflect on their work. Nor need the practice of science prompt questions whose answers would lie beyond the sciences. One can easily be a practicing scientist without pondering why anything exists or any of the "why" type mysteries connected to science, e.g., why the universe is orderly at all or why the universe exists at all. Scientists need not be caught up in the wonder of the that-ness of the world to be great scientists any more than athletes are necessarily led to consider the mind-body mystery when they marvel at what their bodies can do.

Science, Wonder, and Mystery

Scientists seek explanations, and this does involve trying to unmask mysteries in one sense: scientists try to "demystify" nature in the sense of finding the mechanisms underlying the phenomena we see that make things work. To use Max Weber's phrase, scientists are "systematically disenchanting" nature, at least to the extent that the how-ness mysteries to the world are eradicated. Some scientists and nonscientists alike may react by seeing science as revealing how magic tricks are done. Scientists need not subscribe to this approach to practice science, but there is a long tradition of scientists as conquerors of nature. Indeed, modern science was founded on assaulting wonder and mystery, as the writings of René Descartes and Francis Bacon indicate. For many scientists, there is a turn from mystery to an interest in only the controllable: scientists look to the unknown but only to conquer it. "Unknown" in science becomes, not the "unknowable," but simply the "not yet known." The objective is to remove all mysteries related to how our vast and complex universe works, and if all goes well, all such questions will be answered. Thus, in the end, the universe will hold no more mysteries of this type.

However, science itself is neutral to the question of whether mystery surrounds what is discovered, i.e., whether the "more" to be explored is mysterious or comprehensible. The different sciences are attempts to formulate laws for the regularities in the flow of nature that research scientists detect and to explain them in terms of causes producing changes in events. In short, scientists "seek the causes." They attempt to discern the hidden order—the structures of entities, fields, forces, and so forth—behind appearances that is responsible for the invariances and patterns we observe. The workings operating beneath the surface of things have often proven to be quite bizarre, just as the human body has a relatively simple surface but a very complex interior. Scientists may be blasé about their discoveries, or they may feel a shudder at being hit with the unexpected and feel that reality is an overwhelming mystery beyond our control, or they may react somewhere in between. But scientists need not reduce reality only to the known or to the measurable and dispel all mystery from our view of things. The "scientific spirit" does not demand that all of the workings of reality must be ultimately fathomable.

In addition, the wonder scientists experience at *the workings of nature* need not be connected in any way to the sense of *transcendence* in religion. That is, the religious sense of wonder and the scientific "wondering how" at the natural are not the same. One may well have the latter regardless of one's religious convictions or lack thereof; so too, one might have a religious sense of wonder without developing the scientific sense of wondering how things work. Theologians may argue that if we just keep pushing a sense of wonder we will all end up with the philosophical and religious senses of wonder at the that-ness of nature and speculate on its source. But the point is that this goes beyond science, and not all scientists feel compelled to go into the philosophical and religious realm. Scientists who are religious can certainly see their work in a religious context, e.g., seeing scientific work as a way of learning how God made the world work or as a way of glorifying God. So too, religious scientists may take the astonishment and elation of a new discovery as a religious experience. But one need not have a sense of religious wonder to be a scientist—a scientist's wonder can start and end with the how-ness mysteries. Of course, one who is already theistically-inclined will no doubt see science leading to religious and philosophical questions. But scientists may remain unconcerned with these mysteries, or *qua* philosophers they may give non-transcendental answers (which, of course, are as metaphysical as the theists' answers). Scientists may be atheistic, agnostic, or otherwise nonreligious. Nothing in science compels a religious response to what theists see as signs pointing to God. Thus, theologians cannot coopt scientific wonder and make it inherently religious.

How much scientists in fact know about reality is a matter of dispute. Some people believe we now know almost all there is to know about the structures of reality and what we do not know will be quickly filled in. Thus, the end of science is at hand. Of course, the end of science has been repeatedly forecasted in the past only to be proven wrong. But there is reason to think that today the age of new *revolutions* in theory may be at an end, either because we are finally reaching the fundamental laws of nature or because we may have reached the limits of our own capabilities to check empirically between competing theories.[2] The universe and our place in it may have been basically mapped out. This would not mean the literal "end of science" since there still remain many interesting issues to research and revisions to our view of things, but such research would be only a matter of filling in the gaps of our knowledge and not true revolutions in our thinking. Perhaps science in a hundred years will not look all that different from science today.

But other scientists and philosophers think science is just beginning: the universe is infinitely complex and there will be fundamental questions to ask in the future at every level of organization that we do not currently have the background even to conceive. These people believe we know only the tiniest portion of all there is to know of nature's underlying mechanisms. The biologist J. B. S. Haldane famously said that "the universe is not only queerer than we suppose, it is queerer than we *can* suppose." Naturalists can certainly affirm that there are deeper levels of order and creativity to nature than scientists can in practice fathom. The zoologist Richard Dawkins is prepared to believe that the "deep questions" of existence will prove to be far more mysterious than anything the scientists have an inkling of at present. But wondering at the marvels of nature is not the same as admitting there is mystery to reality—Dawkins dismisses "mystery" as "plain insanity or surrealist nonsense."

But even those scientists who want to dispel all mystery from nature can still sense wonder. The poet John Keats may have thought, as he put it in his poem "Lamia," that Newton destroyed all the beauty of the rainbow by "unweaving" it and conquered "all mysteries by rule and line." Ludwig Wittgenstein said something similar, if less poetic: we have to awakened to wonder, but science is "a way of sending [us] to sleep again."[3] However, scientists need not react that way—a rainbow can be as pretty and fascinating even if know how it is arises. Even reductionists such as Richard Dawkins and Carl Sagan speak of their wonder, amazement, and reverence at what they see in nature. To them, science reveals a universe that is much more awesome than any religious myth ever envisioned. Even if all the how-mysteries of their field of study are solved, the sense of awe at something as simple as a blooming flower or as grand as the majesty of the night sky need not be dispelled. As Richard Feynman said of the quantum wave/particle problem, scientists can explain how it works—the mathematics of it is very precise—but they "cannot make the mystery go away by 'explaining' how it works." Wonder does not need this type of mystery—we can still react with marvel at the fact that such things exist, even if we understand how they work. And it is an awe no longer based on ignorance. Edward O. Wilson remarks that "[o]ur sense of wonder grows exponentially: the greater the knowledge, the deeper the mystery."[4] This is a new sense of wonder at the complexity and majesty of the world revealed by science. A biologist walking in the forest certainly experiences a richer, more complex and wondrous place than does a nonscientist.[5] Thus, even in removing how-mysteries, science can increase our sense of awe at reality. And this in turn may lead to a new sense of the philosophical mystery of why such a complex reality is here at all. In sum, science only increases our wonder at the how-ness of nature. It may remove God as a direct causes of events in nature and may answer some or even all "how" questions, but it also increases our wonder and fascination with nature.

Types of Scientific Mysteries

There are two types of mysteries connected specifically to science: the underlying philosophical questions related to why science works at all, and the particular questions at the current edge of each science's field of research. The first type involves "why" questions. Why is there the order of causation (i.e., causes consistently producing the same effects)? In fact, why is there order in nature at all and not total chaos? Moreover, why does reality appear mathematical in structure? And why is our mind able to grasp some structures? Is sense-experience a reliable guide to the underlying structures of reality? The surface of reality must be comprehensible to us to some degree if we are to survive, but why are the inner workings apparently comprehensible to us at all, even if only to a small degree? Reality could be even more mysterious than it is—why isn't it?

As discussed in Chapter 6, religion introduces another type of "why" question related to the meaning of life and why anything exists. Some theologians argue that science leads naturally to the "limit questions" of the philosophical mysteries such as why anything exists at all. If the question of why anything exists were raised by science, that would make the philosophical issue part of science and religious answers part of science. However, one can be a scientist without dealing with that issue at all—nothing in any of the sciences turns on that question and the question existed in philosophy before science did. It is in no way a question raised by science that scientists do not have an answer to. It is not on the "border of science" but a separate philosophical issue; it is not raised by practicing science but by a more universal metaphysical concern. Of course, scientists in their more philosophical moments may well ask that question when thinking about their work, but so may people who have no scientific background or are from pre-scientific cultures. Thus, it is hard for theologians to make that question an actual part of science or to claim that science raises issues it cannot answer. *functionality*

The second type of mystery relates to the how-ness of underlying structures. These are questions within science itself. "How" questions involve identifying what things are and how they work. They may be in the form of "why"—why are some planets gaseous and some solid? why is grass green?—but all scientific questions are about the mechanisms that cause events and can be empirically addressed. That is, scientific questions are answered in terms of identifying what exists in nature and the forces at work in them—the what-ness and how-ness of things. Not all "how" puzzles qualify as mysteries; some are simply the latest unanswered questions. As scientists increase the size of the island of what is known, the shoreline of the unknown also increases leading to new questions. Often the more we learn, the more puzzling reality appears. Nor do we classify every interesting new problem scientists encounter as a "mystery," but only questions of a profound significance to us as humans and that have been repeatedly addressed. How did the universe begin (if it did), and how will it end (if it will)? Why is the universe made of matter and not just energy or an equal amount of antimatter? What are the fundamental components of nature? Why is reality organized into multiple levels of causal interactions, and how do higher levels emerge? How did life begin? Are humans evolved organisms and nothing more? Some of these puzzles may one day prove temporary and be solved, but scientists may not be able to devise theories to answer them all. In the case of the latter, those questions are genuine "how" mysteries that, like the philosophical "why" mysteries of science, will remain a permanent fixture of the human situation.

The Metaphysics of Science

Science's "why" mysteries introduce the issue of the metaphysics in science. The mere mention of metaphysics in connection with science alarms many philosophers, but the issue here is not possible transcendental realities existing outside the scope of science or the possible metaphysical interpretations of science. Rather, the issue is the metaphysical *presuppositions* for the practice of science—i.e., what we believe reality must be like for science to proceed.

The most basic presupposition is simply metaphysical realism—that a world exists independently of our ideas about it. Without a sense of an "objective" world,

,there would no point in doing science. Second is that this world has a causal order. Research scientists look for patterns of causes and effects in reality—Aristotelian "efficient" causation in addition to the "material" causes. Historical sciences—cosmology, evolutionary biology, geology—complicate the picture, as do functional explanations in biology, but they too are based on examining causes and effects occurring today. But no matter how many specific causes scientists find, they cannot prove that all parts of the universe are causally ordered or that what holds here and now holds everywhere and at all times since the Big Bang—that simply has to be assumed until shown otherwise. Science thus presupposes that there is a causal order to be found in reality—not all events in reality occur randomly. (The word "random" is often used ambiguously. It does not necessarily mean *uncaused or indeterminate*—the outcome of a roll of a die is random, but that does not mean it is physically uncaused.) Scientists need not reduce reality into parts or view all of reality mechanically or treat objects as only externally connected, but the practice of science does presuppose a causal order to nature. So too, we assume that we do not merely make events look ordered by artificially selecting only what is convenient for us. Causation itself also has another type of mystery first identified by David Hume: all we see is a regular succession—a "constant conjunction"—and we cannot know if in fact one event "causes" another and if so, how. Something else unknown might be responsible for the events and the correlations. Causation may be no more than a Kantian category we impose on reality and not part of the noumenal realm independent from us.

Causation has an important consequence: it rules out agents with free will as scientific causes since agents do not act in an automatic fashion but choose their actions. This does not mean that agents cannot be causes or that their actions do not have lawful consequences. It only means that scientists cannot force the duplication of their choices as they do with events involving inanimate objects, and so they cannot establish a causal pattern. This rules out God (if he exists) from being a scientific cause: even if a transcendental reality does work miracles in nature and even if we know its character, we still could not predict when, where, or what the god would do; nor can we empirically study how a transcendental reality could bring about effects. Thus, there are no empirical claims we could test. If God had to act automatically and in a consistently lawful manner, his actions could be empirically tested and he could be part of a scientific explanation. However, such a god would not have free will and not be the god of theism. That is, it is freedom of action of a theistic god, not that he transcends the natural realm, that rules him out as a scientific cause — as an agent with free will, he would not have the law-like nature that produces testable claims or constitutes scientific explanations. Any unpredictable interventions in the natural order could not be reproduced. Only causes that behave in a lawful manner can be studied by controlled observation and experiments—i.e., if a scientist has in fact found a causal relation in reality, others will be able to find it in the same way. In sum, only causes that can be duplicated by others can enter into scientific explanations, and this cannot be done with the causes resulting from the free choices of agents, whether natural or not.

Causation and Determinism

A distinction here must be made between determinism and simply causal order. Causation is about lawful conditionals: whenever conditions *a* exist, phenomenon *b* will occur. But determinism is about how the causes in these conditionals arise: the order of the universe is a closed and gapless chain in which every event has a cause. That is, determinism is about how *causes* arose, not the production of *effects*. Like a series of dominos, the first cause makes the second one topple, the second makes the third topple, and so on. The outcome is strictly predictable after the first event occurs unless another cause intervenes. Determinism thus would remove any mystery within the order of inanimate events, although it would introduce the major mystery of why there is a deterministic order in the first place.

But determinism is an extra-scientific, metaphysical claim concerning reality about which science itself is neutral: scientists search for causal order and expect to find a cause, but for this there is no need to presuppose that every event in the universe must have a sufficient prior natural cause. Scientists detect and try to explain regularities of causes and effects within nature, and no long chain of previous causes needs to be postulated. The causal order of "if the same conditions arise, the same effect always occurs" is presupposed, but scientists *qua* scientists do not make any claim about whether all events have causes or the nature of all causes. If an event arises that is not covered in any conditional, then conditionals do not apply and the event is not reproducible and hence is not covered in science. Nor does science require that every cause in the universe be physical or even natural—science makes no pronouncements on this subject. The conditionals of laws presuppose that the antecedent causes will be natural events (since only they can be reproduced at least in principle), but no metaphysical commitment to the universe being a closed and complete chain of causes and effects is necessary to be a scientist—one can practice science perfectly well without getting into that issue. Scientists will, of course, go on looking for natural causes, but they need not demand that such causes be found—that is a step into metaphysics. Newton's physics *per se* does not presuppose a determinism of causes—only the metaphysics of a clockwork world à la Pierre Simon Laplace does. In sum, science only gives the causal skeleton of nature, not a complete picture of everything that might happen in nature.

Indeterminism in quantum-level events or any true chaos on our everyday-level would eliminate determinism as a possible account of the natural world. (But, it must be noted, there are deterministic interpretations of both quantum and chaos theory.) However, the presence of some uncaused events in the natural order, i.e., events with no causes whatsoever, is not a problem for causation. Uncaused events would not rule out repeatable, lawful relations between events. That is, uncaused randomness and determinism are competing claims about the source of the ante-

onals, but they are neutral to the lawfulness of the
___ __e causal conditionals themselves. The fact that an
interpretation of quantum physics in terms of genuine indeterministic events—even
if it is replaced some day—is currently popular shows that science does not require
determinism.

Another type of metaphysics—materialism—is usually added to determinism:
all the conditions in the conditionals are material. All minds, sensations, thoughts,
emotions, and will are either reduced to the material or completely eliminated.
Mathematics is reduced to our ideas, and our ideas are only events in the brain.
Nothing immaterial can be a cause of material events, and thus no immaterial
causes can enter the causal chains of science. In addition, if consciousness did
involve some sort of energy, its introduction would violate the law of the
conservation of energy. Moreover, materialists conclude that this means not only
that immaterial causes could have no place in a scientific picture of reality but that
nature is a closed chain of material occurrences. Materialists may accept that
science cannot prove this conclusion and thus that it remains a metaphysical claim,
but they contend that it is the only position consistent with the practice of science.

However, the distinction between causation and determinism undermines this
conclusion. Science itself has no position on the nature of the causes in its conditionals—a scientific cause can be immaterial as long as it consistently produces the
same effect and is reproducible on demand. Only materialists put restrictions, based
on philosophical assumptions, on what the nature of all causes in the world must
be. How the mind might be a cause may remain a mystery, but as discussed in
Chapter 2 antireductive materialists make consciousness a system-feature of natural
objects and thus can make consciousness part of one causal order.[6] Also scientists
developed the law of conservation of energy based on observation of physical
events, and perhaps the law will remain unviolatible only for the interaction of
inanimate, physical objects. Thus, what the law was designed for can remain intact,
and it is only a metaphysical view of all of reality that would have problems if
consciousness turns out to be a form of "energy" within the law's definition.

Naturalism

The natural sciences are by definition "naturalistic" in one sense: scientists are
interested only in "natural" phenomena, i.e., events occurring within the experienced order of causes and effects and posits within space and time to explain these
events, because only these phenomena can be reproduced and tested.[7] But this does
not mean that science is a form of *naturalism* in the stronger sense that philosophical naturalists intend: that all of reality is reduced to the type of realities posited or
discovered through scientific research—i.e., the spatio-temporal realm and its
forces. The distinction is between a "methodological naturalism" and a more
sweeping "metaphysical naturalism." Scientists are only committed to the former.

But to metaphysical naturalists, science is not about only gaining causal knowledge of nature's workings but is providing a complete picture of all that is real.

Under a realistic interpretation of science, scientists do detect real structures in nature, but this does not mean that scientists are committed to the view that the regularities and the underlying mechanisms posited as causing them are all there is to reality. To be a practicing scientist, one need not claim that science produces the final picture of all of reality. There may be levels of realities in the natural world (e.g., sense-experience, free will, and the rest of consciousness) or transcendental realities that are outside the scope of lawful conditionals and thus outside of science. Science *per se* cannot make any pronouncement one way or the other on whether these are real. We can understand part of reality by means of science without concluding that that is all there is to it. Any decision to deny reality to what does not come through what is only a filter must be a philosophical decision. The scientists' way of approaching reality has proven very useful, but that does not mean reality must be, or even can be, reduced to only what scientists find. Overall, science makes no judgments on whether there is more to reality, the ultimate ontological status of all entities and forces, or the status of the natural order itself—such ultimate matters remain metaphysical and are irrelevant to the practice of science. On such matters, science is simply neutral.

An Ordered Cosmos

Thus, science is metaphysically leaner than many philosophers suppose. But scientists may also make metaphysical assumptions as part of the criteria internal to science for constructing and selecting theories. One is Occam's Razor: nature is constructed with a sense of economy.[8] That is, if two theories explain and predict the same phenomena, scientists are betting that the one postulating fewer types of entities or processes is more likely and not merely that it is more convenient for us to work with. However, deciding what is the simpler theory is not as easy as it sounds—symmetry of properties or a simple equation over a complicated one is a measure of it, but the rest is intuitive. And scientists may pursue a theory they think is more elegant and less contrived even though it currently has problems. The classic example is Nicholas Copernicus: he had to conquer the obvious—that the earth does not seem to move and that the stars do not seem to change position as the earth moves around its alleged orbit—but his theory had notably fewer epicycles and eliminated some of Ptolemy's artifices and thus seemed more elegant. However, sometimes scientists will have to admit that the more complicated theory is probably more likely. No one accepts Thales's theory of water as the one element over modern chemistry even though it is far simpler. Overall, as Alfred North Whitehead said, scientists should "seek simplicity and distrust it."

As noted above, science also presupposes that nature is ordered—a cosmos, not chaos. Simplicity, symmetry, harmony, and unity are all part of this. Scientists cannot prove that all of the world is ordered and not chaotic. The noumenal world-in-itself may not have a causal order, but then we are left with the mystery of why order appears to us at all. Nor can scientists rule out the possibility that we live in an especially well-ordered little corner of a vast and generally disorderly universe, or that forces and constants in nature are different elsewhere in space or have changed over time. For example, it may turn out that the speed of light is slowing down, which would cause physicists and cosmologists to rethink some very basic laws and theories. Scientists simply must assume that the laws they discover are uniform until they are shown otherwise.

A second problem is that disorder may also be programmed into reality: one interpretation of quantum indeterminism of the decay of individual particles is that it is inherent to reality, not merely a problem with our cognitive ability to understand what is going on that level of physical structuring. The current theory of evolution states that the course of evolution depends on chance mutation, even if the mutations are caused by physical events on another level. Indeed, our mind may be able to see only an ordered abstraction of reality. We may be filtering out only a small fragment of reality—the parts that we can order. Perhaps we can see only what our mind can understand: the simplest features of a complex world. Perhaps one role of the mind is to simplify since we would be too overwhelmed to survive if we received too much information from our environment. On the other hand, there may in fact be no mystery to the structures of reality but only order—reality may be penetrable to its core by any rational mind. In any case, realists insist that reality itself is orderly. We are finding genuine order and the mind is not merely imposing a subjective sense of coherent order like a Kantian category for experience on a thoroughly amorphous and chaotic reality or otherwise unconsciously ignoring what does not fit predictions. Science under this realism is revealing more and more beauty of a lawful order existing independently of us that can be discerned, however darkly.

Objectivity

Phenomena within the natural order that are law-like in their actions and relations and are repeatable at least in principle are "objective" in the sense of being presentable for testing by other observers, and only such phenomena are open to a scientific approach to reality. This stance is metaphysical in the sense that scientific practice presumes reality in fact has such a law-like, objective order and that the apparent order is not our creation. Testing involves looking at the interactions of things and isolating things from the total flow of reality. This can involve controlling and quantifying nature and in general treating it like an object to be manipulated. Thus, another metaphysical assumption of science is that approaching realities this way does not distort reality, i.e., some real feature is revealed even though the whole is always more. Scientists also assume that our consciousness does not affect the physical and biological processes being studied in experiments. Paradoxically, what science reveals about the real world is based on our interaction with what is being *studied—interfering* with what occurs—but it is still objective.

"Objectivity" in science can mean different things. It can mean existing independently of our perceptions and beliefs, i.e., metaphysical realism, or what the world would be like if there were no subjectivity in it. Or it can mean "intersubjectivity," i.e., being duplicable by other trained observers. Objectivity does not mean that scientists must be disinterested or dispassionate about their work: scientists can be quite adamant for personal reasons about the truth of their claims and may push hard for their acceptance. Being fully open-minded to all possible theories is obviously very valuable in science, but it is also difficult for anyone and is not required to practice science.

However, scientists cannot be free of a point of view: theories provide scientists with a particular perspective on reality, and science cannot proceed without that. Objectivity itself is a particular point of view invented by the early modern scientists. It has apparently been very successful in finding out about the structures of the world. But it is artificial: it restricts all the realities at work in the mind, and on the other hand it reduces the complexities of realities to manageable portions. Thus, the scientific approach does mean we are not encountering reality as full persons: scientists are omitting the "subjective" experiential elements in favor of the elements that can be checked by others. Thus, consciousness and the qualia of sense-experiences are omitted from the scientific picture. This again raises the issue of whether science even in principle could provide a complete picture of reality.

Why Does Science Work?

But a major question remains: why is the universe intelligible to us at all? It need not have been. Our minds may not have had the capacity to grasp any depth to reality at all. Why can our minds penetrate, at least to a degree, nature's workings? That is, why is objective structure reflected in our subjectivity? Obviously if we are to survive, we must be able to grasp some of the surface of reality—without some reliable knowledge of our environment, we would not last long. And our brain has evolved as part of nature, and so the forces obeying the laws of nature are at work in it. But this does not explain why our minds can grasp at least something of the underlying structures at work deep in reality. Simple organisms function perfectly well without being able to contemplate the workings of nature. Why do we have a mind that can even conceive mysteries? We have a limited mental capacity, and yet we can discern some underlying structures—perhaps all of them. Where does this mental ability come from? Is human reason somehow deeply ingrained into reality? Are we an integral part of the scheme of things? Does our mind reflect "the mind of God," as some theists believe? But if our cognitive capacities are God-given, why do we have optical illusions and why are our intuitions about scales of reality far from the everyday level so far off? Is our capacity to know instead a fallible product of natural evolution, as naturalists believe? That is, is our knowledge just a matter of our evolved mental functions, with technology extending our everyday experience, rather than anything about our minds innately reflecting the structure of reality? Currently reality does not appear to be rational to its core, although the indeterminism we see may be merely the result of our limited capacities to observe and reason and not an ontological feature of reality. That is, reality may in fact be thoroughly rational—finite beings like us simply lack the ability to see it all. But why our minds can discern even some underlying rational structures remains a mystery scientists cannot answer. That science works is, as Paul Davies says, the greatest scientific miracle of all.[10]

Add to this the mystery of why mathematics work. Mathematics provides structuring to the language of modern science—to Galileo, God's "book of nature" is written in the language of mathematics. Why is mathematics such an effective tool in our laws and theories? As Eugene Wigner asked, isn't it a miracle that the musings of mathematicians, not rooted in any facts, turn out to be so much in harmony with the structure of the universe? Mathematics is an artificial language we invented by the free use of our minds—so why does it seem to have an existence of its own? (Of course, why we created it as we did and our intuitions of what is mathematically true are also shaped by our experiences.) If, as formalists argue, what mathematicians "discover" are simply the logical implications of the axioms and rules they themselves define, why does mathematics apply to reality? Unlike our senses, mathematics cannot be easily related to evolution or our survival. Reality corrects some of our intuitions (as with finding Riemannean geometry to be more useful for relativity than Euclidean geometry), but still why does mathematics lead to new insights about reality in realms we cannot directly experience, such as the unpicturable quantum realm? Indeed, why does it seem to be the key for understanding all the structures of reality? Even if the world is ordered, why should the order be mathematical in structure?

The problem with mathematics is not just a matter of the simple numbers and rules of arithmetic that people in various cultures have devised for dealing with everyday life but also calculus, topology, and other advanced forms of mathematics that we have devised. Have we mathematized reality (and thus can see only what our mathematics permit)? Or does reality have a mathematical structure, like some underlying musical harmony? Are the structures of reality mathematical in nature, and do our evolved brains merely reflect them? Are there also structures that are not reflected in mathematical laws?[11] Why is the ratio of the circumference of any circle to its diameter—pi—and the square root of 2 "irrational"? Does the fact that we cannot express them in terms of the ratio of some whole numbers reveal something deep about reality? If mathematics were only a matter of our symbol systems, shouldn't we be able to invent something simpler and neater? Or do irrational numbers merely reflect the strict rules of the mathematical language we invented? And why is pi useful for equations that have nothing to do with circles?

Or why should the orbits of the planets have anything to do with the sections of a cone? Do numbers, including pi and the "imaginary" numbers like the square root of minus one, exist in reality in some way? More generally, why do physicists so often find that the mathematics they need for some new phenomena has already been devised by mathematicians from considerations that have nothing to do with physical phenomena? Or is this circular: we create a language that creates a particular picture of nature, and then we take that picture as confirmation of the language's objectivity.

In sum, are mathematical truths just the necessary products of the logic of our symbol systems, or do they reflect something of reality independent of us? James Jeans thought the universe was the product of a purely mathematical god.[12] Pythagoras held similar ideas concerning a transcendental realm of mathematical order. Theists think they have an advantage here: they claim that God created the world with a mathematical structure and that we can comprehend it because we are created in his image. The mathematician Brian Davies believes the opposite: mathematics is a human cultural creation, and thus our mathematized theories in science are unlikely to survive long.[13] But while mathematics plays a major role in new scientific discoveries, it should be noted that there are still many aspects of nature that have proven resistant to all attempts at mathematization—failures so far outnumber successes—and conversely there are many mathematical ideas that apparently do not have any application to nature. In addition, theists too have a problem: could God make 2 + 2 = 3? If so, then mathematic truths do not have the necessity mathematicians think—God just happens to make 2 + 2 = 4. If not, then even simple mathematical truths are not within God's control and introducing God does not explain their source. (The same question applies more broadly to *logic*.) That a sustaining god would uphold the mathematical realm as he does everything else is not news—the more basic question is whether he creates mathematical truths or whether they have some other source and have a necessity. Thus, introducing God does not solve the most basic mysteries of mathematics.

In any case, no mathematicians think that they can make up anything they want—even if it is just a game whose rules we make up, the rules restrict our moves. Most practicing mathematicians defend a Platonic realism and see themselves as discovering objective truths about the universe. Many feel they are exploring a timeless realm of reality much fuller than anything that mathematicians have yet devised, not anything that is an invention of the human mind. Gödel's theorems (that any consistent mathematical system having a certain minimal power has true propositions that cannot be deduced from the system's axioms) do not help resolve the mystery: do the limits they expose mean that mathematics is merely our language, or can we state true things about reality in mathematical language that we cannot prove? It is hard to see how to resolve the formalist versus realist dispute here—neither side seems right. To Roger Penrose, the connections between three "worlds"—the mathematical, the physical, and the mental—are three profound mysteries. Simply put, why a product of our mind works so well in depicting the

....s a mystery.

Laws and Theories

Most scientists gather and analyze more data, not establish new laws or test new theories. But this does not change the centrality of laws in science. Laws are scientists' attempts to summarize the regularities they discern in the flow of events, preferably with a mathematical syntax. After the nineteenth century, physicists have typically begun to speak of "principles" or simply of "equations" rather than "laws of nature." Laws may involve invariable conjunctions or only a statistical probability, but laws plus initial factual conditions lead to predictions based on the causal order to things. The more scientists can quantify matters and express their findings mathematically, the more precise the predictions. If the predictions are fulfilled enough, laws are seen as detecting a real feature of the structure of reality, especially if the predictions are of novel phenomena. The fundamental mystery surrounding laws arise from the underlying causal *structures* of reality—the fields, forces, and so forth governing events. Planets and stars come and go, but the structures governing their motions do not. They are fixed, or at least that is the working assumption of science until shown otherwise. In what way do structures exist? Do they exist outside the physical universe in some eternal realm like Platonists think numbers do? Is Stephen Hawking correct when he suggests that the equations of a unified theory exist independently from the physical universe and are ontologically prior? They are not the energy and forces that arose in the Big Bang but something else entirely that structure the latter, and apparently they are just as fundamental as the matter and energy they structure.

But attempts to devise laws have another, less metaphysical, problem: laws involve abstractions from the total complex flux of reality. Scientists cannot isolate just the objects they want to study from the rest of reality. Reality is not the simple interaction of isolated items. Scientists must idealize situations in simple terms, e.g., treating the mass of objects as if it were concentrated in one point. It is not merely that we have an incomplete knowledge of the workings of reality but that the knowledge we do have is simplified. Indeed, scientists may be restricting themselves to the simplest possible abstractions of reality—i.e., only those that provide us with some intelligibility in a world of encompassing unintelligibility. Physics seems like such a successful "hard" science precisely because it deals only in certain abstracted aspects of the world. As Bertrand Russell said: "Physics is mathematical not because we know so much about the physical world, but because we know so little; it is only its mathematical properties that we can discover."

In one sense, science is not about the real world: it is only about abstract and idealized systems. Reality in its full, complex grandeur, with its messy interaction of causes, is not mirrored in laws. But Nancy Cartwright overstates this situation by saying that laws "lie."[14] Laws are not "false" or "wrong" just because they work only in the abstract. Ideals by definition do not "get the facts right," but idealized versions of things still may enable scientists to see how the structures isolated for their attention really work. So too, models simplify our picture of reality by isolating some phenomena from a complex whole. The scientific skeleton should not be substituted for the complex real thing, but the "scientific world" is a work of "fiction" only in the sense that we humans have devised science. It does not mean scientists are getting something wrong but only incomplete. Scientists do not necessarily distort what is at work in reality by having to abstract and simplify what they are studying to see the structures at work. A related issue is whether things can in fact be ignored with impunity when devising laws. For example, when calculating the earth's orbit around the sun, astronomers ignore the fact that the solar system as a whole is traveling in different directions; thus, the earth's true line of travel is much more complex than an elliptical orbit. Another issue is that it is not at all obvious that scientists have identified all the relevant factors operating in reality when they propose laws. After all, there may be structures beyond our current understanding and the most fundamental aspects of all the structuring may remain a mystery forever.

Scientists also advance explanations for the lawfully related phenomena they find. Explanations in science suggest the causes responsible for a phenomenon's occurrence. Scientists typically go for deep explanations—explanations fitting a phenomenon into a theoretical framework that brings together a broad range of phenomena (e.g., tides, falling objects, and planetary motion) under one theory of underlying causes, thereby simplifying our account of reality. Theories provide accounts of the underlying structures allegedly at work in the causation—entities, fields, properties, processes, forces, or whatever else scientists postulate. That is, theories explain why the laws are what they are and thus ultimately explain the lawful phenomena. Fundamental theoretical structures are thereby used to give us the sense that we understand how and why a phenomenon occurred, and the world thus appears less complicated and puzzling. Our beliefs are thereby simplified and more unified, and this we take to be a stronger claim to knowledge.

The unification is achieved through the introduction of the bane of anti-realists—"theoretical objects." Instrumentalists would reduce theories to only what they say about the observable world; unobservables are introduced only as a convenient shorthand for connecting observations, and no commitment to their existence is needed. Empiricists, for whom experience is the only source of knowledge, also want to "save the phenomena," but they deny we can have any knowledge of what may be operating behind the appearances; empirical adequacy is all we can hope for in science.[15] Contemporary empiricists do not deny that there is some reality behind appearances, but they see any attempt to specify the reality as an unwarranted metaphysical move. Most famously, Isaac Newton confessed that he did not know the causes of gravity and claimed to frame no hypotheses concerning those causes but only to depict how it worked; still, this did not prevent him from using the concept as a theoretical term to unify and explain different phenomena. But realists argue that scientists are on the right track with regard to the realities and so should be at least tentatively committed to the reality of the entities they postulate.

Theoretical explanation is a matter of our understanding. On the one hand, merely being able to predict something (e.g., eclipses) does not mean we understand its causes and its place in the scheme of things. On the other hand, we may have a sense that we understand why a phenomenon occurred even if we have few causal laws connected with it (e.g., in the historical sciences such as evolution and geology). In the end, a theory is our best current account for how something occurs—with it, we would expect that what actually has occurred should have occurred. Such theorizing is a very human art. As Einstein said, imagination is more important to being a scientist than knowledge. But then are we letting limitations in our imagination dictate what we think the universe must be like? How do scientists know they have identified and understood all the relevant structures?

Our ideas about how nature works become important. This can be on the more metaphysical level of guiding principles (such as the laws of conservation) or on the level of theory. What Stephen Toulmin called "ideals of natural order" operate within theories and dictate what questions need answering. For example, to Aristotle, the question to answer about what happens to an arrow after it leaves the bow was "why does it *keep going*?" but once the idea of inertia was introduced in modern physics the question became "why does it *change* motion?"—what was "natural" and what was puzzling changed with the changes in theory. Explanations thus depend on the questions we ask. Our reasoning becomes essential, and scientists have to worry about our "common sense" distorting the situation. Observations and experiments that conflict with our intuitions provide a control. Nevertheless, at the frontiers of research at the smallest and largest scales of phenomena, our experience has the least control on our speculation.

Models

Theories bring up the issues of models in science. "Model" is sometimes used to refer to a preliminary theory or sometimes interchangeably with "theory." But what is important here is that scientists have to think by means of the familiar to understand even the most exotic scales of phenomena. All of physics, no matter how much it is guided by experiments and mathematical consequences, must proceed by metaphors and analogies to the ordinary for any understanding. Arguably all our thought has a metaphorical component, but new realms of study create a special problem. Theoretical models are more rigorous and systematic than everyday metaphors and analogies, but the fact remains that scientists are using terms designed for the everyday realm to refer to features of another realm. For example, under Einstein's view, gravity does not have any "force" in the ordinary sense of pushing and pulling—things just "roll" along space-time. The same applies to "wave," "particle," "field," and every other term: when a term is extended into a new realm, it both means and does not mean the same as in the ordinary world. Even if scientists come up with unique new terms for features of the new realms (just as "quark" was adopted without analogy from a James Joyce novel), they still must think in terms of the ordinary. Thus, the limit of the applicability of the ordinary in new realms becomes a limit of our thought.

Empiricists can accept that scientists need models in new areas as the only way to suggest how to make new predictions: models make the unintelligible initially intelligible to us, thereby shaping how we see the new realm. But for empiricists, the final theories are only a matter of observational predictions, and thus the models about the underlying workings must be dispensed with. Models offer nothing in the way of explanations; consistency with observations is all that matters. Visual images are necessary to scientific thinking, and thus models may be a useful aid, but they provide no more than hunches—as long as they work, calling them "real" adds nothing to the science. Realists, however, see models as telling us something about the structure of the realm being modeled. Models thus provide the content to the theories' mathematical framework, and the word "model" can be used interchangeably with "theory" (and "model" also makes their claims sound more modest). To critical realists, we have no or final descriptions of the realm, and so no model is an exact model of what is there. But something like the structures envisioned through our metaphors are real. Thus, useful models provide tentative insights into real structures. Something in the ordinary realm must be like some feature identified in the new realm, e.g., electron must have some features in common with orbiting planets to be thought of as "entities" circling the nucleus. But some parts of any model also do *not* apply to the new realm (since they present realities that are from another realm, an exact parallel is unlikely), and there is a real danger that the models can lead scientists in wrong directions.

Thus, the role of metaphors only increases the tentativeness and revisability of theoretical claims. No matter how counterintuitive any experimental results are, no matter how much scientists want to reject common sense, how we think about scientific questions is ultimately limited by our common sense and by our limited range of experience in the everyday realm. Modern physics also defies our common sense, but, as Einstein said, the whole of science is nothing but a refinement of our everyday thinking. Hence, there is a constant tension we cannot get around: reality's counterintuitiveness as revealed by observations, and the necessity of our thinking in our common sense terms. Our understanding of other scales of reality remains limited by the need to see a similarity between what is measured and the ordinary world—only this eases our urge to understand. But in the end, is what makes sense to us still the ultimate criterion of what we accept as real?

What is Matter?

The best place to start looking at the second type of mysteries in science—the

...cience—is with what is obvious to us: matter. This topic was introduced in Chapter 2. And considering that so many philosophers today are materialists, one would think that the nature of matter would be a major topic of analysis, but it is not. Philosophers simply assume that the issue is in the capable hands of physicists and do not bother with it.

But things are not that simple. Two aspects of reality must be distinguished: *structure* and *beingness*. Consider an analogy to language. The rules of the structure of language (the grammar) cannot be reduced to the substance structured (the letters) but represent a different dimension. Whether a string of letters forms a word or a sentence or is just nonsense cannot be determined by examining the letters alone—the other dimension must be examined (the rules of structure). In looking for the material and efficient causes in nature, scientists are like the grammarians: they identify the fundamental building blocks (the "letters" of nature) and the patterns of the causal interactions (the "grammar" of nature), but they are not interested in the nature of the substance the letters are actually embodied in—beingness (the "ink" that the letters are written in). (For a point made below, also note that physicists and chemists are not interested in a total narrative of the text—the history of the universe—but only in the rules of grammar governing the levels of words and sentences within the narrative.) In short, scientists are interested in the structures of reality, not the nature of what is structured.[16]

Physicists have revealed layer upon layer of physical structures in amazing detail, but the nature of what is being structured—*the beingness of things*—remains a mystery. The nature of the "substance" in the phenomena is not of concern, only the rules of their relationships between things—"laws of nature." Physicists, like all scientists, find out things about nature only by the interactions of things. They cannot study bits of matter in isolation. Even "mass" is measured only by the interaction between objects, i.e., the resistance to change in motion, and even defining "mass" as invariant becomes problematic under relativity theory.[17] Interactions can reveal the structures governing the encounters, but they can never reveal anything about the substance since it is equally common to everything. "Mass" figures in physical equations, but the nature of what gives mass "being" is not what is studied, and we cannot expect answers from physicists on this point.

The sheer beingness of things is not an inferred reality behind appearances—we are immediately aware of it. But it brings up a metaphysical issue: what is the nature of the reality in which physical structures are instantiated? Physicists might find this question a muddle since they assume that the end of the issue is whatever is the level of indivisible, impenetrable bits of matter or fields of energy or whatever else that they will conceptualize. For example, what quarks are made of is answered by finding smaller parts until we reach the smallest parts, and that is "pure matter," free of all structure. But the question of what constitutes those bits of matter—what gives them being and makes them real—still remains after whatever scientists find. Even if there is something within the natural universe that gives particles their mass (e.g., the Higgs field), we still have to ask what gives that reality its being. The same problem applies to alternatives to matter. What is "energy"? All we know is energy is "the potential to do work." As the physicist Richard Feynman said, "It is important to realize that in physics today we have no knowledge of what energy *is*." Or what is the nature of the "field" from which matter arises? The "field" may be, as Einstein said, the most important concept introduced into science since Newton's time, but its nature is not specified. Michael Faraday called it simply a "state of mere space," and scientists since then have not been too much more illuminating.

The problem is not that matter is manifested as a "wave" and as a "particle." Whether the lowest level we can discern is particulate or holistic still leaves the question of what constitutes it. That matter is convertible into energy—as Einstein made famous with $E=mc^2$—only complicates the question. That some theories make matter a manifestation of space-time further complicates the situation. Space is no longer seen as a big empty box but as the reality out of which matter arises. Thus, matter loses its status as the primary category for what is "physically" real to the space-time field itself. If matter is just an excited state of space-time—the surface fluctuations on an ocean of energy—then it is not just "extension in space," as Descartes thought, not that that was very helpful to begin with. "Empty" space-time thus has properties and is not actually "nothing." But to see matter meta-phorically as "condensed energy" or "condensed space-time" only pushes the question back one step to "what is space-time?" and adds the mystery of how matter arises from space-time. Energy and space-time are just as "material" as sticks and stones, and the question of what they are made of still remains.

Add to this the problem that matter is the ground for many different physical properties, not to mention life and consciousness. Whether one is a reductive materialist, antireductive materialist, dualist, or idealist, how matter can manifest different properties is a conundrum. Matter in any form may always be subject to physical forces, but this does not mean that physical as opposed to biological factors are "intrinsic" to matter. Or consider another type of property: solidity—the impenetrability, rigidity, and lack of holes of solid objects. It is the very basis of our idea of "matter." The surface of the world feels solid, but this property is manifest only on the everyday level, not on all levels. Atoms and molecules consist of vast reaches of "empty" space punctuated only with a few "particles." That is, something so central to the macro-world does not exist in the micro-world. Physicists and chemists can detail how solidity is produced by chemical bonding, but why reality has the property to produce that effect is not answered. Indeed, why there are different properties and levels of organizations remains a scientific and philosophical mystery (as discussed below).

All in all, what the "stuff" of the world is is unknown.[18] The physicist John Wheeler sees this problem and argues that today "we are no longer satisfied with insights only into particles, or fields of force, or geometry, or even space and time. Today we demand of physics some understanding of existence itself."[19] He then proffers a form of metaphysics—idealism—to explain "existence itself." But no

...tes to the "beingness" of the universe (the that-ness of things), while science is interested the structures (the what-ness and how-ness of things) in the universe. Physicists simply do not deal with "substance" or "matter" but only with certain levels of structure to reality—the "lowest" or "most general" levels common to all known things. If matter is ultimately quantized into small packets, it is still the interactions of the packets that physicists study, not the nature of the "stuff" in the packets. Scientists may uncover more and more layers of organization and perhaps even bare matter free of structure (if such a state exists) but not the beingness of matter.[20] Moreover, as noted above, these structures show more permanence than the shifting configurations of "stuff" that they structure: the universe is constantly changing—all states are impermanent. Thus, the structures have an equal or even greater claim than matter to being called "real." Indeed, one can argue that structure is in fact more fundamental—"matter" is here just to support the "structures."

Thus, as long as the word "is" is connected to matter, even quantum physicists have not changed what the meaning of "is" is—they are identifying the structures underlying surface phenomena, not being. Physics is often called the "science of matter," but a better title would be the "science of the most common level of structures." In any case, it is not the "science of being." The bottom line is that we do not know even where to begin to ask about the nature of matter, despite it being the most familiar "thing" we experience. It is much more of a mystery than we recognize. The universe is made of something about which we know nothing scientifically. It is simply the medium in which structures are embodied. But saying matter is "whatever constitutes the substance of the universe" only labels the mystery. Calling it "matter" reveals our prejudice in metaphysics toward the common sense (here, solidity and weight). Philosophical naturalists simply bypass the issue by defining their position in terms of what scientists study and not in terms of the ontology of what scientists study. But materialists do not see that they are ignoring a substantive philosophical issue and taking a mystery for their banner.

The Smallest Scale of Things

The astronomer Martin Rees notes that there are three great frontiers in science today: the very small, the very big, and the very complex.[21] Consider first the mysteries at the smallest scales of nature. The ancient Greek Democritus postulated that all of reality is made up of different arrangements of indivisible atoms in motion. That view was generally accepted until scientists about a hundred years ago began to discover smaller and smaller parts on a truly mind-bogglingly tiny scale. More importantly, none seemed to be permanent. "Fundamental particles" also bizarrely began to manifest wave-like as well as particle-like characteristics (wave/particle duality).[22] Electrons moved from one orbit to another without ever occupying the space in between (the quantum leap). Scientists also found that they could establish either the momentum or the location of a small-scale particle, but it was impossible to establish both properties with great precision at the same time; the observer can choose to measure one property with incredible precision, but this disrupts the other (the Heisenberg Uncertainty Principle). The apparent independence of particles that have been "entangled" in the past came into question: events appear to influence other events vastly separated in space at faster-than-light speeds, e.g., one particle responds immediately to the change in spin of the other (Bell's Theorem). Some physicists try to resolve the bizarreness of the subatomic realm by suggesting that time and causation can be reversed on that level—of course, this introduces a new mystery. Werner Heisenberg can rightly ask, "Can nature possibly be as absurd as it seemed to us in these atomic experiments?"

Why the quantum realm should be so confusing is as big a mystery to physicists as to anyone else. However, they are learning how it works. Most importantly, the mathematics giving very exact predictions became well-established early on and has remained well-confirmed for eighty years. It should be noted that the mathematics involves the *probability* of a particle being in any of a set of possible states. This is usually labeled "indeterminism" because of the unpredictability or apparent randomness of an individual event, but whether subatomic events are in fact uncaused and thus ontologically indeterminate is another issue. The universe may end up, as Einstein thought, being thoroughly deterministic and yet not totally predictable by us. The act of observation in quantum physics fixes the particle out of a cloud of probabilities. And the theories cannot rule out the possible existence of the "surplus" we do not experience.

Although the mathematics is set, our *understanding* of what is happening on this level remains a matter of controversy. Indeed, it is quantum physics that gives antirealism and the idea that explanations are empty any credibility at all. Empiricists argue that quantum physics is only a predictive tool and should be free of ontological commitments; what occurs behind the scene on that level is a mystery. The apparently *ad hoc* theorizing of much quantum physics also helps their case. We have no intelligible picture of what is going on down there, but we still try to visualize the unvisualizable because that is how we think. Stranger and stranger proposals are put forth. (Niels Bohr once remarked to Wolfgang Pauli, "I know your idea is bizarre; the question is whether it is bizarre enough.") Even replacing two-value (true/false) logic with three values (true/neither/false) has been proposed, but it has not yet proven helpful. No one model in terms of particles or waves makes sense of all the experimental results. Electrons are not literally balls: they exhibit some particle-like features but also dissimilarities. More generally, everything seems to fly in the face of the common sense of our everyday world. The basic problem for constructing models in this realm remains that we must use everyday concepts for a realm that they were not designed for—we must still think in everyday terms even when the experiments and mathematics tell us something else. As Einstein said, even though we know classical mechanics fails as a foundation for all physics, it still occupies the center of all our thinking in physics.

Where exactly the uncertainly in the quantum realm lies has spawned three types of interpretations of the formalism of the mathematical equations. First, some physicists believe that the problem lies only with our present ignorance and that the quantum realm is strictly causal—God does not play dice, as Einstein put it. Some variable that is as yet hidden to us keeps the realm causal. (This option is becoming more popular today.) David Bohm put forth such a view in terms of underlying "pilot waves" guiding old-fashioned Newtonian particles. Some physicists have expressed interest in his theory recently, but most physicists do not like its dualism of waves guiding particles, that the values of hidden variables can never be known, that it requires signals traveling faster than light, and that it is deterministic. Second, some physicists (such as Werner Heisenberg) hold that the indeterminate nature of current theory reflects an objective feature of reality and is not the product of the limitation of human understanding. The "many-worlds" interpretation discussed in the next chapter fits here. Under Niels Bohr's third option, there is a fundamental limitation of our ability to study the quantum realm that we can never overcome, and thus the uncertainties will always remain part of any theory. Under this interpretation, we will never know whether the realm "in itself" is determinate or not—we are permanently foreclosed from complete knowledge even of a particle, and thus we will forever be left with a mystery about the realm's nature.

Many, perhaps most, working physicists are content to accept the mathematical equations of quantum physics and to remain open on the question of the theory's actual physical content as long as the predicted measurements are fulfilled. They are undisturbed by the ontological questions and are happy to just "shut up and calculate!" That they accept this positivist-like view is in itself interesting, but a few things should be noted. None of these interpretations deny a common sense, metaphysical realism: there is some reality independent of our observations that is disturbed by the energy of the detection apparatus involved in the act of observing. However, the problem here is more than the general problem that in science, to quote Heisenberg, "what we observe is not nature [in itself] but nature exposed to our method of questioning"—experimental observation in quantum physics requires that physicists bombard particles with light, and on that scale a quantum of light is so big it disrupts the events it is sent in to probe (unlike using light in measuring, say, the speed of a train). In short, classical physics intrudes on the quantum realm through the measurement procedure. Moreover, how much we can know of the mechanisms at work may remain limited forever, but all physicists agree that the subatomic realm is not unstructured chaos free of some identifiable laws and characteristics. Our acts of observation via light may disrupt the natural flow of things and become part of the reality observed in that sense, but there is still an objectively existing independent state.

The search today continues for a simplicity of fundamental entities to explain the confusing hodgepodge of subatomic particles currently accepted. A level below the components of atoms (quarks and leptons)—is being tested. Quarks successfully account for the properties of dozens of particles. They were theorized to have no external dimensions or internal structure, but that is now being questioned. They too are beginning to look complex. A newer idea is that identical infinitesimal thread-like entities million of light-years long called "superstrings" consisting of ten or more dimensions (three being of our space and one of time) vibrate with different frequencies and produce the different types of observed "particles" and "forces." Thus, the ultimate units of matter and energy are more like vibrating rubber bands than billiard balls. There are currently several completing theories and none of them are particularly elegant. An even more esoteric theory is "M-theory," with "M" standing for "matrix" or "magic"—or "mystery"—that tries to combine string theory with gravity (and hence relativity) by adding another dimension and to show how space-time is woven by the strings. The "strings" in these theories are truly tiny: they are as small in comparison to a proton as a proton is to our solar system. They obviously cannot be visualized since we cannot visualize even four spatial dimensions let alone more, but mathematical depictions are another matter. Under this theory, subatomic "particles" are not entities but "mathematical resonances" of the vibrating strings or a stable vortex in a field of movement. Thus, we will have to view the reality of electrons, not in terms of distinct entities, but in terms of one state of the superstring's vibrations. What a superstring ontologically is is left undefined, but "matter" loses its last claim to solidity. Particles become merely quantized aspects of a string field. And we are then left with the question of what a "field of force" is and how it works.

Theorizing about "space bubbles" and "quantum foam" also raises the whole issue of whether energy fields or particles are more basic: is what physical forces organize smooth or lumpy, continuous or digital—is matter, to use Bertrand Russell's metaphor, a bowl of jelly or a bucket of shot? And why? Some recent theorizing even speaks of "atoms of space-time" with literally nothing in between the tightly-meshed atoms—it only looks continuous because the intermeshing is so tight. However, much of the theorizing is simply spinning out the consequences of our mathematical constructions, not attempts to explain any experimental observations. To some, the current theories sound too contrived to last long. Because the energies required to test these theories exceed any we can achieve on earth, others doubt that it will ever be possible to test superstrings or M-theory to see if they are just consequences of our mathematics or are actually empirically grounded. And even its mathematics does not predict only the variables of our universe but an entire "landscape" of innumerable possibilities. This leads some theorists to ask how this differs from medieval theology or whether physics is becoming, as many postmodernists claim, a branch of aesthetics. The worry is increased because string theory has come out of the experimental side of physics whose members usually express only disdain for philosophizing rather than the more theoretical side whose members are more comfortable with thinking about the underlying structures of things. Some physicists such as Lee Smolin who worry that string theory fails as a science are beginning to look for some testable alternatives. But superstring advocates are confident that at least indirect empirical tests supporting the theory are possible and will be conducted soon.

In any case, at some point scientists will theorize the smallest components or the most basic field we can conceive, and that will be the final frontier in this field. But whether this means scientists then will have reached the most fundamental physical level of organization to reality is still an issue: are there levels of scale beyond our reach or beyond our comprehension even through mathematics? Are there also other fundamental factors at work on these levels that we cannot know? Indeed, we may in fact be vastly ignorant of the workings. Current quantum physics may be like Ptolemaic astronomy: great on predictions, but fundamentally wrong on theory. In addition, the problem of modeling what is further and further from the everyday realm will only get worse. The reality that strikes the physicists will no doubt remain counterintuitive. Of course, there is no reason to believe that concepts devised for the everyday level of reality should have to apply to levels where our experience fails, but to dismiss a theory as "illogical" unless it conforms to our everyday experiences is to try to force reality to conform to our standards.

This problem may forever close the "how" mysteries on the smaller and smaller scales from our comprehension. As Arthur Eddington famously said of an electron, "Something unknown is doing we don't know what." When a physicist of the stature of Richard Feynman says "I think it is safe to say that no one understands quantum mechanics," we know we are in strange territory. And that is significant: while quantum physicists can calculate very precise predictions, they concede that they have very little understanding of what is going on at the lowest levels of physical organization. This means that at the heart of what is usually taken to be the paradigm of science lies serious limitations to our understanding of the fundamental nature of physical phenomena.

The Largest Scale of Things

When we turn to the other end of the spectrum—the largest scales in the universe—things once again are immediately counterintuitive. Einstein's work on relativity points to the problems. The problem is that we travel at speeds that are extremely slow compared to the speed of light. In fact, the speed of light remains constant for all observers no matter how fast they are moving or in what direction—and space and time bend to accommodate this. Indeed, space and time become united in any description of things. Space is not flat (Euclidean) around mass but deformed by the mass. Gravity is a field, a warp in the space-time continuum. Gravity thus is not a matter of magnetic-like "attraction" between objects or indeed any "force"—it is just a matter of objects "rolling" along the curvature of space-time. Thus, planets are moving in as straight a line as space-time permits. Distance and mass become relative to a frame of reference. The idea that time is absolute—some objective quantity in reality that everyone measures the same—has been abandoned: there is no more one absolute "now" than there is one "here." Instead, two clocks would agree only if they are at rest with respect to each other, not if they are moving away from each other. Hence, the "twin paradox": one twin flying in a spaceship at a speed close to the speed of light ages significantly more slowly when compared to the twin on earth. Nor is there an absolute "at rest" in the universe but only relative motions since no one framework is absolute. (This does not mean, however, that "all is relative": the speed of light is constant and magnitudes in intervals of space-time are invariant under transformations between frameworks. The order of causally-unrelated events depends upon the frame of reference used, but the order of succession of causally-related events does not. Thus, results of different observers can be compared. So too, the universe as a whole provides a frame of reference for determining its age. Einstein gave no grounds to question the reality and objectivity of space-time. If he had stuck with calling his theory a "theory of *invariants*," as he called it for the first two years, perhaps fewer thinkers in the twentieth century would have been so enthusiastic about relativism.)

Under one interpretation of relativity, the universe is a four-dimensional block in which our sense of the "flow of time" is merely an illusion our consciousness foists on us: everything exists tenselessly.[23] In Hermann Weyl's words, "the objective world simply *is*; it does not *happen*." There is a before-and-after causal sequence of events on a line through space-time, but there is no additional element to reality of "past," "present," and "future"—no "time." Events are like the still frames of a motion picture: there is nothing in reality corresponding to the motion. The illusion of motion is concocted by the mind of conscious beings to keep us from seeing all things at once. If we could step out of the four-dimensional block and look down upon it, we would not see any flow of events—there would only be the timeless block. The sense of "time" is only in our minds—as Einstein put it, the "distinction between past, present, and future is only an illusion, even if a stubborn one." The past, present, and future all currently exist "timelessly." This is reflected in the fact that there is no "now" in scientific equations, only time as before-and-after succession. (The block universe view does not explain why there is a "now" at all, or how our consciousness is roaming over the block universe.)

However, other physicists disagree. As the physicist Lee Smolin says, "Time is so fundamental in our experience of nature: how could it not be a fundamental part of the world?" That we can conceive reality with time only as a variable indicating a change of states does not mean the passage of time is not real—it merely shows something about how we are able to conceive things, not anything about reality itself. It is like Zeno's paradox of Achilles and the tortoise: just because Zeno showed that we can conceive their race in a way that makes it impossible for Achilles to pass the tortoise does not change the fact that in the real world Achilles would fly by the slow tortoise. Physical equations outside of the second law of thermodynamics are time-symmetrical: they work the same whether the universe runs forward or backward in time, i.e., whether the time-variables are plus or minus. But the aggregate universe has an asymmetry of past and future revealed through the measure of disorder (entropy). This means that there are features of physical reality that basic physical laws do not yet cover.[24] Under this view, the flow of time is real, and the future does not yet exist but is open (i.e., if we have free will, we can control some events and determine their outcome).

In the end, there remain two opposing camps: those who would eliminate time from reality because it is only a subjective illusion created by our minds, and those who find it an absolutely fundamental and irreducible property of the objective world.[25] This itself reveals something fundamental: if physicists can defend such extreme differences today, then it is safe to say that time is still a mystery.

The universe is the totality of what exists, has existed, or will exist in space-time. (This is not as simple as it sounds, as the discussion below of "multiverses" will show.) Observation is not just by the naked eye—radio, infrared, ultraviolet, and x-ray telescopes are used in addition to optical ones. Cosmologists try to give a coherent account of the largest scale structures governing galaxies and of the origin (if there is one), structure, and demise (if there is one) of the universe *in toto*. Relativity is especially relevant to cosmology, the science that involves the largest-scale structures of the universe. But mysteries surround cosmology from beginning to end.

Consider origins. If the current theories of the origin of our universe are on the right track, our universe arose from a "big bang." The name is something of a misnomer since there was no explosion—no noise and nothing blew up—but an expansion that still continues today and of which we all are part.[26] All matter and space present in the universe today was once compressed into a very hot point smaller than an atom with an infinitely dense mass (thus having an infinite space-time curvature). Then it began to expand. It did not expand *into* anything—space is created as it expands. (This is impossible to visualize properly since we see it as an inflating balloon expanding *into* something.) This expansion has been going on, according to current calculations, about 14 billion years—in fact, contrary to all common sense, it seems to be speeding up. After the first billionth of a second or so, cosmologists have a degree of confidence on the subsequent course of events. First, according to "inflation theory" there was an unimaginably rapid release of a tremendous energy in a tiny fraction of a second—nothing material can travel faster than the speed of light, but this does not apply to space itself.[27] Next, expansion slowed as temperatures cooled rapidly from around 100,000,000,000 degrees to an average of just 3 degrees above absolute zero today.

As the cosmic expansion continued, different subatomic entities and forces appeared through natural mechanisms. At first, the four known physical forces operating today were not present. Gravity was the first to separate from the primal unified "superforce." Over time the four known physical forces separated and their values evolved to their current ones. Matter was "congealing" from the energy drawn from the underlying energy-field saturating space. Particles and their antiparticles emerged and collided, thereby annihilating themselves, with perhaps a billionth of the particles surviving to make up the material of the universe today.

More and more differentiations arose. Quarks, subatomic particles, and light elements were formed in the first three minutes. Matter decoupled from radiation around 380,000 years later at which time the universe switched from darkness to light, but radiation dominated the universe for the first billion years, when the weakest force—gravity—started to draw more and more bits of matter into different clumps. Clouds of hot gas condensed into galaxies and then stars, the source of heavier elements, formed hundred of millions of years later. Galaxies and clusters of galaxies became the basic large-scale unit of the universe, with hardly anything between them. There are at least several hundred billion galaxies, each with billions of stars. Our beautiful spinning galaxy—the Milky Way—has over two hundred billion stars. Our star and its solar system, products of some earlier supernova, formed around 4 to 5 billion years ago, with life developing on our planet remarkably soon thereafter. Still, all in all, our planet and everything on it, including ourselves, begin to look very small in the physical scheme of things.

But the Big Bang part of the story only pinpoints an event; it does not itself explain anything. As Alan Guth asks, what "banged," how did it "bang," and what caused it to "bang"? What was the mechanism that triggered it? How did things get from a pre-Big Bang state to the current state of space and time? Why did it start around 14 billion years and not, say, 50 billion years ago? What existed before the Big Bang, and what is it made of? And where did the pre-Bang state come from? After the expansion started, why did different components arise? Why do particles and antiparticles arise out of an underlying energy field? And why didn't they all collide and destroy themselves again? Why did a small excess arise? And why these particular forces and subatomic particles? Did laws govern the process, and, if so, where did they come from? Did the laws and the structures that now govern nature exist prior to the Big Bang? Did either the laws or energy come first, or did they both arise at the same time? How could the laws exist if the material universe did not? Do laws somehow exist outside the realm of time, the way Platonists think numbers are timeless? Such questions might once have been dismissed as merely philosophical queries, but at least some of them are now getting the attention of cosmologists. The beingness of the universe and why there is something rather than nothing will remain philosophical issues beyond scientific scrutiny, but the mechanisms surrounding the Big Bang tantalize cosmologists. For instance, the oscillating-type cosmology of "loop quantum gravity" theory would remove the Big Bang as a singularity (and the ideas of *infinite* density and superstrings) and explains how the Big Bang and the initial inflation of the universe occurred. (Under one version of this theory, before the Big Bang, the universe was in a quantum state that was not even space-like—obviously something we cannot visualize.) What existed before the Big Bang created time was once dismissed as a philosophical mistake—since obviously nothing could exist *before* time began—but the idea has been advanced that there may be a series of Big Bangs and Big Crunches or some other reality preceding the Big Bang, thereby enlarging the time-line.

The searing heat of the Big Bang may well have burned up any possible information about how the Big Bang got started or what substances or forces existed before. There would then be nothing for us now to observe and the question of any previous ground or proto-substance from which our universe and its particular structures arose would remain forever beyond scientists' grasp. Some cosmologists suggest that the universe arose from "essentially nothing." But this does not mean *literally nothing*: there is no matter or space (and hence "nothing" in that sense), but a huge amount of *energy* was somehow already latent in a state called the "primal vacuum." The positive and negative forms of matter/energy may always be perfectly in balance, but this does not explain where this matter/energy initially came from or why it broke into different forms. Thus, the universe is not "self-explanatory," and science does not explain the ultimate creation of things. In addition, the structure for the particles and forces that were to come forth was also somehow written into the vacuum. Current speculation in general on how the Big Bang was initiated is on the order of "somehow an instability arose in a pre-existing quantum vacuum, perhaps spontaneously without a cause." But the possibility of finding an initial impulse to kick the whole thing off appears to be beyond observation and experiment. Thus, even the question of how we get from "nothing" to "something"—the "how" of it all—may remain beyond the purview of science. If so, no scientific account of the actual beginning of the Big Bang will ever be possible and certain scientific mysteries in this area may remain forever.

Another possibility cosmologists have put forth is that of multiple worlds. Each self-contained "universe" is like a bubble percolating from a timeless and spaceless "mother universe," lasting for a time and then expiring. There may be an infinite number of such universes in different dimensions. (The word "dimension" is used here since space exists only within each "universe" and may not intersect other "spaces." The very idea of "multiple universes" requires giving old words new meanings.) Only collectively does the ensemble of such "multiverses" constitute the true Universe with a capital "U." Cosmologists also speculate that one universe may arise from a tiny bit of matter in a "wormhole" that snaps off and becomes a separate universe, generating new matter from an energy-field. Or perhaps when a "black hole" explodes, the residue becomes a new universe with a set of physical laws and constants that are not necessarily at all like ours since the heat of the explosion would destroy all information-transfer from a universe to its offspring. Thus, our local universe alone would self-reproduce zillions of baby universes, and so on, and so on. Our universe may cease, but an infinite number of universes will continue on, constantly popping in and out of existence, and will do so forever. Things get weirder and weirder. The idea has even been suggested that our universe could have been cooked up in the basement of some scientist in another universe who has mastered the physics of black holes, using just a few pounds of matter.

Next consider the question of how our universe will end. As the cosmic expansion continues, eventually all other galaxies will disappear from our view, while nearby galaxies are pulled together by gravity (Andromeda is heading toward us already). What happens eventually? The answer apparently depends on how much matter there is in the universe. If matter exceeds a certain amount, gravity will cause the universe to collapse in on itself in a Big Crunch and our universe will end as it began—as a small point. This in turn leads to speculation about a possible future Big Bang or an indefinite oscillating series of Bangs and Crunches—the Big Bang is just a sudden change in directions, as it were. (Whether the universe is "efficient" enough to bounce indefinitely also raises doubts about the oscillating universe scenario. Of course, another type of energy may be involved for all we know.) But if there is not enough matter, the expansion of the Big Bang will continue forever. A hundred trillion years from now, the last stars burn out. Trillions of trillions of years from now, matter will start to decay, leading to black holes and then to a lifeless, cold and dark age of only subatomic particles expanding forever. Entropy will be at the maximum, and nothing interesting will ever happen again.

Current observations suggest the latter scenario. But some cosmologists have proposed "quintessence"—a new form of the "ether"—that permeates all of the universe and that would supply enough matter for a collapse. In addition, observations of the movements of galactic gas clouds suggest that the gravitational "pull" of vast amounts of "dark matter," i.e., matter that emits no light or other radiation and thus is unobservable (and thus whose nature is a total mystery to us). However, the acceleration of the rate of expansion also suggests a repulsive anti-gravity force that increases with distance and as the average density of matter decreases—"dark energy." This is the successor to Einstein's "cosmological constant" that he postulated to counter the force of gravity back when scientists thought the universe was not expanding. Dark matter apparently constitutes about a quarter of the substance of the universe and dark energy most of the rest, and there may be different varieties of each. Thus, the vast majority of the stuff of the universe remains completely unknown to us; even current theories do not suggest anything about the nature of the dark forces. This also means that visible matter constitutes less than 5% of the universe: what we take to be the totality of our universe is just a ripple on an invisible ocean. It also means that almost all of the universe has no more than minimal physical structuring. Does this mean that the visible universe is no more than "just a bit of pollution," as one physicist put it? But whether the dark material affects things or not, under any of the scenarios carbon-based life will come to an end in our universe: the universe will either collapse upon itself or all that will be left will be the smallest particles of matter or radiation.

Today cosmologists go to wilder and wilder speculations. Their "best guess" speculations are radically underdetermined by observational data. They are gaining more and more empirical information all the time, but many of the new theories have an *ad hoc* flavor and do not lead to any new predictions. The new findings reveal that there is much we do not know about the largest scales of things and their history from beginning to end. Moreover, the universe has repeatedly shown that it does not conform to the intuitions we have developed on our everyday scale of things: conditions at the speed of light, in black holes, under the extreme heat of the early universe, and at very short subatomic distances force us to abandon how we normally see things.

The astronomer John Barrow concludes from all these problems that the astronomers' desire to understand the structure of the universe is doomed—we can merely scratch the surface of what is there. "All the great questions about the nature of the Universe—from its beginning to its end—turn out to be unanswerable. There is a fundamental divide between the part of the Universe we can observe and the entire, possibly infinite, whole." We can expect the universe to be endlessly diverse both throughout space and historically—it is most unlikely to be even roughly the same everywhere. We most likely "inhabit a little island of temperate tranquility amid a vast sea of cosmic complexity, forever beyond our power to observe."[28]

All of this means that cosmologists must be very tentative in their claims. The current state of cosmology is a proliferation of competing theories with no sign of agreement. Some in the field, as with quantum physics, have taken an instrumentalist position and given up the idea of "corresponding with reality"—e.g., Stephen Hawking says he does not know what reality is and is only concerned with a theory predicting the results of measurements. So too, ten years from now the theories may be very different from those today. Or we may be reaching the end of any new information that we humans are capable of gathering, and any speculations that scientists advance will be less and less tied to checking by observational predictions. Just as we cannot see dark matter, we may well also be missing major features of structure—perhaps structures that will eliminate the need for postulating "dark matter" or "dark energy." Scientists may never get well-established theories of the largest structures of the universe. Cosmology may return to being a matter of only metaphysical musing. At the least, reality in conditions far removed from the everyday is looking less and less the way we thought it would—it is more bizarre than we did imagine and perhaps can. As cosmologist Joseph Silk says: "Humility in the face of the persistent great unknowns is the true philosophy that modern physics has to offer."

Theories of Everything

One topic brings together both the smallest and largest scale of things: theories of everything. A major quest of all scientific theorizing is to bring more and more phenomena under one theory. Newton explained various phenomena—tides, the path of falling apples, and the orbit of planets—all as the result of the same force. Unification can also be done by uniting apparently diverse forces under one theory. Michael Faraday showed that electricity and magnetism are forms of the same force. James Clerk Maxwell used the concept of "fields" to unify electricity, magnetism, and light. Einstein's last obsession was to produce a unified field theory linking electromagnetism and his theory of gravity (general relativity). The quest continues today to unify all the currently recognized physical forces—the "strong" nuclear force holding the nuclei of atoms together, the "weak" nuclear force responsible for some forms of nuclear decay, electromagnetism, and gravity—under the equations of one umbrella theory.

If unifying the laws of the very small (quantum physics) and the very large (general relativity) is successful, all the forces would be seen as different manifestations of one single force at work in reality—just as there is only one substance to nature according to naturalists, so too there is only one force. Progress has been made toward unifying electromagnetism with the weaker nuclear force, but gravity has proven especially resistant. The basic problem is that the perceived indeterminacy of quantum-level events cannot be combined with the determinacy that reigns in relativity theory. In addition, quantum non-locality apparently requires an absolute simultaneity that cannot be squared with relativity. In sum, cosmology needs quantum physics to understand what happens to large masses in tiny spaces, such as at the Big Bang and in black holes, but the two levels of theory do not appear to fit together in any way. Either quantum physics or general relativity will have to be revised. Most physicists believe the latter will have to submit to the former (they are working on string theory). Others argue that quantum physics, not general relativity, needs the fundamental change, however subtle (they are working on loop gravity theory).

If all physical forces are finally unified, the result will be a "theory of every-thing"—a "TOE." It would provide a comprehensive physical explanation of all phenomena with a minimum of natural forces and constants by establishing a law that requires the physical forces to have the values they have. Today a leading candidate for a TOE is M-theory because it fits gravity into the picture. It predicts the existence of particles carrying the gravitational force—"gravitons"—as simply a superstring's simplest mode of vibration. In any case, any TOE will have to combine the smallest scale of things with the largest under one theory. The quest for simplicity and unity within physics will end: all the laws governing the fundamental physical forces in nature would be derived from one theory. All the mathematics of laws would be deducible from one fundamental equation. Physicists will be able to put one elegant master equation encompassing all of the physical forces of nature on a T-shirt.

To TOE enthusiasts, a TOE is the theory that will end the quest dating from the days of ancient myths to make sense of things to us. (It should be noted that not all physicists are excited about the prospect of a TOE. Freeman Dyson, for one, would be disappointed if the whole of physical reality could be depicted by a finite set of equations—he "would feel that the Creator had been uncharacteristically lacking in imagination.") But truly to be a theory of *everything*, the TOE must encompass not only laws but also the seemingly arbitrary and contingent values of physical constants in nature, e.g., the speed of light. And that is just the beginning: a TOE must explain why the universe has the particular laws and features it has and not others. Moreover, how do any laws arise? Where did they come from and what are the mechanisms that produce laws? If physical forces and constants evolve over time, why? Is there any reason why they should *not* change over time? Why are there only three spatial dimensions "uncurled" from superstrings and not more? Indeed, a TOE must explain the more basic scientific mystery of why nature has superstrings or whatever are the most basic elements that scientists discover. Why do superstrings vibrate they do? Why are there six types of quarks and not, say, ten? Why do the particular subatomic particles that exist exist? And why do they have the mass, charge, and other properties they do? And why are there only the types of physical forces emerging from the underlying TOE force that there are? And why do they generate the particular degree of strength each has? Why are there 92 natural elements and not, say, 57 or 314? And why is the universe as complex as it is? Why, as Paul Davies asks, do we have a set of laws that drove the featureless gases of the Big Bang toward life and consciousness? Why didn't the universe remain simpler, e.g., having nothing more complex than quarks? Indeed, why didn't the universe remain in its initial state of perfect symmetry? In the end, why wasn't the universe governed by Newtonian laws rather than relativity and quantum physics? *of evolution*

In short, why does the universe have the general features it has and not others? And then there is the question of why the Big Bang occurred at all. And this in turn is leads to the initial conditions of the Big Bang and to what happened before the Big Bang. Indeed, doesn't a TOE also have to explain why there is any space-time at all? Answering these questions is obviously a tall order, but nothing less constitutes a truly total explanation rather than being only an equation for predictions. There can be no unexplained brute facts left standing for a theory actually to be a "theory of everything."

Since scientists may be reaching the extremes of empirical support available to humans, there is a very real possibility that science will never be able to test any TOE or supply evidence for one. Of course, there may in fact not be one underlying force to reality and thus simply no TOE. Also, must there be only one such equation? How do they know that there is not another possible consistent theory?[29] Steven Weinberg says that we can already imagine sets of laws of nature that, as far as we can tell, are completely consistent mathematically and yet do not describe nature as we observe it.[30] Is any variation of the variables possible, thus permitting a role for a designer? As Einstein asked, did God have a choice in creating the world as it is? The initial conditions, forces, and entities of our universe—its overall structure—may be arbitrary, as some multiple-worlds theories hold. Or reality may not have an ultimate rational structure but be permanently mysterious. Then no simple equation would ever be forthcoming, nor would any simple, compelling picture of things. Or there may be such a structure to reality but our particular minds may not be able to grasp it or conceptualize it into a theory, and the structure of reality would then still remain a mystery to us.

And even if we are able to comprehend the fundamental structure, why do we think we have mastered all the preliminary steps at this particular time in our history? Why do we think we have assembled all the pieces to the puzzle now? Some scientists argue that the laws we have devised may not be fundamental at all but only reflections of some deeper forces that are as yet unknown. As we learn more about the smallest and largest scale of things, might not more pieces appear? For example, is "dark matter" made of some currently unknown type of particle, or does "dark energy" operate by some new force that we need to know about for a complete picture of things? Or perhaps there are larger-scale forces governing galaxies or clusters of galaxies—we would need to find them all before we could create any final unification. Any TOE we devise today is likely a candidate for extinction as more data emerges. Our place on our particular little planet may limit how much we can ever know of the nature of our universe. So too, the more extreme conditions of reality may well remain unknown. Progress in quantum physics and cosmology may also come to an end because of the limitations of our technological abilities before we have all the pieces needed to make a meaningful stab at completing the puzzle.

Theorists tout TOEs as if all the complexities of reality are like the beautiful intricate patterns produced by simple algorithms in fractals such as the Mandelbrot set—things look complicated but really are the product of a few rules.[31] Perhaps a few simple, recursive causal rules might also underlie all the extremely complex phenomena of nature. Moreover, some argue that once we identify a TOE we can replay the universe and the end result would be the same: it all will occur once again in a deterministic way from the Big Bang. But there are problems with this idea. First, if causation and determinism are not the same things (as discussed above), then it is one thing to identify the laws through a TOE and another thing to accept a determinism of causes. There may well be objective indeterminacies on the quantum level. If so, this would nullify any determinism of all of reality. There are also the issues discussed below of random events and human free will. Because of this contingency of causes in the lawful conditionals, scientists may identify the fundamental mechanisms at work in reality and still hold that not all events are determined. If so, no laws by themselves can predict the course of events; hence, actual events would lie beyond a TOE. Thus, if we reran the entire evolution of the universe, the laws and initial conditions may turn out to be exactly the same, but the course of events may be very different. There may not be a planet earth in our solar system or life or any humans, let alone the particular ones who have existed, or any of an innumerable other contingent occurrences. In sum, even if there are fundamental forces, the particular course of the universe may not be inevitable and predictable. History may not be encoded in the Big Bang.

A related problem is that the complexity of reality appears to involve different levels of physical and *nonphysical* organization. Life, consciousness, and everything else in the everyday world would all have to be explained by a true TOE. Whether consciousness is a causal feature of reality or is just an epiphenomenon, its presence still has to be explained. A TOE may encompass all the forces at work in the *physical* level of organization, but this does not mean that it encompasses all the structures of reality. Many laws to reality may not be incorporated. For example, Mendel's laws of genetics may not be derivable from the laws of quantum physics even in principle. That there are different sublevels of physical organization (e.g., the quark level versus atoms) also would have to be accounted for. The problem of how higher levels of organization emerge also remains, especially if we accept reductionism. A TOE would have to explain emergence and all of the known levels in terms of laws present from the beginning. In the end, it may turn out that a TOE is a theory of only some sublevel of the physical level, not of all the workings of reality—a "Theory of (One Level of) Everything."

Most basically, a TOE cannot itself be a brute, contingent fact but must show why it itself was instantiated in reality. Whatever equation TOE theorists come up with, the TOE would have to explain why that equation is in force and not another. But whether scientists can ever do that is a real issue. It is not obvious that any law can explain itself, e.g., electromagnetism cannot explain why the universe is set up to allow electromagnetism to operate in the first place. Can anything be self-explanatory, or must it be simply a brute fact incapable of any explanation? Any TOE may leave a basic why-mystery untouched. In addition, TOEs cannot touch the basic philosophical mystery of the that-ness of reality—why there is anything at all—or the basic scientific why-mysteries of why there is causation in nature or indeed why the universe is lawful at all. In short, TOE theorists will never be able to claim, in Stephen Hawking's metaphor, "to know the mind of God."

Complexity

The last area of current scientific mysteries is the complexity of things. Physical and biological scientists typically deal in idealizations—e.g., how ideal gases would react in a true vacuum—and abstractions that isolate selected factors from the rest of nature. But the actual world is not that simple: nothing exists in isolation but only in a mix of relationships and interactions with other parts of nature. Today the great diversity of things and the intricate interplay of all of the parts of reality are being given more study. Some scientists even in physics and mathematics are taking a more "holistic" rather than "reductive" approach to phenomena, looking at the "non-linear dynamics" of things.[32]

This complexity shows itself in the universe both in the development of more and more levels of organization and in greater and greater diversity within each level. The universe started out as radiation and then as a mass of whatever is the simplest of entities, but it became more and more complex—in fact, it seems programmed to integrate into more and more complex types of things. New levels of complexity become a recurring feature of the history of the universe. On the inanimate side of things, why did reality not remain in the simplest state after the Big Bang? Or why didn't reality stop with quarks? Why did things go from quarks to stars and planets and all the esoteric things populating galaxies? When hydrogen atoms are crushed together by gravity in stars, why are heavier elements created rather than piles of crushed parts? And, on the animate side of things, why did more and more complex forms of reproducing organisms evolve? Why did complex organisms with specialized organs evolve? Why didn't life remain on the single-cell level? One-celled organisms, after all, can adapt to an amazing variety of environments. Natural selection and gene-mutation certainly do not explain why the very possibility of greater and greater complexity existed in the first place. And why did conscious beings evolve? Indeed, why did life appear at all? In sum, why does reality appear continuously creative? If anything, the law of entropy that scientists have proposed suggests that the universe would tend toward disorganized chaos, not produce islands of greater and greater organized complexity.

The interest in complexity began with the somewhat misnamed "chaos theory." Chaos theorists brought attention to the sensitivity of changes to initial conditions. Their great illustration is that the flap of a butterfly's wings in China may set off atmospheric disturbances that produce a storm in New York City. But they do not argue that reality is truly *chaotic*, i.e., free of a causal order. The problem they identified is that we lack the capacity to isolate and measure all the factors involved in the dynamic interaction of things above the level of the simplest interactions. Scientists can identify the different laws at work, but when it comes to the actual intricate interactions of the real world, they lack the capacity to proceed so cleanly: different causal laws will be interacting, and neat factual conditions will not be isolatable. We cannot predict weather a week from now. Where exactly a leaf falling from a tree will land may be impossible to predict precisely, given the odd shapes and surfaces of each leaf and the intricacies of wind currents. Even the interaction of just three objects in each other's field of gravity is very hard to predict even using a computer. Overall, there is unpredictability (and hence a loss of certainty) on the everyday level of reality because of the inherent messiness of the total interactions of things in reality.

But all physical events may in fact still be strictly deterministic. In the case of the three-bodies example, there is a strict deterministic chain of causes causing new causes even though we are not able to predict where the bodies will end up because we cannot penetrate the chain. Thus, there is a big difference between the *epistemic* problem that chaos theorists have identified with our inability to know and measure the actual dynamic nature of reality (predictability) and the *ontological* claim that reality is in fact chaotic (acausal or indeterminate). Just because we cannot identify all the conditions that are actually at work in a given process does not say anything one way or the other about whether the process is causal or whether reality is a deterministic chain of causes and effects. In sum, unpredictability only points to a limitation in our current scientific abilities—something that a greater refinement of computer technology may be able to cure—not necessarily an ontological fact.

The more recent "complexity" theorists in fact put great weight on the role of causal order in the dynamics of reality.[33] They see reality's complexity as the result of an intricate interplay of causal order and mixtures of contingent factual conditions in which the causal conditionals produce effects. A strict determinism is still possible. For all we know, everything may be the aggregate of subatomic processes —it is simply that scientists cannot solve equations for anything more complex than a collection of a few molecules. However, it is also possible that the initial conditions may contain random elements and thus that there is no great deterministic chain to reality but only discrete causal conditionals operating on sets of conditions. If that is the case, reality is an interplay of strict causation and an openness with regard to causal conditions. This would mean that the history of the actual factual conditions is necessary to complete the explanation of all of reality.

Complexity theorists are not trying to supplant older sciences; they are dealing with a new problem, not a new approach to old problems. (Physicists and chemists show little professional interest in the history of the world: they are content to identify the laws at work in the world and ignore the course of "contingent facts" that constitutes the conditions in lawful conditionals.) They focus on how order arises in the complexities of the world. They believe all things in nature are driven to become ordered. They speak of the "self-organization" of things "spontaneously" arising from "chaotic" conditions into ordered patterns, but this is a misnomer.[34] "Self-organization" occurs under the direction of natural laws. No magic is involved. When these theorists speak of a pile of iron filings "self-organizing" into order under the influence of a magnet, they do not mean that the filings literally "self-organize" themselves—a *force* (magnetism) is at work ordering them. Water does not "spontaneously" freeze for no reason—it does so only under specific physical forces and conditions. So too, with simple computer programs that end up producing order rather than chaos: we program in the restrictions in advance—there is nothing "spontaneous" about the order at all. All the idea of "self-organization" means is that no supernatural organizing force or agent is at work in nature—all the forces of ordered complexity are natural. Order is "free" only in that it arises naturally from the laws of physics and chemistry.

Emergence

One of the strangest things about how reality is organized is that when energy is fed into certain unstable systems, more of the same or greater chaos is not created—instead, a new "level" of order emerges.[35] There is a hierarchy leading from whatever are the entities on the simplest physical level to subatomic particles to atoms (the smallest units having chemical properties) to molecules to larger bodies ending with stars and other entities populating the universe, and, more importantly for us, also to ecosystems with living and conscious beings. Each level is simple compared to the complexity of its underlying base, and in turn it becomes a base for higher levels of phenomena. That is, each "higher" level in the hierarchy has its own simplicity and order compared to the complexity operating "below," but more complexity is also added to reality.[36]

This means something in reality enables higher levels of complexity to emerge. The lower levels are a necessary base for the higher levels to emerge, but whether they *produce* them is an issue. That is, there is a leap between levels—not a gradual accumulation of effects from the base's properties—and nothing occurring on the lower levels would predict higher levels of new phenomena emerging. "More" comes from "less," not ontologically but structurally. The properties of a cell are distinct from the properties of its components on the atomic level. The higher levels of an entity have their own properties that are not reducible to the sum of the properties of the lower; they also have their own order of causal interaction. Thus, atoms are still real "entities" because of their causal interactions and effects, even though they are not distinct "particles" that are ontologically independent of their bases. They need not have to be able to modify their bases to be considered real: they still *do* something their components either individually or collectively cannot, and thus they are causal entities in their own right. There are organizational, "holistic" properties on each truly emergent level that are real in their own right even if base-conditions are necessary for their appearance. So too, there are laws and explanations about activity on each level that are unique to that level. We can neither conceptualize nor explain such biological phenomena as reproduction and adaptation in terms of the forces of physics. The higher-level activity does not violate the laws of the lower levels but remains neutral to those laws—from a physical point of view, nothing new is occurring in a living organism. Still, even complete knowledge of the lower levels will not explain the operation of the novel properties on the higher levels or why the new properties emerged at all. Thus, no theory of the parts can predict those properties or otherwise explain the whole.[37] In sum, each level introduces a new, unforeseeable type of organization to reality. This shows a creativity to reality one would not expect from the sciences as now constituted.

Reality's interconnections function more like an integrated system than the strictly external interactions of a machine. "Gaia" theorists treat the entire earth holistically, studying the symbiotic interactions between living beings and their environment (the earth, the seas, and the atmosphere), such as how living beings chemically regulate the composition of their environment to promote their own survival; the planet as a whole is seen as self-organizing and self-regulating. The idea of "systems" emphasizes the relational nature of reality and that reality cannot be fully understood in terms of its parts alone.[38] Primacy must be given to the whole over the parts in order to understand the whole itself. But the same idea of levels applies to non-systems. For example, a bag of radio parts does not constitute a working radio, even though they are materially the same—the structural element of a radio cannot be found by analyzing the parts. Nor will knowledge of the parts alone explain what the radio's output is.[39] Indeed, why the parts are what they are is explainable only by the whole. Relations within the whole create new properties, and those relations cannot be understood by studying the nature of the parts individually. Nor will the study of the parts and the mechanisms at work in a whole answer why reality made the whole exist and thus will not explain the entire situation. We need to know the parts to understand the whole, but we also need to know the whole to understand how the parts function as they do. Thus, understanding the base is necessary but not sufficient for understanding the higher-level order of things. This also raises questions: Do parts exist only because of the whole? Is the whole primary, and are the parts just means to an end?

Life and consciousness are the two levels most often seen as emergent, but various physical sublevels are also just as novel and emergent. Antireductionists consider the higher level's properties more than the sum of the properties of its base: the higher level properties are causal, and not "illusory" in any sense; they cannot be explained away by showing the role that the parts within the whole. Biological processes are not simply the consequence of physical and chemical forces and the universe's initial conditions—new laws of nature become operational once lower level conditions become complex enough. Or consider something as simple as water. Wetness is not a property of its chemical components—oxygen and hydrogen—or their sum. If we pour water on most fires, it will put them out, but if we pour its components in their gaseous form on the fire, it will cause an explosion. Why do the higher levels of new properties emerge? And why are chemicals made to bond? Reductionists think that understanding the parts is enough to explain the higher level phenomena, but this approach misses the simple fact that a new level of phenomena emerges—indeed, on the lowest levels, the microstates of all phenomena never change at all but remain the same. Even though the properties of water are dependent on the bonding of its components, and even if the properties are epiphenomena and not causal features of reality, that reality is organized to produce higher and higher levels of new phenomena still requires explaining. Reductionists are looking *downward* and trying to explain the properties of wholes in terms of the properties and interactions of their parts. Antireductionists look in the other direction and do not see any *upward* explanations of how the parts can be responsible for the emergence of wholes—i.e., forces governing the interactions of parts do not produce higher-level phenomena, and the question is how the parts actually relate to the wholes. To them, the reductionist approach oversimplifies and distorts the full complexity of reality. Nature is more than just "matter in motion"—it is "matter in levels of organization."

This process of integration and emergence has occurred all the way from the smallest bits of matter up to the most complex configurations in the universe. The Pauli Exclusion Principle governing the arrangement of electrons in an atom cannot be derived from the study of one electron alone. In fact, electrons seem to lose their individual identity when they become part of an atom: to understand an atom, we have to look at the properties of the functioning whole system, not the components. Objects are not just a pile of quarks and people are not just a pile of DNA molecules—they are structured wholes. To say that a chair is 99.99999% "empty space" ignores the role of structure. Each level has a reality not described by quantum physicists. As Gilbert Ryle noted, particle physicists do not describe tables and chairs at all. Even the states of matter—gaseous, liquid, plasma, and solid—are level-properties that do not make sense when discussing the components, which do not have those properties. Solidity and all the other higher level properties are still real factors in the world. Even if higher level properties are somehow completely caused by lower level events (e.g., solidity by the rigidity of bonded chemicals), they are not properties that quarks and their interactions possess, and we have to account for why reality is organized to produce them.

The lower bases carry the information operating on the higher levels but do not otherwise affect it. It is like the chemicals of the ink and paper of this book carrying the words: no analysis of the chemicals will detect the meaning of the words. Similarly, the chemicals underlying DNA do not contain the biological "information" written in DNA—the sequence of molecules is irrelevant to the underlying chemistry. The lawfulness of the bases are necessary for the higher levels to emerge, but that does not mean they produce them. Not only does the emergence of new properties undermine the idea of a lower-level reduction, it also suggests that more forces are at work than just the lower-level ones and randomness. In any case, the universe must evolve sufficient complexities for the higher levels of organization to manifest themselves. That is, the forces necessary for the appearance of water or life may have been present from the beginning of the Big Bang, but water and life could not appear until their physical bases developed. Higher sublevels of complex forms of life on earth depend on the presence of a certain amount of oxygen—something that is merely "accidental" from the point of view of physicists and chemists. So too, it may have taken billions of years of reactions cooking within stars before the elements for any life could appear, but that does not mean that the biological level of organization is any less real than inanimate levels. Under antireductionism, the only alternative to the forces making life possible being programed into reality from the Big Bang is that they were somehow added later. That would be just as mysterious, if not more so.

To antireductionists, there is no one level responsible for everything—no one way the world "really is."[40] Instead, the world exists in a different way on each level of organization. Each phenomenon has multiple levels, and a pluralism of points of view (and hence sciences) is needed to have a complete scientific picture of what reality consists of and how it works. No one scientific account will depict the "real" causes or the "real" explanation. This does not violate Occam's Razor: a complex world simply requires more than one account. The scientific accounts do not compete or contradict each other but are supplements, each revealing something different about reality and answering different questions. Physics encompasses the broadest levels—a physical level of organization apparently operates in everything. But this does not mean a subatomic particle is "more real" than a living cell—fewer levels of organization simply are operating in the former. How structures exist and how matter and structure are connected may remain mysteries. Each "self-organizing" level has an equal claim to reality and is not reducible to other levels. Thus, no science is more "fundamental" than any other, even if physicists study the broadest and "deepest" levels that are the bases for all phenomena.

The principal problem of the connection of levels—how emergence occurs—is not yet covered by the sciences. Chemists and biochemists can explain the "how" of chemical bonding very well, but they cannot explain the "how" of the process of the emergence of higher levels of effects, let alone why reality is organized with the propensity to produce higher and higher levels of order. What is the source of nature's power of creativity in creating higher levels of organization? Theists may invoke a designer god, but scientists look for something within nature that is producing both the structures and the upward thrust of new levels of organization. A new science of emergence would be needed. A science of systems and complexity is not enough.

Perhaps a new type of natural force or field accounting for emergence may be found. We are accustomed by scientific analysis to think in terms of a whole being built by its parts, but perhaps wholes are not created by the properties of their parts but by other forces at work in nature. That the universe as a whole may be a system working on its parts through uniform forces (thereby producing a coherence and unity to the whole) is the most extreme view, but more moderate is the idea that a currently unknown natural forces comparable to the physical ones might be at work. The idea of a life-force—a distinct "vitalistic" *substance* responsible for life—has long been banished from biology, but the idea of a vitalistic *structure* still remains feasible.[41] The idea of biofields—biotic "form-shaping fields"—has been around since the 1920's, with Rupert Sheldrake's ideas of "formative causation" and "morphic resonances" being the latest incarnation.[42] The four forces that physicists currently recognized do not seem, individually or collectively, related to emergence. Each force would only produce more of the same order and if anything would prevent higher levels of order from arising. For example, gravity would only lump more and more matter together and would do nothing toward organizing it, let alone make higher levels emerge. Nor would the interaction of all the known physical forces produce higher levels. In short, the basic laws of physics cannot establish how the connections leading to higher levels of complexity arise. As the physicist Paul Davies says, the forces of physics merely shuffle information around, not create new information.[43] Throwing in the need for random events to explain the history of the universe does not help but only hurts the possibility of explaining the emergence of more and more levels of organization.

Reductionists of the structural stripe disagree with all of this. They must admit structuring is part of reality—the universe is not a chaotic pile of quarks blown apart at the Big Bang. But they contend there is no hierarchy of levels nor any discontinuity between, for example, the nonliving and living but only one level of structuring having greater and greater degrees of strictly mechanical interactions. There are no nonphysical processes that are distinct from physical ones—there is nothing more to biological processes than the interactions of macromolecules. So too, all chemical data about elements are explainable by physical data about their atomic structure. Everything we take to be real in the everyday world is simply an aberration in the universe resulting from ever increasing complexities of interactions—small eddies of order in a sea of chaos in a universe that is winding down under the control of the law of entropy. Reductionists must concede that different degrees of complexity have their own concepts and laws—concepts about a pile of sand cannot be derived from concepts about a particle of sand, even if the properties of the pile is simply the sum of properties of the parts. But structural reductionists argue that forces at work in the lowest level are all that are needed to produce higher-level phenomena. All the work is really done by the parts, and thus ultimately by only the smallest physical parts—all wholes are merely powerless epiphenomena. Thus, all truths about the wholes are ultimately deducible from the collective truths about the parts.

To reductionists, there is no mystery of emergence because there is no emergence—only more and more complex interactions of the simple parts. A TOE will explain it all. Reductionists can accept the existence of random events at different points in the history of the universe that would keep a determinism from existing; thus, no exact track of the contingent events may be deducible from a TOE. Still, no "emergent forces" are at work in nature—nothing more than physics will ultimately needed to explain emergence or evolution. Understanding how the parts work will eventually explain it all. Reductionists see physics engulfing other sciences one by one.[44] In the end, the scientific account of it all will be remarkably simple, with mechanical explanations in terms of basic physics. All the diversity and complexity of reality arise from the laws governing the smallest particles, like a simple computer program generating the intricate fractal designs. The only reason there appears to be different "levels" operating with different laws is the increasingly complex mixture of things operating under the basic physical forces that is occurring as the universe evolves.

But why reality travels an upward path of more and more complexity remains a mystery to reductionists. And this assumes there is in fact a foundational level of organization. (The fundamental structure may, like fractals, generate an endless depth of new and complex levels "all the way down." Many physicists have no problem with that idea—as Freeman Dyson says, the universe may be infinite in all directions. Thus, there may be a fundamental law but no ontologically fundamental level.) Explaining how the properties of water arise through chemical bonds removes one "how" mystery but not the mystery of why reality is being organized into layers to begin with. Thus, reductionists have a long way to go to explain what is happening. Even the first step in their research program of explaining only the "how" questions—i.e., establishing correlations of higher and lower phenomena that would permit predictions—seems problematic at present. After that, they will have to create a complete chain of explanation that leads down to the smallest physical sublevel. Reductionists then will have to deal with a central feature of reality: why is there the emergence of higher and higher levels of organization at all? Why didn't the universe remain on the quantum level, and why didn't the complexities of interactions remain of one type? In short, why and how did the non-quantum world emerge? The structural reductionists' quest is to find the smallest parts of reality and the most general physical forces and then to explain all of reality in those terms—but this in itself cannot explain the presence of the upward drive in complexity. *How*, and equally important *why*, does a pile of interacting inanimate material apparently operating only under physical forces become reproducing and exhibit the other indices of life? And why then do some of the piles become aware of themselves and exhibit the other indices of consciousness? Evolution may explain the how-ness of the development of life and consciousness once they arose, but why is reality organized to permit the level of life to appear to begin with? Structural antireductionists, who accept the reality and irreducibility of different levels of organization and many of whom emphasize emergence, also as yet have no explanation for the upward-oriented organization of the universe, but at least they accept the need to recognize more mysteries as central to the how-ness of reality.

But whether one is a reductionist or an antireductionist, the mystery of emergence is undeniable at present. There is a general trend toward more and more orders of organization. Reality appears to be an organization-generating system. Emergence appears as a central feature of reality. In fact, for all we know, more levels of organization may emerge in the future—some thinkers suggest a global "collective consciousness." But the nature of whatever occurs will be as unpredictable as life is from quarks. Why is the universe so creative in this way?[45] Even if there are just random changes, why is the universe organized into levels and why do more and more levels appear? Scientists may be able to explain how higher levels of complexity emerge, but why is reality set up to permit it? Our inability to answer these questions reveals the limited nature of our current sciences. The greater and greater complexity and the emergence of new levels reveal that science presents only an abstract view of one aspect of the natural order. And whether we will ever be able to penetrate the mechanisms behind emergence is at present an open question. In any case, this phenomenon shows that reality is more creative and more mysterious than our current sciences would lead us to suspect.

Life

Two higher levels of obvious importance to us deserve special mention: life and consciousness. What exactly "life" is—what distinguishes a living person from a dead body—is harder to define that one might think. The processes of self-reproduction, growth, and self-repair are central. Living beings exhibit an autonomy not exhibited by inanimate objects. (This leads to the problem of intentionality of action and the need for a different type of explanation for human actions than just inanimate causation. More generally, whether an explanation in terms of goals to be achieved—teleological explanations—is needed in the animate world and perhaps still in the inanimate world is also an issue.) But overall, scientists are not in agreement on what constitutes "living." Biologists debate whether macromolecules or viruses are "alive." Crystals can "seed" other material to make more crystals, but nothing characteristic of life ever appears. Nevertheless, despite the definitional problem, life appears as a separate level of complex phenomena, and how nature made this monumental leap is the basic mystery underlying biology.

Strides are being made in finding a natural explanation for the origin of life—e.g., self-reproducing molecules have been created in experiments. But an even more fundamental mystery is how ingrained into the universe life really is. The appearance of the biological and mental levels of reality seems to be as natural as the appearance of atoms and molecules. But reductionists believe life arose only as an accidental product of physical forces and random events—the inanimate is "more real," and life is just an amazing fluke of nature.[46] But to antireductionists, life is a feature of the universe on an equal footing with the physical: it is as much built into the overall structure of reality as quarks and atoms. We are a normal feature of the universe and not a freak occurrence. To reductionists, life should be rare in the universe; it will occur "spontaneously" under the right base-conditions, but random events are necessary for these conditions to evolve and thus the right conditions only occur by chance; complex beings crossing the threshold to self-consciousness would be an even rarer accident. To antireductionists, life should be fairly common throughout the universe since some structures in reality as funda-mental as the physical and chemical structures are responsible for its appearance. To some antireductionists, the universe is not only "bio-friendly"[47]—life is absolutely central to the universe's design. (The Anthropic Principle will be discussed in the next chapter.)

The problem that reductionists have is that the laws of chemical bonding alone will not explain the emergence of life. The bonding is completely closed on its own level, and it cannot lead to anything new in terms of higher levels of phenomena. Why are the laws such that life will appear? Chemistry simply does not appear to account for the origin of the existence of a biological level of "information." Reductionists must explain by the mechanics of bonding or other physical or chemical processes how molecules obeying only chemical laws could become arranged into entities that maintain and reproduce themselves. Conversely, unless there is top-down causation, events on the biological level do not interfere with the causal completeness of the mechanisms at work on the chemical base-level.

Biologists admit that they do not yet know how life began. Indeed, it is one of the biggest puzzles of science. Neo-Darwinian theory does not purport to explain it but only explains changes after it arose. A cell is more complex than any inanimate entity of the same size. Even the simplest unicellar animal is so complex that how it arose perplexes reductionists and antireductionists alike. All sorts of conditions had to appear for life to arise, requiring millions of years of evolution on earth. Most scientists agree that the original template for life (whether it is the DNA or RNA chemicals in proteins) was so complex that it could not have appeared by pure chance—natural forces had to play a role in restricting the events surrounding how it came into existence. But life on earth still appears so miraculous that the reductionist Francis Crick suggested that it began by being seeded from space. (Of course, this only pushes the question of the ultimate origin of life back one step to how the life of the seeds arose.) Certain chemical bases are needed, but how life arises from them is unknown. What exactly the necessary bases are may not be known until we find life on other planets. For example, is carbon necessary, or are non-carbon-based life-forms possible? Scientists have been able to create some self-replicating amino acid molecules, but even if scientists can generate life in a laboratory, this does not mean that they have duplicated how nature did it any more than the airplanes we build duplicate how nature made birds. The historical question of the emergence of life may never in fact be answered. Moreover, even if scientists succeed in cooking up life from a prebiotic molecular broth, this would not resolve the issue of whether there are biotic structures at work or not: we would still not know whether physical forces are producing life or whether we merely have assembled the necessary base-conditions—the "building blocks of life"—for the biotic structures to kick in. Nor would it change the fact that, as Karl Popper said, life represents something utterly new in the universe.

Conscious Beings

Conscious beings present an even greater mystery. There is even less agreement on the scientific characterization of consciousness than with life. For beings with consciousness to emerge, the biological bases must become more and more complex. But a new level of organization again appears to emerge from the bases, and how all the various aspects of consciousness emerge is at least as mysterious as the emergence of life. Sensory sensitivity to our environment may be fairly easily explainable in biological terms (like plants being sensitive to light), but other forms of consciousness become harder and harder to explain. The subjective "feel" of the redness of an apple or abstract thought or the sense of self does not seem possible to explicate in physical terms of sensitivity to electromagnetic vibrations or even in biological terms of the actions of neurons. The possibility of a social cause in the emergence of consciousness further complicates the picture.

Materialists have no explanation for why or how consciousness appeared, and it is difficult to see how a physical theory could explain it. As David Chalmers puts the problem, how can something as unconscious as matter produce something so immaterial as consciousness? The problem is not merely that at present we do not know the mechanisms that give rise to consciousness—rather, we cannot imagine how *any* mechanisms *could* give rise to it. Thoughts and sense-experience take up no space and have no weight or any other physical properties; when we see the greenness of grass, neurons in our brain do not turn green. But reductionists reject any idea that the mental is anything but something physical. In Crick's words, we are "no more than the behavior of a vast assembly of nerve cells and their associated molecules."[48] Like all animals, we are just machines with only inanimate physical properties. Physics eventually will explain everything. Matter evolves into certain configurations governed only by the physical forces, and purely random events cause life and mental faculties to arise; these higher-level physical phenomena disappear when the configurations become disorganized at death. Moreover, there is no causal agent called the "self"—in fact, the entire mental level has no causal power over matter but is merely a powerless epiphenomenon. The sense of "self" is an illusion caused by the interaction of epiphenomenal mental activity.[49] Indeed, some reductionists dismiss all mental phenomena as purely illusory: there are only "objective" events occurring in the brain and no "subjective" side to things at all. Computers will eventually be able to duplicate everything "human," and then there will be nothing interesting left to say on the topic.

Antireductionists, on the other hand, accept the mental level of organization as being as real as any other level. Life and the mind are regular features of reality, reflecting structures that are as natural and inherent in the fabric of reality as the physical ones. The physical and biological levels are necessary bases for the appearance of the conscious level, but they are no "more real" or "more basic" or otherwise privileged.[50] There are "explanatory gaps" between the inanimate and the living and between the non-conscious and the conscious that will make any accounts of the bases of consciousness inadequate to explain the rise of the new level and the activity on that level. To antireductive naturalists, the emergence of consciousness may remain mysterious to us forever, but it is just as ordinary as the emergence of different physical sublevels.[51] The new level is an objective part of the structural order of things even if we experience it only subjectively. Our subjective experience is not reducible or unreal but is as much a feature of reality as what is experienced. Similarly, the idea that the mind has a causal power in the scheme of natural structures seems only natural to antireductionists: a person is a natural system, and our mind is a perfectly ordinary part of the causal structure of things. How the mind emerges and then interacts with the brain remains a mystery—e.g., is it like a field? But in any case, consciousness is as real as the brain. And most importantly, this means that we have free will: lower-level bases do not determine human actions, and a determinism of inanimate objects is rendered impossible. Our conscious decisions control the consciousness that has a causal power in our actions. There may be no distinct "self," but the mental components of a self are there and have causal power in nature. Indeed, all the structures of the universe that are familiar to us meet in personhood, which suggests that we are in fact at the center of reality structurally, if not geographically. At a minimum, it means that we are quite normal in the scheme of things. Pound for pound our brain is the most complex bit of organized matter we know on any scale, and that only adds to the significance of a person.

One consequence of both complexity and emergence is the centrality of the whole. If we start with the primacy of the person and the lived body, as William James and phenomenologists assert, we may learn something more important about the nature of reality than focusing on the possible connections of the parts or the interactions of structures. The idea of human free will and control seems obvious to us. To deny the reality these faculties or to dismiss consciousness as an epiphenomenon is difficult to maintain: why would they have evolved or exist in any way if they did not have some value in our survival? Accepting them as having power in the world is more in keeping with the nature of a dynamic and creative reality growing in complexity than its denial.

Because of the complexity of any actual phenomenon, we may never be able to identify all the factors at work in an event and thus we may never be able to tell if consciousness can be reduced. It then comes down to a decision in metaphysics, and there is no scientific reason to adopt the reductionist point of view. First, as discussed above, scientists can readily accept that there is more to reality than just what they study. Second, reductionism is not the "scientific view" of reality but a philosophy. The *analysis* of how a phenomenon's parts work is part of science, but *reducing the reality* of a phenomenon to those parts is not. Science itself does not compel the conclusion that macroscopic objects are no more than the effects of microscopic processes. Just because we know how the parts work does not mean that we know why the whole formed. Third, the metaphysics of reductionism may guide research, and scientists do try to unify forces within overarching theories, but they are not committed to coercing a unification of causes from different levels if the factors on different levels are found to be distinct. Scientists have no problem working with distinct theories and models, and any forced attempt at unifying theories of different levels of phenomena is a metaphysical enterprise, not an actual scientific task. Fourth, reductionists reject consciousness because they believe mental causes would invalidate the scientific picture of reality. But, as discussed above, science is a matter of causal conditionals and not the nature of the causes. Nothing in physics rules out that there are nonphysical causes at work in reality except the more metaphysical principles about how reality operates that lie outside of actual causal conditionals, e.g., the law of conservation of energy discussed earlier. Fifth, reality, not metaphysics, should dictate what scientists study: if after studying the issue we cannot but accept consciousness as a cause of events, then scientists are stuck with it—scientists have to explain consciousness, not dismiss it out of hand on metaphysical grounds.

It may well be that consciousness will remain impervious to scientific scrutiny, even if it is only an ordinary natural product, simply because we are organisms that have not evolved to understand it.[52] Furthermore, we may never know whether conscious beings are an almost impossible miracle in the universe or a common occurrence. It may be that consciousness is a fundamental feature of reality programmed into the universe, i.e., the universe has written into its forces and laws a way of knowing itself. Or maybe not. As Bertrand Russell says, we simply have no idea of life's relation to the universe: it may be the final climax, an accidental and unimportant by-product, or just a disease affecting matter in its old age.[53] We are left with a basic mystery.

Biological Evolution

The evolution of life on earth is a different type of complexity than emergence: emergence deals with the appearance of new levels of organization (e.g., life emerging from chemical bases) while evolution is about the development of new complex configurations within one level alone. It is another part of the overall complexification occurring in the universe. Evolution shows on one scale how change and the appearance of greater and greater complexity both appear to be a central feature of reality. It also reveals a historical dimension to reality—that time moves in only one direction—that physics and chemistry omit. Its own mysteries involve how life blossomed from one cell to the incredible diversity of mind-boggling complex multicelled organisms today.

The evolution of diverse life-forms on this one tiny planet is indeed remarkable. Life on earth appeared surprisingly fast (in cosmological terms) after the sun and the planets were formed. All life appears to have arisen from the same DNA template. All plants, fungi, and animals appear to have descended from this material. Thus, all forms of life today are related. No one knows how replication began, but evolution took off after life got started. Life remained in simple one-celled forms for two billion years. Relatively simple multicelled life-forms existed for the next billion years or so, transforming an atmosphere that would be toxic to modern forms of life into an atmosphere permitting complex forms of life to appear. After plants generated enough free oxygen in the atmosphere and when the other physical and chemical base-conditions were right, there then appears to have been a sudden explosion in the Cambrian period of different complex plant, fungus, and animal species that gave rise to the major phyla existing today. (Since the fossil record is so slight, how much evolution of simpler parts for these species occurred before this explosion is an issue.) The earth has transformed over and over again throughout history, and life has adapted to the changes. And it must also noted that simple and even complex forms of life seem to thrive in very extreme conditions of heat and cold in the thin membrane on earth that permits life.

The convoluted nature of evolution is hard to miss: life is a sprawling bush with intertwining branches going in every direction, and with innumerable stunted limbs. There appears to be no "progress toward a goal" other than achieving a mushrooming diversity and complexity. Today there are a hundred thousand known species of fungi, about a quarter of million species of plants, and over a million species of animals—nine hundred species of bats alone. With such a plentitude of diverse species, it is hard to see humans as the crown of creation in the sense that all of reality was merely leading up to us. Over the past few million years, over a dozen different hominoid species have existed—many at the same time—and have gone extinct. *Homo sapiens* have been around for perhaps a few hundred thousand years. We are just one more branch of life whose chief distinctions are our highly social nature and the development of our level of consciousness. Our ability to develop language has allowed us to create symbolic images of the world that enable us to react against our environment and to pass on information as other species have not, thus giving a Lamarckian twist to cultural evolution and thus indirectly to further human biological evolution. Our linguistic ability may have had an effect on our subsequent physical evolution (including the size and organization of our brain) that separated us more from other primates, even though we are closely related— other primates lack the neural capacity to handle abstract thought. None of these things—consciousness, self-awareness, language, social organization—are unique to humans. (Edward O. Wilson sees colonies of ants and bees as mindless "superorganisms," with each individual being a cell that operates as an automaton; the rules they operate by even give the appearance of intelligence. Humans differ by actually having intelligence.) Our degree of each feature does separate us from other organisms, but our continuity with the rest of life, past and present, is at least as important to what we are.[54]

That life has evolved slowly over billions of years from one-cell beings to the vast diversity we see today is not disputed by anyone who takes science seriously. There is a biological continuity from common ancestors, but organisms have not remained stagnant over time: they have adapted as their environments have changed and they have grown more complex. This is a fact if scientists have established any fact at all. However, the *explanation* of how evolution has occurred is disputed. Charles Darwin and Alfred Russel Wallace proposed a natural mechanism for the changes: natural selection.[55] Members of a specie that have survival-related advantages over other members in their environment tend to thrive and to produce more offspring that survive. This idea has now been combined with modern genetics: changes occur randomly in the genes (e.g., radiation may alter a gene), and these are then passed on in a lawful manner to the offspring through inheritance. Thus, both randomness and laws are involved. (Again, randomness does not mean *uncaused.* The meteorite that destroyed the dinosaurs may have been the result only of causal forces, but it was a contingent event that is not deducible from laws alone. So too with the other four global events that destroyed huge percentages of the existing life-forms. Because of the random element, if the history of life were rerun, the outcome would be very different—forms of life may be very different, with none of them conscious. According to the biologist Stephen Jay Gould, if the history of life on earth were rerun a million times, chances are that it would never again give rise to mammals, let alone any mammals with the intelligence of human beings. Indeed, if the history of the earth were rerun, life itself may not arise at all, unless some biotic structures to reality are also involved that assures the presence of life. This makes the study of life on earth very much a historical science of contingent events.) Genetic mutations that are advantageous in adapting to the environment or to environmental changes lead to more offspring surviving. Any adaptative advantages acquired by actions during an organism's life are not passed on genetically (Lamarck's version of evolution) and so are ultimately irrelevant to the future of evolution. If an organism fits its environmental niche and its environment does not change, it need not change. Some species (e.g., sharks) have been around since the age of the dinosaurs because they are well-adapted and their particular environments have not changed with regard to their needs. Species can also devolve if their environment permits simpler entities (e.g., fish living in caves losing functioning eyes).

But neo-Darwinism does not explain the general trends toward greater diversity and greater complexity. Nothing about natural selection, any more than the underlying chemistry, would predict the very possibility of a continuous increase in diversity and complexity, even if complexity somehow enhances survivability. The development both of divergent lines and of greater complexity is consistent with natural selection, but why would they occur? Even if adapting to changes in the environment is responsible for the appearance of new species, why do new degrees of *complexity* develop? Natural selection may effectively weed out the unfit, but how does it create something new? When an environment changes, shouldn't a species die out? Why adapt? Or why do distinct new species arise rather than just better adaptations of the existing species? Even if an accumulation of small changes over long periods of time can create a whole new species, why is there a drive to create new species at all? More generally, why didn't life remain on the one-cell level? Wouldn't one-celled animals be more easily adaptable to changes in the environment? Conversely, if complexity is so valuable, why aren't all species now complex? Why are there any single-celled species left?

Molecular biochemists in the field of evolutionary developmental embryology—"Evo Devo"—are making major strides in explaining how a simple cell develops into a complete adult by focusing on how the same genes are switched on and off at various stages in an organism's development from the embryo to adulthood in different species. They are also studying the development of life-forms such as the Cambrian explosion and more generally how evolution works with genetic tools to build increasingly complex organs. But why did cells begin to have specialized functions? Why would single-celled organisms begin to fuse together? If we want to argue that, say, consciousness simply arose to help species adapt and thrive, we can still ask why reality is organized to permit such new levels of complex phenomenon to appear.[56] That we are now capable of thoughts and scientific theories that go far beyond what is necessary to survive also presents a problem for the idea that consciousness merely evolved for our survival. But more generally, why would changes in the environment cause beings to acquire greater and greater complexity? Growth in size and complexity without structure should lead to instability and collapse. And even if stability were to result, it is not clear how randomness and the transmission of genetic changes alone could produce greater types of complexity rather than simply more of the same type of complexity adapting to the environmental changes.

Some biologists today are moving beyond neo-Darwinian reductionism and arguing that evolution is a multileveled process that involves group-selection and broad types of cooperation.[57] Some antireductionists are arguing that basic biotic structures must be at work in the origin and evolution of life. Nature has programmed an urge for organisms to survive or to adapt to their environment and also for life to attain greater complexity. Some rules must be programmed into reality to generate novelty rather than chaos. As Paul Davies says, "[t]his systematic advance in organized complexity is so striking it has the appearance of a law of nature."[58] Although this idea involves no role for a transcendental designer but only natural forces, it is still basically non-Darwinian (since it limits the role for randomness in the course of evolution). These antireductionists accept that natural selection plays some role in evolution, but they think that neo-Darwinism does not provide the full picture and that some more complete theory incorporating natural selection and genetic mutation will one day be forthcoming when scientists know more about the workings of life. This may mean that a true revolution is needed in biological thought. Davies, for one, thinks a blend of molecular Darwinism and laws of organizational complexity could offer a way forward.[59] If Darwin was the Newton of biology, we may still be waiting for its Einstein to explain the emergence of more and more complex organs and species.

But these antireductionists must confront one big problem concerning consciousness: if consciousness is programmed into reality or if consciousness is so valuable for survival, why has evolution produced so little of our level of development over so long a period of time? Stalwart reductionist defenders of neo-Darwinism such as Richard Dawkins think that random mutation and natural selection operating on a genetic level are the only mechanisms at work in evolution and thus neo-Darwinism explains the history of life. Reductionists believe that Darwin solved all the grand mysteries of life—all that is left is for biologists to fill in the details and for chemists and physicists to reduce biology.[60] All we have to do is discover the basic physical laws at work—the actual history of life on earth is a history of contingent events and thus is, as the chemist and reductionist Peter Atkins says, "scientifically unimportant"[61] since scientists only identify and explain the laws and forces at work in nature.

However, mysteries both of why and how remain surrounding both the origin and development of life. Why and how did the universe get built in such a way that replication was possible and a trend to diversity and higher levels of complexity is so prominent? How did even microbes emerge, and why did higher levels of biological complexity evolve? How do we reconcile the two trends that the universe exhibits: toward chaos through entropy and toward more and more stable organized complexity? Are as yet unknown forces at work? We know less about the basics of yet another familiar phenomenon—life—than might be supposed. And in the end, biology may be more central in our overall understanding of the nature of the universe than the more abstract sciences of physics and chemistry.

The Limits of Science

Thus, to summarize science's relation to mystery succinctly: science does not touch the basic metaphysical why-mysteries, it only increases our sense of awe and wonder at the extent and workings of nature, and, while defeating how-mysteries, to date it has only revealed deeper how-mysteries. Scientists begin with a sense of wonder at both the ordinary and the exotic features of nature. But what drives science is a relentless curiosity that tries to turn mysteries of how nature operates into problems to be solved. It is a process of demystication of how-mysteries. But the limits of science must be recognized. The limited scope of what scientists can achieve is both why science is so successful and why it is not a comprehensive account of reality. Even if complexity theorists get past the hurdle that scientists deal in abstractions and idealizations of what is real, and not with the full complexity of nature's interactions, limits persist.

First, it must be remembered that science is very much a human enterprise. While science in recent decades has become more and more rigorous on the empirical side through the use of computers and other new technology, it is also becoming clear that it is less exact on the theoretical side than logical positivists thought. Science cannot be reduced simply to the correlations of phenomena or predictions of instrument readings—it involves what we think are the causes underlying the phenomena. Science is a matter of explaining and our understanding. As discussed in the last chapter, that scientists can get some predictions right and others wrong suggests that they must be on to something with a successful theory. Still, it must be remembered that reality responds only to the questions we ask, and predictive success is no guarantee that we are asking the questions that will reveal the true causal structure behind reality. We can see why empiricists mistrust scientists' claims about the nature of reality apart from the causal patterns scientists detect. Observation guides science, but the feedback to our questions from observations does not always end the matter—our guesses at what is going on will always be central.

Thus, much depends on the limitations of what we can conceive. How reliable are our intuitions shaped by our everyday experiences? Our brains have evolved for minds that work in the everyday world, and reality constantly shows scientists that our common sense is useless in the more exotic conditions of the quantum and cosmological levels of reality—there the universe appears bizarre, perhaps even more so than we could ever imagine. This does not mean that all scientific mysteries must be beyond our ken, but the essential role of the familiar through models limits science. We may resist the urge to accept as real only what we can comprehend, but our ideas of simplicity and elegance and what finally makes sense to us become a criterion of what we accept as real because we cannot do otherwise. As the physicist John Wheeler says, "[w]hat we call 'reality' consists of an elaborate papier-mâché construction of imagination and theory filled in between a few iron posts of observation." But contrary to postmodernism, there are those firm "iron posts." Theories are only our guesses at the workings of reality, but it is hard to deny that some are well-established. If our minds can reflect something of nature because it has a base in the evolved brain, this does suggest that scientists can attain at least some genuine knowledge of something about reality's causal structure. The universe does not appear to be a totally unstructured mystery, even though it remains a constant source of resistance to our attempts at understanding.

Second, science is limited only to the search for the causal skeleton of reality, and everyone agrees that we do not yet know all there is to know scientifically even about that, and we must accept that we may never know all there is to know about such workings. The question is whether there are ontological or epistemic mysteries that will remain forever. The world is comprehensible to us to a degree, but it may well not be totally comprehensible to its core. It may be that what is rational to us about reality is only a very limited aspect of it all—scientists may be abstracting out only the easiest parts that we can control but missing many more fundamental causal factors at work. We have no reason to conclude that what we can experience directly or indirectly or can conceive must be all there is to reality. There may well be depths we cannot comprehend or even conceive—the intellectual analog to "dark matter." Newton likened himself to a small boy on the seashore diverting himself by finding a smoother pebble or a prettier seashell than most while the great ocean of truth lies undiscovered behind him. That there is more to reality than the scientific abstractions is not a problem for science, but it is a problem if there are more to the structures of reality than science currently recognizes. Even if reality is rational through and through, we may never be able to fathom all of it, and thus we may be stuck with epistemic mysteries.

As the astronomer John Barrow concludes, "[t]here is no reason why the most fundamental aspects of the laws of nature should be within the grasp of human minds, which evolved for quite different purposes, or why those laws should have testable consequences at the moderate energies and temperatures that necessarily characterize life-supporting planetary environments."[62] In fact, why should an organism with the particular mental capacities evolution has provided us up to this point be able to know all there is to know about how nature works? What makes us think we now have a perfect mind that can know everything? As Darwin asked, why would anyone trust in the convictions of "a monkey's mind"? Only a portion of reality is accessible to our senses, and this is a reason to keep a tight check of testability on our theorizing. We only see a fraction of the electromagnetic spectrum, now expanded by technology to the ultraviolet and infrared. Would our science differ if we saw another range, as some animals do? Also remember our limited location in the universe: even if we could reason perfectly, any limitation of data limits what we can imagine. There may be a lot we could learn if we could travel further. With all these limitations, it is only to be expected that we may never be privy to the ultimate secrets of how nature operates—indeed, it is arrogant to think otherwise.

Einstein once remarked that there is no mystery in science that does not point to another mystery beyond it. In fundamental research, this has certainly been true so far. Indeed, contrary to what the logical positivists might have expected, as some puzzles are allayed, the presence of greater mysteries has only become clearer. Each new finding has only opened up new horizons that we previously did not even suspect existed. And, unless reality is truly transparent, at some point we will be up against the limits of what our mental and technological abilities can find out about nature or even conceive about the causes. Perhaps we are at the end of major evolutionary changes in humans. Perhaps the era of great scientific discoveries is coming to an end. Perhaps not. It may be that in the future humans or their successors or machines or something else will learn much more of the how-ness of reality.[63] But we have to admit, however reluctantly, that we now know less about the fundamental how-ness mysteries than we usually like to suppose and that we may never conquer them all. In fact, scientists' repeated failures in their effort to close the openness of reality should make them especially aware of the mysterious quality of reality's structures. In addition, the basic philosophical problems related to both matter and structures—why there is anything at all, why there is order, where any laws of nature come from—will remain untouched. In such circumstances, we need less self-assurance and a little more humility.

As discussed at the beginning of this chapter, the vastness, subtlety, and complexity of the fullness of reality brings a sense of awe and mystery to scientific knowledge. Most scientists in their workaday life do not look at the big picture and ignore mystery—they focus on only gathering data or conquering the next research puzzle, not on the deeper how-puzzles underlying their work. But in their more reflective moments they may be struck with how amazing it is that there is, for example, something as simple as a blooming flower. Even if we knew everything about the mechanics of how a flower blooms, the awe and sense of mystery are not decreased—Richard Feynman and others argue that the excitement, awe, and mystery is in fact *increased*, since the mystery is no longer based on ignorance. The same is true of all scientific findings, even the most familiar—that the earth moves, that stars are trillions of miles away, and so forth. Science thus does not necessarily lessen the mystery of the known. It can deepen our sense of mystery by revealing that there is more that we do not know. The philosopher's disease here would only work to stifle that sense by finding closure where there is none.

Scientists can split the atom and detect events lasting a billionth of a second, but our scientific knowledge, to use William James's phrase, is only a drop in a sea of ignorance. Three great revolutions occurred in physics and astronomy in the twentieth century: quantum physics, relativity, and the discovery that the universe is expanding. But one of the most profound philosophical discoveries of that century also came from science: we now realize that we may never know all there is to know and that uncertainty surrounds much that we do know. If nothing else, uncertainty about natural events is currently a necessary part of our most basic physics—quantum physics—and hence uncertainty is "inherent in the nature of things," as Paul Dirac said. Thus, as Jacob Bronowski pointed out, we now know that physics' goal of giving us an exact picture of the material world appears unattainable. Mystery remains at the center of our knowledge. Our sense of mystery of the how-ness of reality should be greater now than at any time since the rise of modern science. According to Einstein, "the real nature of things—that we shall never know, never." Max Planck would add that "science cannot solve the ultimate mystery of nature. And that is because in the last analysis, we ourselves are part of nature and therefore part the mystery that we are trying to solve." Even so, he did not think that this stopped science from being "the pursuit of the unknowable" —indeed, he thought that is its chief attraction. We may like to concentrate on the part of reality that we know, but that it is surrounded by a sea of mystery should not be forgotten.

Notes

1. See Hepburn 1984.

2. See Horgan 1996.

3. *Culture and Value*, trans. by Peter Winch (Oxford: Blackwell, 1980), pp. 5. He added that wisdom is like "cold, grey ash, covering up the glowing embers of life" (p. 56).

4. Edward O. Wilson, *Biophilia* (Cambridge: Harvard University Press, 1984), p. 10.

5. Ford 2007, p. 240.

6. See Jones 2000, chapter 3.

7. Whether mathematical realities should be included in a naturalist's ontology is controversial. Consciousness also presents a problem since naturalists want to consider it a purely natural product, but how to study it empirically is difficult to specify even in principle—scientists can identify the neurological bases, but they still cannot figure out how or why consciousness arises or how to study consciousness itself. This leads many materialists to reduce the mind to the body or to deny the reality of consciousness altogether. Nevertheless, it is considered part of the natural order even if scientists can never study it.

8. Whether the universe is actually constructed with economy might be open to question. For example, the DNA sequences of living beings have redundancies and back-up systems.

9. See Rescher 1984, pp. 83-98.

10. Davies 1992, p. 20.

11. Many other questions can also be asked. Does the fact that autistic savants can calculate mathematical results with lightning speed without any idea of how they arrive at the results tell us anything about the grounding of mathematics? Are they tapping into some real feature of reality, or are they showing that mathematics is just a matter of the human mind? How about mathematicians like Srinivasa Ramanujan for whom novel and difficult theorems seemed "to flow from his brain effortlessly"? Was he discovering something about reality or just reflecting his mind?

12. Jeans 1928.

13. Davies 2003, p. v. "Whatever some over-enthusiastic physicists might claim, there is much which is beyond our grasp, and which will probably remain so" (p. vi). He gives consciousness and the true nature of quantum particles as examples.

14. Nancy Cartwright, *How the Laws of Physics Lie* (Oxford: Clarendon Press, 1983).

15. See van Fraassen 1980, 1989.

16. The conflicts between religion and science can also be framed in terms of this analogy: in general the religious are concerned with the meaning of the text (the "why" questions related to the meaning and purpose of the world and our place in it), while scientists are concerned with only the vocabulary (the "what-ness" of reality) and the grammar (the "how" questions of nature's works). But occasionally scientists find vocabulary or a rule of grammar that precludes a reading that the religious want to give nature (e.g., that the earth is not the center of the universe or that humans are evolved), and so the religious want to tell scientists what the vocabulary or rules of grammar *must be*—hence the conflicts.

17. Under current theory some particles are charged with force (bosons) and some not (quarks and leptons), but the ontological dichotomy of being and structure remains.

18. On the issue of "matter," see Young 1965, McMullin 1978, and Trusted 1999.

19. John A. Wheeler, "Law Without Law." In John Wheeler and Wojciech Zurek, eds., *Quantum Theory and Measurement* (Princeton: Princeton University Press, 1983), p. 210.

20. If matter is just for the embodiment of structures, there may be no "bare matter." Or if structuring is equally fundamental to matter, again there may be no matter without structure.

21. Rees 1999, p. 176.

22. All wave phenomena (e.g., light) also have particle effects, and all particles (e.g., electrons or protons) also have wave effects. This situation is not as paradoxical as it is usually portrayed. Of course, nothing can be both a wave and a particle, and nothing subatomic is. It is wrong to say that "light can be seen as composed of either waves or particles." In itself, a photon or electron is neither a wave nor a particle but a reality that manifests such properties in different settings that physicists create by their experiments. That is, whatever is there behaves like a particle under some conditions and behaves likes a wave under other conditions that physicists create. The subatomic realities "in themselves" are never revealed in experiments. When subjected to one experimental set-up, what is real reveals wave-like properties, and when subjected to another set-up, it reveals particle-like properties. The subatomic reality consistently appears wave-like under certain experiments and consistently appears particle-like under others, and we cannot observe both properties simultaneously. Each point of view is limited and partial. In short, if we do one thing to the reality that is there, wave-properties appear; if we do another, particle-properties. What we see is always a combination of the reality that is there and our intruding experimental procedures. Thus, what appears depends on different combinations of things, including what we do; if we did not interfere, these properties may not exist, and the reality is unaffected. We make what is there into a "wave" or "particle" by shooting photons of light into the mix. But this fact has none of the internal inconsistency of a paradox. That x (what is there) plus y (one set of conditions) does not equal x plus z (another set of conditions) is not a paradox. Not only is this a logically consistent idea, it is not difficult to understand once we realize that a photon of light is comparably powerful on the subatomic scale: we cannot "illuminate" something without affecting it and fixing some properties—the act of observation becomes part of the quantum event. In addition, the mathematics of quantum physics produces a consistent account that permits predicting probabilities very precisely. Nevertheless, the situation does point to the fact that we do not know the subatomic realities in themselves. Our lack of understanding leads to the "contradictions" in the limited aspects we do know. The underlying reality-in-itself that produces the different appearances when approached

differently remains a mystery.

    23. See Barbour 1999.

    24. Sean Carroll says that "[t]he arrow of time is arguably the most blatant feature of the universe that cosmologists are currently at an utter loss to explain." "The Cosmic Origins of Time's Arrow," *Scientific American* 298 (June 2008): 49-50. This relates to the broader mystery of why the entropy of the universe was low at first and then evolved to high-entropy states. Carroll argues that either the asymmetry of time is a blunt feature of the universe that escapes explanation or it points to deeper workings of time and space, hinting that the existence of a much larger space-time of a multiverse having universes in which time runs backward (thereby making the total universe time-symmetric).

    25. Fagg 1995, p. 250.

    26. Fred Hoyle, an opponent of the theory, created the term "Big Bang" as a term of derision. He, Thomas Gold, and Herman Bondi proposed a "steady state" theory in which "little bangs" of hydrogen atoms continuously occur, created by some unknown mechanism in a pre-existing space to keep space uniform as it expanded. (The same basic mysteries surrounding origins thus apply in this theory, except for the centrality of one historical event.) This theory was simple and elegant, but empirical evidence concerning the cosmic microwave background radiation showed that the universe in the past did not look like the universe of today. This strongly suggests that the Big Bang theory is more accurate, although Hoyle was never totally convinced and a small group of dissident cosmologists today have advanced a modified form of the steady state theory.

    27. Superstring theory is currently being applied to cosmology, producing an alternative to inflation theory. Another alternative is the ekpyrotic theory in which different universes are seen as different three-dimensional membranes floating in a higher dimensional super-space; occasionally two such membranes collide, and our Big Bang was the result of such a collision. There may be empirical ways to distinguish the theories related to fluctuations of temperatures in space. If so, this would lead someday to testing the existence of other universes.

    28. Barrow 1998, p. 250.

    29. Another major problem any TOE theorist will encounter is Gödel's theorems. Any TOE will be a mathematical equation or simple set of equations. However, Gödel's theorems hold that any mathematical system that has at least the power of elementary arithmetic will not be able to prove such consistency and completeness. If a TOE is such a system, there will be laws within the TOE that cannot be proven to be true or false. There will also be an inexhaustible number of possible consistent laws. Thus, unless a TOE has the formal logical structure of axioms and theorems simpler than arithmetic, then Gödel's theorem precludes the possibility of a complete theory of nature.

    30. Weinberg 2001.

    31. A question today is whether physics should approach reality as being structured through a computer program or through a set of mathematical equations. Some physicists think they should be looking for the software programs guiding the world; others think that symmetries revealed by studying equations—the "laws" of nature—is the way to go.

    32. This is usually described as science treating reality as "linear" when in fact it is "nonlinear." Before complexity theory, science made no pronouncements on the total interactions occurring in reality. Whether even complexity theorists will have to deal in idealized versions of complex situations raises a question for that science.

    33. See Lewin 1992.

34. See Kauffman 1993.

35. See Haldane 1996 and Jones 2000, chapter 10.

36. Three concepts must be distinguished here: the *emergence* of higher levels of organization, the *evolution* of more complexity within one level, and the *emanation* of being. The evolution of higher life-forms may be a matter of emergence, but it is not a matter of the emanation of being. "Emanation" is a concept in religious and metaphysical systems about how "levels of being" arise from a ground of being, typically in reverse order from a materialist perspective: spirit, consciousness, life, and finally matter. (More on this in Chapter 7.) The idea of "degrees of being" has fallen out of philosophy today—only whether life and consciousness are reducible to matter remains an issue.

37. For an example of a physicist arguing that physics ought to be rethought in terms of emergence rather than reduction, see Laughlin 2005. In his terms, the "phases of matter" —solid, liquid, gaseous—are "organizational phenomena." He argues that we should accept common sense and leave behind the practice of trivializing the organizational wonders of nature and accept that organization is important in itself—in some cases, it is even the most important thing (pp. 218-19). This rethinking would also have the effect of ending the idea of the absolute power of mathematics since emergence may not be formulatable that way (p. 209).

38. "General Systems Theory" as a philosophical school never gained wide support, and "systems" are emphasized by New Age advocates. But today even mainstream biologists are taking the idea of "systems biology" seriously, along with help from computer specialists and physicists, to try to determine how whole organisms work, e.g., how an embryo gets from a few cells to a full human with cells organized into specialized functions, or, indeed, how the genetic code assembles proteins into the embryo, or why humans have so few genes.

39. See Michael Polanyi, "Life's Irreducible Structure," *Science* 160 (21 June 1968): 1308-12.

40. Goodman 1960, 1978.

41. At least so argues the philosopher David L. Hull in his *Philosophy of Biological Science* (Englewood Cliffs, N.J.: Prentice-Hall, 1974), pp. 128-29.

42. Sheldrake 1988. For Sheldrake, laws are more like nature's *habits* than fixed *forces*, at least in the field of biology.

43. Davies 1999, p. 259.

44. Reduction is a matter of *explanations*—e.g., explaining biological phenomena in terms of chemical interactions. To look at this as "theory reduction" may be misleading since any biological theories are not so much "reduced" as superceded.

45. Theists may want to bring in design and God at this point. But it must be remembered that scientists will be looking for repeatable, natural causes for complexity and emergence. A strict causal order of forces combined with random natural events also remains a definite possibility. Thus, the idea that there may be other natural forces at work in the process of emergence annoys both reductionists who think physical forces will eventually explain everything and those theists who want to preserve a role for a god in designing the universe for life and consciousness. Both sides want the only alternatives to be reductionism and theism—because they each believe they will win that contest—with no third option (nonphysical but natural forces producing emergence) available.

46. "Lifeless matter" is a misnomer since there could not be any "living matter." That is, unless panpsychists are correct, in which case all matter is conscious in some way, part of reality is organic and part is inorganic because of the *structures* involved, and the dispute

involves what structures of organization are real.

47. Davies 1999.

48. Crick 1994, p. 3.

49. Dennett 1991.

50. Just as some thinkers have suggested that there are biotic fields structuring the emergence and development of life, some have suggested that there are "consciousness fields" with archetypal structures and transpersonal connections at work in our mental life.

51. See McGinn 1999.

52. See Chalmers 1996.

53. Bertrand Russell, *Religion and Science* (Oxford: Oxford University Press, 1997), pp. 218-19.

54. Some biologists argue that, if natural selection alone is at work, the shear quantity of people and the intermingling of groups today may mean that human evolution is now over—no significant innovations may be incorporatable into such a vast gene pool. Also unless the birth canal can get bigger, our brain cannot get much bigger. But that there may be other natural forces at work besides natural selection and physical forces cannot be ruled out. Nor do we know what the effect of genetic engineering or other human attempts to affect evolution might have.

55. The concept of "natural selection" is metaphoric since there is no agent, blind or otherwise, who does any "selecting." The concept is about impersonal changes in the genetic makeup of organisms. So too, the concept of a "selfish gene" is metaphoric: genes do not have the ability to decide to act in their self-interest; they also do not act individually or in isolation of the actions of other genes—the whole metaphor misleads by making evolution look more deliberate, individualistic, and competitive than it really is.

56. The arising of complexity in consciousness brings up another mystery: the origin of language. Those who advocate a universal grammar (e.g., Noam Chomsky) argue that Darwinian notions of evolution probably will never explain how our ability to use language developed.

57. See, e.g., Peter A. Corning, *Holistic Darwinism: Synergy, Cybernetics, and the Bioeconomics of Evolution* (Chicago: University of Chicago Press, 2005).

58. Davies 1999, p. 266. See also Kaufmann 1993 on order-generating effects rather than natural selection as the reason why genes tend to settle into recurring patterns.

59. Ibid. He adds that a "complexity law" might act to create information (p. 259).

60. Richard Dawkins, *The Blind Watchmaker* (New York: Norton, 1986), p. ix.

61. Peter Atkins, *Creation Revisited* (San Francisco: W. H. Freemans, 1992), p. 3.

62. Barrow 1997, pp. 213-14. Barrow delineates four limits to science: existential (i.e., our research may be directed at discovering theoretical structures that do not in fact exist), conceptual (if a deep theory of everything exists, how confident should we be about our ability to comprehend it?), technological (limits to observations), and fundamental (seeing the built-in limits of any mature theory, e.g., the Heisenberg Uncertainty Principle). See Dewdney 2004 for limitations related to mathematics.

63. Scientists have not always been the best prognosticators of the future of science. See Duncan and Weston-Smith 1977 for examples of how leading scientists did not always foresee what the central problems would be in their fields in the next decade. But scientists typically express an eternal optimism—that solutions lie just around the corner and in ten years these puzzles will all be solved.

— 5 —

Explaining the Universe

Whence did this creation arise? Whence did it come?
Who knows for certain? Who here shall declare it?
The gods were born after the creation of this world.
Then who can know how it arose?
None know from where this creation has arisen,
And whether he who surveys all from the highest heaven has
produced it or has not produced it.
He alone knows—or maybe even he does not know.

— Rig Veda X.129.6-7

Yājñavalkya: So ... all of this world is woven back and forth on
the Brahmā-worlds.
Gārgī: On what then are the worlds of Brahmā woven back and forth?
Yājñavalkya: Don't ask too many questions, Gārgī, or your head will
shatter apart! You are asking too many questions about a deity about
whom one should not ask too many questions.

— Brihadāranyka Upanishad III.6

Scientific inquiries into the Big Bang lead to a host of how-mysteries. How did the Big Bang get started? Why did it occur when it did? Where did it come from? What reality did it arise out of? If the Big Bang is a "quantum fluctuation," what is fluctuating and where did that come from? And how did the fluctuation get started? Within our universe, a second set of questions arise. Why does the universe possess the particular properties it has? Why does it have the creativity it has? Why are the laws of our universe the way they are? Could there have been alternative sets of laws? In fact, why is there any order at all? Where did the laws come from? In sum, why this world and not some other? This inquiry involves both the ultimate philosophical why-mystery—in Gottfried Leibniz's classic formulation, "why is there something rather than nothing?"—and the search for ultimate explanations in the realm of how-mysteries. Today the search for such ultimate explanations is taken with a seriousness, even by scientists, not seen since the Middle Ages.

"Why is There Something Rather Than Nothing?"

The physicist John Archibald Wheeler once said, "I know of no great thinker of any land or era who does not regard existence as the mystery of mysteries." But the question "why is there something rather than nothing at all?" is actually very odd. More fully, Leibniz, reflecting the philosopher's demand for answers, stated, "Why is there something rather than nothing? For nothing is both simpler and easier than something. Moreover, assuming that things must exist, there must be a reason why they exist thus and not otherwise." Milton Munitz refines Leibniz's formulation to "Is there a reason for the existence of the world?"[1] It is a more neutral way of framing the question than, for example, "Where did whatever exists (including God) come from?" That question requires an answer in terms of causes or origins. Friedrich Schelling's question "Why is there not nothing at all?" is hard to think about since we end up trying to form a visual image of "nothing at all," which is impossible. What is meant is the total absence of any thing, but there is the danger here of our mind reifying "nothing" into a reality—a big, black silence, empty of all things. So too we cannot realistically imagine that we do not exist: we can, of course, imagine a world in which we were not born, but *we* will be thinking this—what we cannot imagine is reality without that. Thus, the Cartesian "I think therefore I am" comes back to mess up the process.

The Christian theologian Richard Swinburne argues that "the most natural state of affairs of the existing world and even of God is *not* to exist *at all*!" That is, the most natural state of affairs would be the empty set, as it were. That would obviously be the *simplest* state of affairs. But it is hard to see how to evaluate whether that is the "*most natural*" state or even what exactly that means.[2] The philosopher Behe Rundle tries to dismiss what he sees as philosophy's central question by applying an "ordinary language" analysis and arguing that we cannot make sense of the idea that "there might have been nothing"—there must be something.[3] For example, we cannot conceive of a universe coming into being or going out of being—it always comes out of something or something is left. We are always left at least with the setting from which we ask the question and so never with true nothingness. Thus, he argues that there is no mystery why there is something rather than nothing: we cannot actually imagine there being nothing. Ordinary usage of the word "nothing" thus shows that "there could be nothing" is not a genuine, intelligible claim—it can only be applied to parts and not to the whole of reality. But just because any idea of "nothing" that we can possibly form makes it into a "something" does not mean that the idea is unintelligible. Certainly just having a word for it does not make into a something. It is impossible to *visualize* true "nothingness," but it is not clear why we cannot *conceive* the removal of all somethings that leaves nothing including the big, black space the things were in. In science, we can conceive of space as an inflating balloon expanding into nothing, but we cannot visualize that nothing. Similarly, here we can conceive the opposite process: the removal of all things. In short, it might have been the case that nothing ever existed, even if we cannot visualize the situation.

The Cosmological Argument

But can we at least explain *why* some universe exists? Theists rely on a simple intuition: everything that exists must have a cause, or why else would it exist? They present "cosmological" arguments in two forms: (1) everything that begins to exist or changes needs a cause; and (2) even if the universe is eternal, the contingency of things within the universe requires a non-contingent explanation. Thus, we can go horizontally backward in time through a series of causes and effects to the beginning of change, or vertically through a relation of dependence—either way, we need an explanation that by definition requires no further explanation, and that is a "self-existent" reality that does not depend upon anything else for its existence. No other alternative is imaginable and thus possible.

Consider the first form of the argument. Every change requires a cause, or else why would the change occur? But once we get on this track where can we stop? When we think about it, a great chain of causes producing effects must exist. This chain must either be infinitely long or have a beginning. The first alternative is an infinite regress and the second has a cause that produced itself and so requires no further cause—Aristotle's "unmoved mover." The first alternative (an infinite regress) is patently absurd because the chain of events leading to the present could never get started if it extends infinitely backward in time and had no first cause. We simply cannot conceive that reality is constructed that way. Thus, the second alternative (a "self-producing" and "self-existent" cause) must be the case even though we have trouble understanding how such a reality could exist. This self-existent cause must contrast with the universe or it too would be part of the chain that requires an explanation. That is, we can explain why, for example, one object is hot by pointing to its being heated by another hot object, but the ultimate explanation of heat itself is not a chain of hot objects but something nonthermal (the motion of heatless molecules). So too with the universe: the source of the changing natural realm (including any noumenal realm) must be unlike what occurs within it—it must be uncaused, changeless, and timeless.

Moreover, as the second form of the Cosmological Argument holds, even if there is an infinite regress of causes, we would still ask why this chain of contingent events exists at all. This leads to the conclusion that there is a sustaining source, i.e., a reality that by definition lies beyond the natural realm and that continuously sustains that realm in the sense that if it withdrew its ontological support we would disappear like characters in a dream when the dreamer wakes up. This sustainer is a "necessary" reality, not in the logical sense of the Ontological Argument, but in a metaphysical sense: without it, contingent events could not exist. The source is thus still a type of "cause," but it is not a cause in the chain of natural efficient, material, or other causes and effects within the natural order;

rather, the source underlies and holds the natural world in existence at every moment. Therefore, even if the natural realm is eternal, such a sustaining source would still be needed to continue its existence.

Limitations of the Cosmological Argument

Note four things about the Cosmological Arguments. First, theists think it is an extrapolation into metaphysics of the scientific search for causes, i.e., the search for a cause of the universe as a whole. However, scientists look for causes for why particular events within the natural order occur, but nothing comparable is possible for seeking a reason for the entire universe. The problem is that we cannot examine the alleged cause or empirically test various alternatives. Thus, advocates of the Cosmological Argument may think they have taken their lead from science, but it is not *scientific* in nature or method. Thus, they cannot invoke science as the justification for their quest. Nor are they pursuing science to its "logical conclusion." The reality postulated in the Cosmological Argument is not the ultimate unifying force in science: it explains *all of existence*, while science is about the structures determining *one particular set of affairs*—the entity it posits is the first cause but is unlike any natural cause (or it would be in need of an explanation too) or a cause sustaining all of the natural realm; either way, it cannot explain why one state of affairs within the natural realm is the case and not another—it is the source of being fro all phenomena and thus equal for all. The argument does not "complete" science but is an entirely different type of explanation than the empirical mission of science. While a desire to understand drives both the Cosmological Argument and science, the argument evolved, not out of science to know *how* things work, but out of a prior, more basic metaphysical drive to understand *why* things exist. Science *per se* in no way points beyond the natural order to a possible transcendental source of being or order, although theists who are already thinking in terms of such a source will frame science that way. For all that science can say, being and order may have ultimate causes internal to the natural universe itself.

Second, not only does the argument not explain any scientific structures, it also does not explain the nature of beingness or otherwise help us understand it. What is the stuff of the universe? Is it part of God, or as a creation is it something completely different? But either way, what is the being of God? The argument offers no information on that. Moreover, in trying to answer a why-mystery, the argument introduces new how-mysteries. How did the changeless creator image and then create the universe? How does he sustain its being? The argument simply does not address such mysteries.

Third, the argument leads at best to an uncreated and timeless source that created and now sustains the universe—not to a full-blown loving, caring *theistic god*. The argument's "creator" is just the name we give to whatever caused the universe or is not dependent upon any other reality. No other properties are

No

warranted from the argument itself. It is simply the reality that by definition is the source of our world. It need not be more than whatever it would take to create this universe since all we may safely infer in a cause is only the minimum necessary to produce the effect. It does not warrant inferring an *omnipotent, omniscient,* or *all--loving* being or one that is *personal* in nature. Theologians cannot simply call the source "God" and then automatically load in all the other properties of a theistic god that they want. (This sleight of hand also occurs in many discussions of the Teleological Argument: an abstract ordering principle is labeled "God" to name the mystery, and then later in the discussion it is magically treated as a full theistic god without further argument.) At best, the argument gets to a *deistic reality,* i.e., a reality that establishes the universe and its laws and then lets it go as a self-regulating and self-sufficient reality. Even if we argue that such a source may require the intentionality and will of a person to explain the choice to create, getting to an authentic theistic god—a loving person who cares about reality and intervenes in the natural order to help humans and with whom we can have a personal relationship—requires a very different argument. (Thus, deism can have an *antireligious* effect: it offers a new option of affirming a transcendental reality but one without any "living god" necessary to a religious life. Its transcendental reality is merely a posit to answer a metaphysical question.) Theists will see creation as an ontological function of a theistic god whom they believe in on grounds of revelation. But the argument by itself does not justify that. The reality posited in the argument is merely the product of a line of reasoning—the paradigm of the "god of the philosophers," not the "god of religion."

Fourth, the argument is not tied to the science of cosmology or the theory of the Big Bang. (Nor as discussed in the last chapter does any science force us into religious "limit questions," although the religious are free to see such implications.) The first form of the argument does require some beginning in time to the universe, but the argument was advanced before modern cosmological science and is not tied to any particular theory, such as space-time starting as a point. Under the second form, even if there is no such beginning, the question still remains of why there is any universe at all. As Thomas Aquinas realized, it is conceivable that created reality is eternal, i.e., without a temporal beginning, although the universe would still be totally dependent upon God. For example, if the universe is an infinite oscillating series of Big Bangs and Big Crunches, one can still ask where this series came from.

Thus, we must distinguish two concepts here: any particular physical event of *origination* (the first event in the present world) and *creation* (the act of giving being to the natural realm regardless of how it arose)—i.e., a particular historical event versus ontological causation and dependence. The science of the universe's origin—if scientists can ever study what went before our Big Bang—may explain origination, but it is neutral to the question of creation. In effect, the Big Bang theory is only about the process of the expansion of our universe, not *creation* in the literal sense—the Big Bang did not cause the universe to exist but is only the first event "in" it. Scientists have theories of how the Big Bang may have occurred by arising through purely natural mechanisms from an underlying energy field without the need of a creator god.[4] This may well explain how the Big Bang occurred, but it does not make the universe "self-creating": it does not explain where the underlying energy field came from. The "spontaneous" appearance of matter out of the space-time plenum or however the Big Bang arose is not creation "out of nothing" but only the development of whatever already existed, i.e., it is only a change from one state of existence to another. As Paul Davies asks, where did the energy of the Big Bang that went into generating all the matter and radiation we now see in the universe come from?[5]

This issue of creation is simply not something addressed by scientists *qua* scientists. Science is not designed to address the metaphysical issue of why there is anything at all or how reality came into existence since it is a matter that cannot be empirically tested. No TOE or other scientific theory can explain why a universe exists—why there is something here to be structured by that theory—no matter what cosmologists say about a "self-generating" theory bringing about its own existence. Some physicists argue that there is a "purely naturalistic explanation" for why there is something rather than nothing, but their arguments turn out to be only that matter ("something") comes from a "primal vacuum" ("essentially nothing") or from an earlier universe and not literally from *nothing*. This "nothing" is just the state of a pre-existing primal vacuum free of "something" (matter). But there is no explanation for where the primal space-time vacuum and its energy came from. Physicists are not explaining at all why there is anything. Energy is not nothing. Space-time without matter is not nothing. The physicists are merely equating "something" with *matter*—i.e., equating "something" with something we can see rather than with "anything at all"—and thus are making a bolder claim than they actually defend.[6] It is a matter of the transition of one state into another, not of why is there anything to convert into "something" in the first place. Nothing means literally *nothing*, not just energy or space-time free of matter. Nor is nothing something that is free of structures, or something that can be stable or unstable—*it is not anything*. These thinkers thus are confusing the philosophical question with a scientific one.

Stephen Hawking presents a cosmology that removes the need for a transcendental creator: the universe is a closed whole with no boundaries in space or time—there is thus literally no "place" for a creator god to act and nothing for him to do.[7] But he realizes that the question of whether God created it all remains open.[8] He thinks that it is "perfectly consistent with all we know to say that there was a Being who was responsible for the laws of physics." But Hawking is an atheist, and he adds that "it could be misleading to call such a Being 'God,' because this term is normally understood to have personal connotations which are not present in the laws of physics."[9] Still, he can ask what is it that "breathes fire into the equations" and makes a universe for physicists to describe?[10] The universe may be eternal both backward and forward in time, but the ultimate "why" question still remains open.

Whether there is a transcendental "ground of being" giving rise to and sustaining the reality of the natural world is simply not addressed by science.

The Unconvinced Naturalists

The naturalists' counter position to the Cosmological Argument is that there is no reason to postulate a transcendental reality as the cause or sustainer of the natural world—any notion of a "source" or "first cause" is a transcendental notion that is simply not necessary. The natural world—the world open to scientific study—is all there is. Asking where the universe came from is like asking what physically holds it up: it might sound like a legitimate question, but in fact it is not. The world is eternally going through an infinite number of Big Bangs and Big Crunches, or our world "spontaneously" arose out of an eternal natural underlying reality and expires while other worlds arise, or (to use John Wheeler's expression) the universe is a "self-excited circuit," or it simply has no explanation. Thus, there is no "creation" and thus no need for a creator god initiating the origination of events in the world.[11] Indeed, as David Hume asked, how could there be a cause of something eternal? As advocates of the first version of the Cosmological Argument assert, only what *begins* to exist needs a cause, but the universe never began to exist. But how could *reality itself* come into existence? By definition, there would be no prior cause or condition. The universe is eternal and simply exists. A Big Bang, if it occurred, is a product of purely natural forces; no outside, transcendental power source was needed to switch it on. To naturalists, this is more in keeping with science than the idea of a unique singularity of origination in which the laws of nature are suspended—in fact, how the universe could arise suddenly is unimaginable to them since they believe that the natural universe is all there is.

Naturalists also reject the second form of the argument. They see no need to postulate a sustainer of the natural world: we can stop with the universe as the only ontologically "necessary" reality for the contingent events occurring within the world.[12] The universe is just a brute fact, i.e., something that happens to exist and we must simply accept it with no hope of any possible further explanations. The physicist Edward Tyron may have been a little too glib when he said that "Perhaps the universe is just one of those things that happens now and again" (like the uncaused decay in nuclei), but it may be that the universe in fact has no further explanation and we just have to accept that we are left with a mystery.

The Problem of an Infinite Regress

Naturalists can poke enough holes in the Cosmological Argument to deflate any confidence we might have in concluding that it answers the mystery of our existence. Consider first the idea of an "infinite regress" of changes and their causes going back eternally. Theists want to argue that this idea is absurd since by definition there would then be no first cause to get the chain of causes and effects started. To theists, this idea would render the world "intellectually unsatisfying," "unintelligible," "incomprehensible," or immune to a "rational explanation." But naturalists reply "so what?"—why must the universe be comprehensible to us? Where do we get the arrogance to think that the universe must be comprehensible to beings like us? Contrary to what theists following Aquinas might think, merely because something is not conceivable to us does not mean it cannot exist. So what if the alternative is unimaginable to us? Our thought do not control what can be real. Why must things be intelligible to us through and through? Is there any reason to believe that except that we want it that way? Obviously just because we want something to be so does not make it so. Are we going to fabricate some answer that is "satisfying" to us regardless of the nature of reality? Why is it "unbearable" to theists to think otherwise? Theists like to borrow the existentialists' term "absurd" to make this situation seem irrational, but there is nothing logically contradictory about the possibility that the universe "just is" with no further explanation or that we cannot know the ultimate explanation if there is one.

Add to this the problem that only our *a priori* reasoning about the impossibility of an infinite regress reaches that conclusion. At best, we are applying our limited experience to the issue, and there is no guarantee that this will lead to the correct conclusion. Our experiences obviously can never reveal a first cause, nor is there any way scientifically to test the alleged complete rationality of reality. The history of science is a history of *a priori* intuitions that proven false. For example, Johannes Kepler had *a priori* arguments based on geometry of "the harmony of the spheres" for why there are precisely six planets. That nature is constantly surprising scientists reveals the danger of such reasoning in any factual matter. The mathematics of *infinity* also often conflicts with our intuitions of what should be the case (e.g., in the infinite whole number system, there are as many even numbers as there are numbers in the entire list, not half), and that is directly applicable to the situation with an infinite regress. However, if there is an infinite chain of causes and effects, there would be no uncaused events—every event would have a prior cause even if there is no uncaused first cause at the beginning. The causal chain is just there—an uncaused brute fact—but every event has a cause even if our minds cannot imagine a chain without a first cause. That is, there is no first event in an eternal universe: to say that the universe is eternal is to say that it had *no beginning*, not that it had a beginning *infinitely long ago*. Nor is an eternal universe "self-created"—it simply exists uncaused from an eternal past. Only once we introduce the question "why?" do we get into the question-begging mindset of looking for origins and causes. Thinking the universe must have created itself or have popped into existence at some point in the past or otherwise be *caused* is simply the wrong way of looking at the situation, even if this way is natural to a theist.

Here naturalists reverse positions with theists on what is absurd: the idea of an infinite regress is easier to wrap our mind around than the absurd idea of a "self-created" reality. We cannot visualize the complete chain, but we can in fact comprehend the idea of an infinite chain without a beginning: the natural numbering system, with its infinite number of negative numbers leading up to zero and then an infinite number of positive numbers. There is no first number in the series and yet every number has a predecessor. No first number is needed to get the chain started. (The other way to look at it is as a chain going backward with zero as the starting point, but this does not change the fact that we can look at the other way.) How can we rule out that this is not physically possible of reality? What makes it impossible other than our sense of what should be the case? "Reason"—which only means our own reasoning—does not "demand" otherwise. It is certainly not a logical impossibility, as the number series shows. A first cause would be conceptually neater, but there is no way around the fact that this may only reflect our way of thinking. That we have trouble with the idea of an infinite regress may say more about the limits of our imagination than anything about reality.

"Self-Existence"

The theists' alternative to infinite regress is a "self-explanatory" reality. That is, reflection on the logic of the chain of causes and effects dictates that we ultimately must reach a reality that is its own cause and the source of its own existence—it is "self-caused" and "self-existent." Thereby, every other entry in the chain and the chain itself are both explained. Only by also explaining itself could it explain "all there is" since there obviously could be no cause outside "all there is." So, even though it may well be counterintuitive—we think a cause must already exist to cause something, and so nothing can be "self-caused"—but a self-caused reality must be postulated. If we ask for an explanation of that reality, this simply shows that we have not grasped the concept: the self-explanatory source is by definition whatever reality cannot have a prior cause; thus, it explains itself. Asking for its cause would be like asking why a circle is round.

Why would we think such a reality exists? Because the only alternative is the infinite regress. But as just discussed, naturalists raise very solid reasons for accepting the possibility of that alternative. And even theists admit that what a "self-existent, uncaused cause" is is incomprehensible.[13] How could something be its own cause? How could something create itself? Theists must say "Well, it must be its own cause or we would have to go back indefinitely, and that makes no sense to us." It is merely the posit required to stop the reasoning of the Cosmological Argument—its nature, how it exists, and how it arose defy any attempt at understanding. All we know is that by definition it must be unlike anything we know (or else it would be part of the chain that is being explained). Theists do not deny the mystery of it all—after all, God is a mystery. They say this at least makes creation (if not the creator) ultimately comprehensible and thus is more in keeping with the spirit of science than just accepting the universe as a brute fact.

Nevertheless, theists must admit that the idea that God's existence is "self-explanatory" is unintelligible. We can say the words "She's a married bachelor," but the sentence is meaningless. So too, we can say the words "self-explanatory" and "self-caused" to end the Cosmological Argument, but they still do not mean anything. It is not just that our minds cannot conceive *how* a reality could make itself—the words contradict themselves. If we grasp what "cause" means, then we realize that nothing can be the cause of itself. Nothing can somehow generate itself from nothing; nothing can be induced to self-create. In fact, since the whole idea of "self-causation" or "self-creation" is self-contradictory, it is a *reductio ad absurdum* of the argument and thus sufficient reason to reject it. To have to speak of "self-causation" is just an admission that we do not know anything about what is going on regarding the existence of the universe. Even if one interprets the concept of an "uncaused cause" to mean only that something does not depend for its existence on anything else or is eternal, we still do not have a clue how or why it exists. We are left with the mystery of its existence, which is fine with theists, but this cannot be taken to mean that it has explained anything.

The question "what created God?" may sound like the question of a child who has not grasped the concept of the explanatorily "necessary reality" of the Cosmological Argument, but even if there is a first entry in the chain of causes that by definition is uncaused, one can legitimately ask why it exists. However, the existence of any first cause or ultimate explainer remains a mystery. The idea that something can come from nothing is absurd—indeed, even if were possible, the whole idea of how we can get from nothing to something would be saturated with mystery. But if the axiom "from nothing, nothing comes" applies to worlds, then it applies equally to gods. (Also note that the axiom presupposes that the universe or God had an *origin*—this way of thinking begs the question against eternally existing realities.) It is not just a problem of how a creating reality could arise but why it exists at all. Why is there a god rather than nothing? Even if a god is eternal, we can still legitimately ask why that reality is there. A creator may not need an originating cause, but it still needs an explanation for why it exists in the first place. Merely saying "it is eternal or uncaused" does not explain anything as to why it is there—simply stating *how long* something has existed does not explain *why* it exists. Theists themselves raise this same issue with respect to the idea of an eternal universe. If it is "incredible" or "unbelievable" that the universe should just have happened, then the same is true of God. God's existence is, in Alfred North Whitehead's phrase, "the ultimate irrationality."

More generally, we cannot answer "why does something exist rather than nothing?" by citing *anything*—any "something" is always part of the totality (all there is) to be explained. Any answer we can conceive will always be from within the true totality of reality to be explained. Thus, we cannot formulate a real answer in those terms since the solution would always be part of what we are asking about. Obviously, we cannot expect any explanation to be causal: no explanation would be possible since there is nothing outside the totality of "all there is" to be the cause. Nor, as just discussed, can anything explain its own existence. Thus, the totality of what is real will always remain unexplained. In the end, for the ultimate question of existence we are left with no final answer but only an unexplainable brute fact, whether it is the natural realm or a transcendental source. Either way, why there is the totality of what is real remains a mystery

In sum, even accepting the Cosmological Argument's type of reasoning, we are inevitably left with a why-mystery. We cannot cut off the questioning by means of the bizarre concept of "self-creation" without admitting our ignorance of how "self-creation" or "self-causation" is possible. God rather than the universe itself becomes the brute fact with no further explanation. But how does a new brute fact provide a complete explanation of the universe? Simply placing the brute fact outside the natural world in a transcendental realm certainly is not more "scientific" than accepting the universe as the brute fact. Scientists too must accept brute facts in the end—scientific explanations must stop somewhere. The necessity of positing a creator to end the regress may sound compelling to someone who has a prior commitment to a creator god, but the idea that the universe is a brute fact is not absurd. How is our understanding advanced by adding another layer of mystery?

Also notice that we cannot make the world and science "intelligible" by invoking something we admit is beyond our comprehension. Creation is not rendered rational if its source is "the ultimate irrationality"—the ultimate explanation is not complete, and creation is still grounded in mystery. This means that the Cosmological Argument does not fulfill what theists see as its major advantage.

The Principle of Sufficient Reason *emanationism*

The Cosmological Argument is based on the "principle of sufficient reason": for every event, there is an adequate cause or reason. It is taken to be a principle of science or even a necessity of thought. But there are problems here too. First and foremost, theists have to adjust it with a "self-caused" reality to avoid an infinite regress. This does give a cause for the initial event in the natural world, but as just discussed it is an absurd move. It also means that they have to twist the intuition underlying the Cosmological Argument by claiming something can cause itself.[14]

Second, in quantum physics the decay of a radioactive atom's nucleus appears to be literally uncaused. It is an example of a beginning without a cause. Physicists can predict probabilities extremely accurately, but they cannot say anything about the behavior of any given individual nucleus; hence, they will never be able to predict when a given atom will decay—even God cannot know that. A complete causal picture is thus impossible. Currently the most widely accept theory is that this is not just a matter of our present ignorance but we can never know if this in fact an objective feature of reality. Thus, for all we know, the existence of some realities have no causal explanations, no "sufficient reason" for existing.[15] *no*

Thus, the principle of sufficient reason may well not hold in the physical world. It is another bit of our common-sense intuitions that scientists have rejected. But if the principle of sufficient reason need not apply to the parts, how can we be certain it applies to the whole? Indeed, why should we think reality as a whole must have an explanation? What reason do we have but our own intuitions that have now been shown to be empirically questionable? Moreover, even if our current theories in physics are eventually replaced with deterministic theories, that scientists currently accept these theories shows that this principle is not in fact necessary to doing science, let alone an *a priori* truth or a logically necessary category of thought—science has ended any intuitive obviousness the idea might have held. And even if looking for causes were in fact a presupposition of human thought, this is not grounds to conclude that reality must comply to our presupposition. In sum, we should forego speculating *a priori* about the ultimate explanation of reality based on this principle—we may well have no idea what we are talking about.

The Natural Universe

Even if we accept the reasoning of the Cosmological Argument, naturalists also ask why we cannot simply stop with the natural universe itself. If we must accept something as "self-caused," the universe as a whole will do as well as a transcendental reality. It too may be beginningless and thus eternal and without an origin—it would then be uncreated, just as the posited creator is supposed to be. That all of its events are contingent upon other events in the universe does not mean the universe *in toto* cannot be an eternal chain of such change (as discussed above) or that it requires a cause. Our particular universe may not be eternal but only one of the infinitely many "universes" that bubble up out of an eternal self-contained and preexisting "field" and are eventually totally reabsorbed in a process that may go on for eternity. If so, this field is as changeless as any transcendental reality, even though all events in a universe are changing. The eternity of the universe may seem unimaginable, but eternity is as much a mystery for theists in the form of the eternity of God throughout time (or, alternatively, God's timeless existence outside of the realm of time, if time is only a property of creation).

Theists also have a dilemma with whether God has free will or not. If God is personal and freely decides to create, he *changes* with his free choice to create. He is thus not changeless, and theists then have a problem: according to the first type of Cosmological Argument—that all things that change need a cause—this god becomes the same sort of reality that needs explaining. Theists thus must make another *ad hoc* adjustment to the principle of sufficient reason to fit their prior commitment by allowing God's contingent decisions both to create and to structure the universe the way he has and yet exempting him from the Cosmological Argument—but then the argument implodes. But on the other hand, if the transcendent is a deistic reality that creates without making decisions, then theists would have to explain how the creation of a universe is dictated by the reality's nature, why the universe has the particular structures it has, and how a deterministic reality could possibly be personally concerned with humans, e.g., how could such an unchanging reality choose to respond to our prayers?

Finally, the events are "contingent" within the universe in the sense that they dependent upon other events within the universe. But there can be an infinite chain of such contingent events. Scientists have no problem with that idea. This counters any need for there to be at least one non-contingent ("necessary") being to explain this contingency and thus the universe's conditionality. Moreover, theists too have to accept one fact that is still contingent in the metaphysical sense—the very existence of God—and naturalists thus do not see how invoking the idea of a transcendental reality helps to clarify our situation.

A Sustainer

In short, to naturalists, the space-time continuum as the ultimate brute fact will do just fine. This leads to their rejection of the second type of Cosmological Argument: they see no need to posit a sustainer of the universe to maintain its being. To theists, it is obvious that God is both the creator and continuing sustainer and that we are dependent on him for our existence. There may be nothing left for a *creator* to do as Hawking says, but the role of a *sustainer* giving being to the universe will always remain even if it is eternal. That is, a sustainer is not a causal agent within nature (e.g., causing the Big Bang), but it could withdraw its power and we would cease to exist. Theists are not troubled by the accompanying how-mysteries: even if the universe is eternal, how did the universe arise from the sustainer? And how exactly does a transcendental reality "sustain" the world?

To naturalists, the theists' response adds nothing toward helping us understand why anything exists but is only a product of our demand to explain. Nothing empirical suggests the need for a transcendental reality to provide "being" to the universe—only constantly looking for causes no matter what would make us think otherwise. The universe is eternal (since it never came into existence) and self-existent (since it is in no way dependent upon anything else). And if it is asked how the universe can boot-strap its own existence, naturalists reply that the same can equally be asked of God. What sustains the universe? Well, what sustains God? How can the universe be self-existent? Well, how can God? Where did the universe come from? Well, where did God come from? How could a god be born? Well, how could any reality come from nowhere? Could the universe cease to be? Well, could God, and if not, why not? No outside force could work on it. Naturalists may be accused of simply accepting that the universe exists and not explaining its existence, but then again theists do the same for God. Even if the idea of "self-existence" could eliminate the question "What sustains God?" it does not eliminate the question of why there is a god and thus why anything exists at all. All in all, whatever can be used to justify a transcendental sustainer can be applied equally to the universe itself. How then does invoking God help our understanding at all? A creator god may seem simpler (since it is timeless, spaceless, and immaterial), but it complicates the overall picture by introducing both a new level to reality transcending the natural order and new how-mysteries. In sum, if the theists argue that we must accept a self-existent reality, why can't we stop with the universe?

In addition, we can still legitimately ask what sustains the sustainer. Whatever reason the world needs to exist would apply equally to a sustainer god. If God gives our universe its being, then what gives God his being? How does God keep existing? Using the ploy of "self-sustaining" works no better than "self-creation." Saying that "he is self-sustaining" or "he exists by his own nature" does not further our understanding any more than saying "he just does!" It is the same as the theists' accusation that naturalists say the universe "just is." And once again, if something can be self-sustaining, why can't it be "the space-time continuum" or whatever is the primordial stuff out of which matter and energy arise? This would be as much as a sustainer underlying the changing events in the universe as we would need.

Moreover, just as there could be an infinite chain of causes and no first cause, what keeps there from being an infinite hierarchy of such sustainers? We have trouble with the old idea that a flat earth is held up by four elephants who stand on the back of a turtle and this turtle stands on the back of another turtle, and after that "it's turtles all the way down." We think there must be a bedrock for the turtles to stand on—a transcendental source. We want foundations; that is how our minds think. But to naturalists there is no need even for the first turtle: an eternal chain of contingent and impermanent events does not need some necessary and permanent reality underlying it to give it existence. And why invoke something timeless and spaceless that is beyond our comprehension on *a priori* grounds alone when it too would need an explanation for existing? It is a product of simply perpetually asking "why?" and trying to visualize the situation. How does introducing a new mystery solve our current mystery? Contra David Hume and Bertrand Russell, the chain of causes does not explain itself—nothing explains its own existence. But this applies equally to God. Either way, existence—whether it is the existence of a transcendental reality or of the natural universe—has to be ultimately accepted as "just is." Taking the universe as a brute fact is not logically different from taking a sustainer as the ultimate brute fact, and introducing a hierarchy of levels of reality only creates a new befuddlement.

Even leaving aside the idea of how a transcendental reality could sustain itself, the whole notion of "sustaining" the natural universe is hard to fathom. How does a changeless, immaterial, and spaceless reality sustain the material, changing world of time and space? Naturalists see no reason for asserting that divine action is constantly required to keep us from going out of existence. It is obviously not a testable empirical claim. It must involve some sort of metaphysical "energy" whose only purpose is to keep the universe from going out of existence at any moment. (By definition, it must be unlike natural energy: it supports the existence of natural energy and thus is not a form it.[16]) But this "power of being" is as much a mystery

emanation

when we are talking about God's own existence as the universe's.

And introducing *creation* only presents new "how" issues. How does the timeless generate the temporal? How does it create anything out of nothing? In fact, how does the timeless and bodiless come up with the idea of a material world with beings with bodies in it—indeed, how does a timeless reality think at all? Saying that we are "thoughts in the mind of God" does not get us very far and makes God *No* a thinker subject to changes and thus in need of an explanation too. The relation between the transcendent and the worldly cannot be studied scientifically (since we cannot present the transcendent for examining different possibilities), nor is it clear even how to address it. Without being able to see the relationship, any concepts from within the created side will not apply. We cannot even call it "causation" since the entire idea of "causation" is applicable only between events within the universe and thus not to the question of the existence of the universe itself. The universe is not an effect in a chain of causes or under the Cosmological Argument it would need explaining too. That is, causation was created when the universe was created. Invoking a creator does not simplify the picture but only greatly complicates it unless there is some explanation of these how-mysteries, and that seems foreclosed forever.

Thus, in the end the theists' position only generates a slew of new mysteries: why does God exist, how does he exist, how does he create, and how does he sustain the natural world. And since the same why-mysteries end up applying to a transcendental reality, bringing a god into the picture in no way helps our understanding of the mysteries of existence but only complicates the picture.

Can the Dispute be Resolved?

In sum, the Cosmological Argument is appealing only if we can rule out the possibility of an infinite regress of causes or if we can show that the universe requires a sustainer, and we cannot do either. Of course, we may want closure and to stop asking why, but we cannot do that just by invoking a "self-existent" reality. The theists' inference to God as the best explanation also presupposes there is an ultimate explanation for the universe rather than it being simply a brute fact. But how do we know that reality must be comprehensible to us and have an explanation? Theists invoke God as a neat solution, but it is not obvious that our chain of questions can end with a definitional stop since the question of whether God exists still remains. Nor is obvious that the universe will not fit the bill. Either way, we are left with the mystery of something that just is—either God or the universe is a brute fact with no further explanation. In sum, whichever is the "self-sustaining" reality, we can ask why it exists. Neither answer defeats the fundamental mystery. *?*

There is no empirical way to test explanations of the basic existence of reality. Nor does either side violate the scientific picture of the world either by adding a transcendental explanation of the universe as a whole or by accepting the universe

[handwritten top margin: no empirical evidence of uncaused physical objects; esp. in Copenhagen. God is not a physical object]

as an unexplainable brute fact. Determining which side is "more reasonable" or "more plausible" is not possible since one's basic metaphysical commitments will determine what one accepts as a "convincing argument," "obvious," or as what "stands to reason." For example, naturalists will invoke Occam's Razor for the simpler naturalistic account (since it only has one substance), but theists will argue that this is question-begging because it is absurd to think the universe could come into being from nothing—a spontaneous uncaused happening. Nothing can create itself or be its own explanation, but naturalists do not think the universe came into existence at all. And then again, theists have no answer to why God exists. Theists argue that the idea of an infinite regress is inconceivable, but they have no trouble invoking the absurd idea of a "self-existent, self-explanatory reality." Perhaps a transcendental reality does exist, but the force of the Cosmological Argument itself has been destroyed once the presuppositions are seen to be questionable.

And naturalists do have some points in their favor. Certainly the theists' concept of a "self-caused god" is the strangest concept in this book. Theists cannot merely pronounce that the Cosmological Argument requires it. But we are not forced to accept its posit—the alternative of an infinite regress and an eternal universe is still viable. Moreover, if theists expect to explain anything, they cannot simply say the concept is beyond human understanding. If theists cannot explain both why God exists and how God creates, they must admit that they have not really added anything to our understanding but have only added another layer of mystery that greatly complicates the picture by adding an unobservable, untestable reality that transcends the natural realm. Why move one step beyond the natural world just to arrive at the same mystery? Also, if we must speculate on ultimate explanations, it is better to keep the speculation as close to our actual experiences as possible—and we do know that the universe exists—and as far away as possible from purely *a priori* reasoning that has proven so wrong in the past and can offer no guidance here. Arguments about whether the universe is eternal or not may well be tainted by theological prejudices disguised as *a priori* necessities.[17]

Also notice again the way the question of why anything exists is normally phrased—"where did the universe come from?" This way of phrasing the issue is question-begging: it presupposes that the answer will be in the form of origins. Even asking "why does anything exist?" puts us in the mindset that there must be a cause. Perhaps that is simply the wrong way of looking at things. Perhaps the universe is eternal and uncreated and there is no reason behind it. This possibility cannot be ruled out simply because that is not how we habitually look at the issue. To theists, it is inconceivable that there is not some reason for the universe—the existence of the universe requires a creator. But since the days of Zeno's paradoxes), we have known that we can conceive what is obviously true in ways that make them seem utterly impossible. The idea of an infinite regress of causes may do the same for an eternal universe. Theists have no problem with the idea of an eternally existing reality—they just think it is a transcendental reality and not the universe. But to naturalists, the idea that the universe existed forever seems

[handwritten left margin: confuses material & non-material]

[handwritten bottom: unscientific]

perfectly acceptable, even if our mind may seem to demand some explanation for it. Their idea eliminates the singularity of any creation at all and brings everything under the scope of natural laws and thus seems more rational to them. Of course, naturalists cannot explain why the universe exists, but then again theists cannot explain why God exists.

Finally, theists have one remaining big problem: one mystery does not explain another. Mystery is not an explanation of anything. It is just that—a mystery. We cannot say "It is a mystery beyond our understanding" and then claim to have understood or ultimately explained anything it is invoked to explain. Claiming something is a mystery is only admitting our ignorance. At most, it is a place-holder—an x identifying a problem. Thus, when theists invoke what they admit is a mystery, they cannot be said to have explained the existence of anything. How can God be an answer if we do not understand what God is, how he exists, or how he works? Under these circumstances, there is no difference between saying "Because God willed it" and saying "it's magic!"

At best, the Cosmological Argument only permits its advocates to say "there is a reason for the universe" (since its own presuppositions entail that) and nothing whatsoever about what that reason is. It is a paradigm of the philosopher's disease: there must be a reason for anything, even when we cannot fathom what that answer would be. Every question must have an answer, and theists simply define the answer here to be "God." This may quiet our urge to understand but only in the way that saying in exasperation "God knows!" stops our questioning. Even if the Cosmological Argument could succeed in establishing a transcendental reality, all that is warranted is an unobservable reality with no other features than once having had the power sufficient to make the universe and now to sustain it. It is a metaphysical x that by definition is unlike anything it creates. It is ineffable in that sense: we can name the mystery, but we cannot otherwise understand it. But this makes the x useless as a devise for our ultimate understanding—having to invoke the unintelligible does not make the universe's existence finally intelligible.

The Intractable Why-Mystery Remains

Advocates of the Cosmological Argument think they have successfully incorporated our urge to explain everything while also successfully curtailing the questioning by the artifice of "self-creation" rather than pushing the questioning further. But their intuition that there must be a cause has not been prevalent in all cultures. The ancient Greeks had no problem with the idea that the universe is eternal. Classical Buddhists and Hindus too believe the universe is eternal—to them, it is the idea of a creator that makes no sense. (The Buddhist rejection of the idea of a creator god is simple: such a god must be either immutable or not—if it is immutable, it cannot change and thus is unable to decide or act to create; if it is not immutable, it is within the realm of time and thus did not create the realm of time, and so the ultimate origin of the universe is not addressed.) But the idea that we can follow time back and ask about the beginning is how we now think. We do have trouble with the idea of something popping into existence out of nothing—how can we exist without a creator?—but theists cannot explain where God came from and only hope the Cosmological Argument cuts that question off as illegitimate.

Still, the theists' rejection of the naturalists' alternative of an infinite regress and an eternal universe must rest on a prior commitment to theism, since the same problems arise for the existence of God. When forced to choose between absurdities, they choose the absurdity that supports their beliefs and claim that it is "obviously" true since the alternative is absurd. The Cosmological Argument does quiet our questioning somewhat by pushing the ultimate mystery out of the natural order into a realm transcending the natural realm. That may be satisfying to many who are searching for a transcendental meaning to the world. Nevertheless, the fact remains that this solution—an "uncaused, self-created cause"—may only be a product of our own quest to quiet our lack of understanding and not reflect reality at all. And again, it introduces a new dimension with its own "how" mysteries and one big "why" mystery but now in a larger form and separated from the natural world we know. In sum, the Cosmological Argument may tell us more about our speculative imagination—and its limitations—than anything about the ultimate nature of the universe.

In the end, the Cosmological Argument brings us no closer to answering the basic question of why should anything exist at all, and it is safe to say that this basic mystery will never be solved: there is no way beings within the universe can know the relevant information. Even if we knew all any being within the universe could know of reality, the basic mystery of why anything exists would still remain. Any reality postulated to end the Cosmological Argument needs explaining as much as any reality—arguably, a first cause or a sustainer actually needs *more* of an explanation than what follows it. Certainly, even advocates of the argument must admit that the wonder of why God exists is not removed by the argument. Where does the being of God or the universe ultimately come from? Scientists may solve all the how-mysteries of nature, but that would not illuminate the mystery of why the universe exists. Science simply cannot touch this basic why-mystery. The contingency of the ground of reality will always provoke the question of why it is there—as discussed above, nothing is an explanation of itself. In short, for whatever explanation we propose, we always must ask why it exists. The basic mystery will just not go away.

Indeed, any answer we can create appears to be logically impossible: the only two options seem to be an infinite regress and a self-caused reality, both of which are absurd. This leads many philosophers to reject the question as meaningless and to conclude there thus was no mystery in a legitimate sense about the presence of the universe in the first place. The ideas of "turtles all the way down" and of an "uncaused floor" for them to stand on are both absurd, and this shows we cannot even conceive the situation properly to frame the right question. Kant dismissed the cosmological dilemma as an unanswerable antinomy—both options must be rejected since we have no experiences in this area to give content to the concepts, and "concepts without precepts are blind." In other words, extrapolating beyond our experience can only lead to conflicting conclusions that our reasoning cannot resolve. We may still have a feeling of curiosity and wonder, but this only represents our inability to formulate a proper question, not anything about reality—we are in a fog, but that says nothing about reality but only about our muddled minds. The atheist J. C. C. Smart thinks the question "Why does the cosmos exist at all?" is unanswerable and so should be set aside, but that anything should exist at all nevertheless strikes him as a matter of "the deepest awe."[18] Other philosophers accept much of this point but argue that the why-question is still intelligible and thus legitimate even if we can never be sure there is an answer.[19] Milton Munitz concludes that the existence of the world is incomprehensible and cannot be adequately explained in any way, although the accompanying feelings of astonishment, awe, and perplexity will persist and not be removable.[20]

Assuming the question is legitimate, all we can do is formulate one of two answers—the one entailed by a Cosmological Argument or a naturalistic alternative. Does the world remain the unexplainable brute fact, or does God? Pick your poison. Either way, we end up with a mystery. Is there no reason for all this, or an unknowable reason? We keep explaining until we are satisfied, but all explanations have to stop somewhere, and finally something has to be taken for granted as just the way things are. Where we stop may be, as Paul Davies says, a matter of "taste."[21] Like Davies, one may prefer the explanation of a transcendental source if one prefers to keep the mystery at least outside the natural realm or has a need to think in terms of a personal creator. Or one may have less confidence in our speculative imagination and prefer to keep matters closer to what is experienced and keep the mystery within the natural universe itself. But either way, reality is not rational "all the way down" but has a nonrational brute fact: to naturalists, the universe "just is" with no further explanation; to theists, this mystery lies with a "self-created" god. In the end, one must choose the mystery one accepts, or we can remain agnostic, if we can live leaving this question open—being agnostic in fact may be the most intellectually honest option. All we know for certain is that the mystery of existence will always be an open question, however we deal with it: beings within the universe simply are not in a position to know why it exists.

The Teleological Argument

If the sheer that-ness of reality remains a mystery, what about the how-ness? That is, when we turn from the question "why is there anything at all?" to "why this particular world?", is there an ultimate explanation? The structures of the universe and their origin need explaining just as much as existence itself. No one can deny that there is an incredibly intricate and complex order to reality—in fact, that the world is ordered at all is no less strange than that it exists. Order is manifest both in the causal orderliness to things that we observe and in the extraordinary creativity of the continuing appearance of new inorganic and organic forms in the evolution of the universe. This order could not have come from a chaotic "explosion" of the Big Bang. And that raises three areas of questions. First, why is there any order at all? Why is the universe structured with laws of nature in the first place? If the universe started out in a totally chaotic state, where did order come from? Second, why is the universe structured the particular way it is? That is, why not a different universe with a different order?[22] Are the laws arbitrary? To use Einstein's phrasing, did God have a choice in how he created the universe? Third, where did this structure come from? Is there an intelligence behind the world? Did the laws governing nature's evolution exist before the physical universe? Is the alleged ground of being also the ground of order? Can theists cite the structure of reality as evidence of a theistic god (a loving, involved, personal being), or at best does it point to a deistic reality? Or can we forego postulating a transcendent architect altogether—i.e., can the universe itself somehow self-organize?

All of this leads to the issue of the Teleological Argument. The classic formulation is from William Paley. If we found a watch on the ground with no further explanation, wouldn't we infer a watchmaker? Wouldn't we infer from its complexity an intelligent, personal being—a designer—at work and not a mere accident? Doesn't this apply by analogy to the order we find in the world? How else can we explain the intricacy of, for example, the eye or the rest of the coordinated functions within a human body, or the fit of animals in their environments? After Hume, it is difficult to maintain a simple inference from empirical phenomena to the existence of a designer, but this argument can still be read as an "inference to the best explanation"—i.e., that some causal explanation is needed and a designer is the best explanation available. However, after Darwin this is hard to maintain. Nevertheless, evolution in the Neo-Darwinian form is still structured: some lawful natural forces (including those producing the mutations) are at work. How do naturalists explain the presence of a lawful order at all? Why do laws continue to hold and the forces they depict not vary? Theists can either look for examples of biological and inorganic phenomena that scientists have not yet explained, or they can strip the Teleological Argument to its basic form and claim that a transcendental cause is needed to explain the presence of any order at all in the world.

What exactly is meant by "design" can lead to a confusion: it can mean that reality has some overarching "purpose" or "goal" or "final state" to be achieved, or it can mean something smaller in scope and more concrete—laws. The design was either programmed in from the outset or events are guided along the way by an active designer. However, physicists and biologists currently reject any idea that nature is somehow guided by Aristotelian "final causes"—thus, transcendental teleological explanations are rejected. The only explanation of a whole is in terms of how its parts interact—Aristotelean "efficient" and "material" causes. Thus, evolutionary biologists can accept that parts contribute to a working whole or have a "purpose" for survival. (Some use the word "teleonomy" for this to avoid confusion with the grander purposes of "teleology.") But no more "purpose" than immediate survival is involved in biological evolution. Living beings react only to the present problems of survival, and evolution does not somehow plan ahead for some future goal that it was programmed to aim for. In sum, the only "design" scientists currently accept is the idea of a lawful causal order.

But even if we reject any grand teleological design, where do the laws governing the parts come from? Why and how were the laws written into the fabric of the universe? Theists cannot see how all this could possibly exist without a plan to it all. In particular, why wouldn't the universe all be unconscious matter if life and the mind were not integral parts of reality already planned from before the Big Bang? To naturalists, our changing universe is evolving greater and greater organized complex forms naturally. But theists see such development as a "story" or "journey" of an "unfinished universe" that must be heading toward a happy ending. They claim that the observed order of nature shows that the creation of life—and especially conscious beings with free will—is obviously part of that purpose. Life is not an evolutionary accident in a purposeless, meaningless universe. Rather life and conscious beings were programmed in from the very start and thus figure fundamentally in a grand scheme of things. Why else would life appear at all? Why wouldn't the universe have remained simpler?

Moreover, Christians since Gottfried Leibniz have argued that ours is in fact "the best of all possible worlds" (since God can envision all possible worlds and must always create what he sees as best) or at least that it is better that this world exists than not (since more good than bad is generated by the world existing). That life or the universe itself may come to an end does not negate there being a purpose: the universe may accomplish its purpose and then cease to exist. Something need not last forever to be valuable. No one says a Beethoven symphony is valueless even though it comes to an end, and the same applies to universes.

The Anthropic Principle

Theists can also point to the truly phenomenal "fine-tuning" of the natural properties that led to life. For example, astronomers calculate that if the Big Bang's expansion velocity differed by only one billionth less or more, it would be either too slow (and gravity would have collapsed the universe) or too fast (and the universe would have flown apart without planets forming). What is even most impressive is the apparent fine-tuning of the properties of submicroscopic particles, physical constants, and physical forces to the incredibly exacting degree that just permits life to appear in the universe: if any of half a dozen constants in the delicate configuration of fundamental variables were off by less than a tiny fraction of one percent, life would not be possible. Calculating precisely the odds of the exact configuration needed for life arising by sheer chance is impossible, but they are so far north of a trillion to one that pure chance as the cause is hard to accept. In Derek Parfit's analogy, it is like hitting a bull's eye in a distant galaxy. The more the physicist Freeman Dyson studies the universe's architecture and detail, the more evidence he finds "that the universe in some sense must have known that we were coming."[23] It begins to look as if we were meant to be here.

This is generally called the "Anthropic Principle," but that is too anthropocentric a term. At best, the intricate arrangement of natural forces leads only to the conditions of complex life, not humans. Why consciousness also arose is not explained. Thus, a "biopic principle" would be a more accurate label of the alleged principle since it does not privilege conscious life over, say, bacteria or trilobites. In its weakest form, the principle is simply that what we can expect to observe about the universe must be restricted by the conditions necessary for the presence of observers. This is obviously true: we would not be here to ask about the universe if the universe were not compatible with life. Advocates of a stronger form of the principle claim that the universe is arranged so that *conscious* life *must* appear at some point in its history. An even stronger form is the claim that consciousness must develop within the universe and then will never die out. By giving the universe an overarching purpose in which conscious beings must appear, advocates of the stronger versions necessarily give the universe a transcendental significance. Conscious beings thus would not be in any way inferior ontologically to the matter that makes them up. If Copernicus kicked us out of the physical center of the universe and Darwin reduced us to just another animal, this principle firmly reinstates us as the center of the whole scheme of the universe. And, theists would add, this principle confirms that we are the purpose of the universe.

The Naturalists' Response

Not all Anthropic Principle theorists are theists. Some treat it from a naturalistic perspective with no transcendental design implications. Nor is everyone impressed by these claims to purpose. Naturalists point out that elaborate order can be generated by random events on a computer. They must admit, however, that some simple rules are always involved—i.e., there is "design" programmed into the computer process at the beginning by intelligent beings even if no further intervention is needed and subsequent events evolve by chance.[24] But naturalists distinguish simple causal order from any sense of a purpose imposed by a transcendental reality. It is possible to refer to goal-directed activity in, for example, plants turning toward sunlight, but naturalists see no evidence of an overarching goal to nature or any teleological causes in nature. The universe looks just as it should if there is no overall plan for it. Life does not serve any purpose—it just is. It, like consciousness, is the result of natural forces not out to accomplish any further goal. Humans are not the goal of the universe, nor is anything else.

The illusion of "design" and "purpose" in nature is only the result of natural ᴺᴼ
selection or natural forces. We humans used to see rain and lightning as the actions
of an intelligent agent—acts of God—and "design" is simply this idea returning in
a more sophisticated form and on a larger scale. For anything that *looks* made, we
see intentionality and thus think in terms of a creator—"who made it?"—but this
is only the result of our way of thinking, not how things really are. One is reminded
of the old joke that God made our legs just long enough to reach the ground, not
too long and not too short. If one looks at reality already thinking within a design
framework (as theists do), things do look that way, but obviously the length of our
legs was not designed in that manner, and the same with alleged "design"—once
we are free of the "design" frame of mind, natural explanations are acceptable.

Do naturalists have any explanation for the order of nature, or is it just a brute
fact like the existence of the universe? Doesn't the intricate orderliness demand a
designer? Can we accept that the complexity exhibited in the universe just
happens? Sheer chance cannot be logically ruled out, but even the most diehard
naturalists have trouble with the idea that order happens just by blind chance—the
odds of the degree of order that produces conscious beings arising by mere chance
intuitively feel impossibly infinitesimal. Thus, we feel justified in looking for a
reason. Nevertheless, the orderliness of nature and the specific laws of nature may
simply be a brute fact. After all, why should the more natural state of the universe
be chaos rather than order? Things may tend to break down, but why is it any
stranger that there are uniform, constant forces at work in nature organizing things
than that anything exists at all? Forces, after all, are just as natural as substance.
Such forces may be another brute fact just like the existence of the universe. Brute
facts may be mysterious, but there is nothing we can do to explain them.

Some theologians (e.g., Richard Swinburne) argue that accepting the laws of
nature as a brute fact renders science "ultimately absurd" or "unintelligible"—we
need an architect or legislator of those laws, to explain why science works or to
make the practice of science rational. But theists too must ground science on a
rationally unexplainable, brute fact: *God's unconstrained will*. There is no explanation of that will. If God's act of will is forced by some necessity beyond his
control, then there is no need to invoke God to explain the laws of nature; but if the
laws are the result of God's free choice, then there is no ultimate rational explanation for why they are what they are except the brute fact that God chose them that
way. Even if there is a general Anthropic Principle for life and consciousness (and
a TOE may end that), there is still no explanation for specific laws and forces.
Thus, there is still no ultimate explanation of the laws of nature for theists but just
recourse to another brute fact. In theism the mystery is pushed back on step, but the
alleged "ultimate absurdity" remains. Indeed, theists now only add a new layer of
a transcendental mystery and new how-mysteries that complicates the picture.

Science cannot itself explain why a natural order exists in the first place. The
ancient Greeks and the Chinese thought an eternal natural order was inherent in the
universe without its source being a major issue. Science *per se* can neither rule out

the possibility of a designer or other transcendental source of laws nor confirm it. But we can still ask three basic "why" questions: why there is any order at all, why this particular order, and why it leads to such complexity and creativity in nature. Answering the first question is difficult. Perhaps there are timeless Platonic ideas behind the order, if impersonal transcendental principles can generate life and consciousness. Scientists are only now just beginning to tackle the third question. No natural mechanisms for these features have yet been found to explain why reality is set up that way. But physicists are addressing the second question. Someday we may have a TOE that will show that no other set of values could be physically possible or in fact that only one set of laws is logically and mathematically self-consistent; there will be no free variables. If so, there would be no role for a designer: any universe could not help but have the same basic structures and laws. Our particular order would be inevitable, and this would explain the apparently "finely-tuned" aspects of the universe at the hands of a designer. Nothing is "self-generating" or "self-explanatory," but since things would have to be the way they are, there would be no need for a further explanation: the laws would have a necessity that eliminates any contingency that would have to be explained. Only the question of why there is order at all would be left unanswered. In sum, such a TOE would eliminate the appearance of "design" highlighted by the Anthropic Principle because no other combination of factors, even slightly different, would be possible—there are no choices for a creator to make, and so no creator is necessary because none could do otherwise.

However, many physicists believe the universe could have had other physical laws. And short of showing the necessity of a specific TOE, the question why any one particular set of equations is instantiated in the universe would still remain. Steven Weinberg, for one, believes that when physicists have gone as far as they can go, we will not have a completely satisfying picture of the world because we will still be left with the question why this TOE and why the world is described by quantum mechanics rather than, say, Newtonian mechanics. Thus, he concedes "there seems to be an irreducible mystery that science will not eliminate."[25] Moreover, the basic mystery of why there is any order at all would still not be explained by any scientific theory—no law of nature could explain why there are laws of nature in the first place.

Evolution

Even if science does not rule out a designer of nature's basic laws, naturalists still have four criticisms of the idea of transcendental design. First is evolution. It is far from clear that biologists yet have a full grasp on all the mechanisms at work in evolution. Indeed, the weakest point in naturalists' account of order is the emergence of new biological phenomena—life itself, complex forms of life, and consciousness. It is hard to account for the appearance of qualitatively new phenomena such as life and consciousness by sheer randomness alone among inanimate objects. To theists, all naturalists have to date are a conviction and unfulfilled promises. So too, while natural selection and random mutation are certainly at work in the evolution of life, it still may be difficult to explain all the emerging complex forms of life solely from those two factors. Naturalists try to explain the appearance of something that is apparently only useful later in terms of present needs. For example, to explain why feathers are present in the fossil record before flight evolved, paleontologists set forth other uses of feathers—e.g., to keep an animal dry or warm or for sexual attraction. Only later were the feathers utilized for flight once that ability unexpectedly arose. But this process, even in conjunction with natural selection, cannot explain why life-forms are getting more and more complex in the first place. That some type of natural biotic law may be at work in biology cannot be ruled out.   *emancthonism*

Those theists who accept science see a design to the course of evolution in one of two ways: deistic evolution (a transcendental reality set up the initial conditions and the laws of nature governing evolution and then let the universe run, including permitting random events, without any predetermined goal; this may include laws for life and consciousness to appear), or theistic evolution (a loving personal being not only set up the laws and conditions but also continues to intervene to create specific species or otherwise to guide the course of evolution to the goal that free conscious beings would eventually appear; or he intervenes to add a soul to humans once the goal of evolution—conscious beings—has appeared). Some randomness of events may be part of God's plan, but the emergence of intelligent beings with free will somewhere in the universe is guaranteed. In short, evolution is how God creates humans. Thus, current biology is wrong: events in evolution are not truly random; the transcendent somehow imposes some constraints. Scientists describe the laws governing events occurring in the hardware without regard to the software, as it were, guiding the actions to a goal, and we may not be able to see the program running the hardware because we are events inside the "computer." *enough.*

However, anyone who surveys the history of life on earth would find it hard to see signs of such an overarching design. At best, there is only a general plan for the blind emergence of complex life-forms, greater "information" process capabilities, and biodiversity through random events. The actual course of life on earth is messy and has none of the elegant patterns we see on the astronomical and quantum scales of things. Overall, the history of life looks much more like a series of haphazard events, and it is hard to see it as the unfolding of a divine will. Some theists make a virtue out of this and claim God is an improvisor, explorer, tinkerer, or experimenter playing dice, not a planner who maps out final results beforehand. He sets up the rules of the game that assures life and conscious beings will appear, but he does not determine how the exact course of the game plays out. Contingency becomes important. This makes for a more interesting universe than one with a preordained blueprint merely being fulfilled over time. But how much "design" these theists can claim beyond basic natural laws is not clear. Did God plan the course of the meteorite whose impact killed off the dinosaurs 65 million years ago in order to make way for mammals? If so, why did he bother making dinosaurs in the first place? And why did they survive for over a hundred million years before that? The same applies to the contingent events that caused four prior mass extinctions in which paleontologists estimate that as much as 95% of the existing species were wiped out. Today biologists estimate that perhaps 98% of the species that have ever existed are now extinct. If God is a tinkerer or experimenter, he fails almost all the time (or has some purpose for it all that is unknown to us).

If in fact intelligent life is the objective of creation, we can also ask why there is apparently so little of it in the vast universe and why it took so incredibly long to develop. Why did human-level intelligence develop in only one small strand of beings on earth? And why does so much life get along perfectly well without it? Why would a god take billions and billions of years just to set the stage for us? Think of the incredible amount of time when nothing more occurred than chemicals cooking in the interior of stars, followed by billions of years of simple life-forms on earth making oxygen for our atmosphere. Think of the billions of years in which waves crashed against the shore while myriads of animals and plants came and went. Think of the millions of years of sentient animals eating other animals, reproducing, and then dying from famine or drought or disease or being eaten. Think of Cro-Magnon man and a dozen other hominid species coming and dying out over the last five million years. Think of their short lives, their fear, and the physical suffering they endured before modern medicine. Why would a designer need to do any of that if this universe is nothing more than a stage for modern humans? Does all of that mean nothing to a god?

In addition, if we look at the entire universe and the evolution of life on this planet, it is hard to see humans as the purpose. Certainly no direct and simple line of evolution leads to us—our evolution looks as convoluted as every other species'. We are no more the center of life on earth than the earth is the center of the universe. Even if intelligent life lasts a trillion years, the first ten billion or so years without it is still a problem. Why would a designer take billions of years before even life appeared, let alone beings with consciousness? One can understand why theistic fundamentalists want to deny both evolution and the idea of a universe billions of years old: the history of life makes any "designer" look like an incompetent demiurge. If intelligent life were the goal all along, this is waste of truly cosmic proportions. As Bertrand Russell said, "[i]f it is the purpose of the Cosmos to evolve mind, we must regard it as rather incompetent in having produced so little in such a long time." And, as he asks, are we humans really such a great result for an omnipotent and omniscient designer to brag about after billions of years of work?[26]

Theists who say that evolution proves that God is "patient" and "plans well in advance" simply are not taking seriously the unimaginable time it took evolution to get to sentient beings. Why would a designer take the evolutionary route and spend billions of years to get to us if it is only about us?[27] Why all the diverse pathways, bizarre variations, extinctions, and dead ends? Moreover, the beings that had survived look extremely badly designed. Some adaptions are described by biologists as "abysmally stupid." Why, for example, do humans have such fragile spines? Why toothaches? The human eye certainly was not designed by a designer using Occam's Razor. Why in general the convoluted bodies and useless organs? Useless organs that exist in a human body are interesting from an evolutionary point of view (since they trace our evolutionary history), but very strange if we were designed by a competent designer. Where is the evidence of planning rather than just letting diversity and complexity flourish haphazardly? And why would a designer just happen to make us look like we evolved naturally without a plan?

And why believe that humans are the focus of everything in such a universe? Before Darwin, people could say that birds and flowers were put here merely for our enjoyment and all the oddities were just for our amusement, but after Darwin no one can say this. Flowers bloomed and died for millions and millions of years with no one to see them. If cockroaches were put here only to annoy us, what were they doing for the other three hundred million years of their existence? How can we think of ourselves as anything but another branch of the tangled bush of life that has evolved and will either die out or evolve into something else? *Menitude*

More generally, can we possibly know if there is any purpose to the world by examining the universe? How could we detect any "final causation"? We can only see a trend toward greater and greater complexity and diversity—what Freeman Dyson calls "the principle of maximum diversity." No further goal appears to be involved since the diverse evolutionary lines are constantly branching left and right all over the place and dying out. Beings having the capacity to be aware of the universe and to contemplate the mysteries of reality may be part of the scheme of things, but how could that be all of it? Nothing, such as a simple and direct evolutionary line leading to us, suggests that the universe has a special interest in humans. The universe will no doubt go on creating even though we humans have now appeared. Will something fundamentally different appear in the future? All the extinctions in the past, which appear as so many dead ends if we were the goal, must also be explained. The apparent role of random events must also have a purpose. We must also explain why there is so much inanimate diversity in the universe. Many believe on religious grounds that we on this one little planet are the only intelligent life in this mind-bogglingly vast universe. Some biologists, such as Ernst Mayr, also believe this on the grounds that conscious life was so incredibly improbable even on earth: if there is no biotic principles at work in reality, the chances of life even on earth if its history were rerun are nearly zero. If this it turns out to be true, it would only compound the problem: there are billions of galaxies each with billions of stars, and if we are the only intelligent life we begin to look to naturalists like either a fluke or the product of an incredibly wasteful designer. It is hard to see us as the central purpose, let alone the only purpose, of all this.

Natural Suffering

Add to this a second problem connected to evolution: natural suffering, i.e., the pain that occurs to every sentient being merely by being alive. Such suffering has been endured for well over a hundred million years by sentient animals and now by humans. Modern technology can lessen the scope of suffering, but this does not change the basic problem. Theists, however, tend to focus only on moral evil (i.e., the evil acts that humans commit) since they see the universe as only about humans.[28] Theologians may have a plausible counter for that problem: God has given us free will to choose or reject him, and we have misused it in other regards.[29] But natural evil is another matter.[30]

Some pain is, of course, biologically useful (e.g., pain from sticking our hand in a fire to stop us from further physical damage). But what of the evils such as earthquakes, poisonous snakes, miscarriages and stillbirths, cancer, and Alzheimer's Disease? And if pain is here to protect us from further injury, why are there "silent killers" like heart disease and high blood pressure? If there is a designer, didn't he design the AIDS virus? Or did he merely design the laws that made AIDS possible and then let it happen? Either way, doesn't he have some moral responsibility for it? Arguably, to create something interesting, this realm by definition must be less than the timeless perfection of the transcendent—hence, there must be "evil." Nor could we expect a real life to be easy and unchallenging. To quote Yogi Berra, "if the universe was perfect, it wouldn't be." There have to be obstacles in life to overcome and the possibility of harming others for a genuine morality to be possible. And theists can also argue that our physical suffering is a test that God is subjecting us to. In this way, some evil is necessary to achieve a "greater good," such as, to use John Keats's phrase, "soul-making."[31] Today some of the suffering gives us occasions to respond to our own suffering with patience and humility and to respond to others' suffering with acts of compassion. But how, for example, can Alzheimer's Disease help with anyone's soul-making when it leaves people wasting away unable to recognize even family and friends?

Theologians cannot focus only on human suffering of the last few thousand years and ignore the agony that sentient animals had to endure as they struggled to survive for perhaps hundreds of millions of years before humans came along. Theists cannot simply dismiss animal suffering as irrelevant—i.e., the universe is all about humans, and so the suffering of animals does not matter. Sentient animals are capable of experiencing suffering, and that is enough to make them the subject of a moral agent's concern; thus, if we are to conceive of the creator as moral, theologians must find some answer. If God is a loving parent, wouldn't he try to conform his creatures' suffering rather than remain silent? (Theists could deny the autonomy of the natural realm and argue that in instances of animal suffering that do not impact upon human soul-making God does unbeknownst to us in fact intervene in the natural order and stops the suffering—he only lets humans and animals suffer when it helps us.) If a greater good is being accomplished, we do not see it in the world, and this raises the problem of how we could know if there really is a greater good achieved. As it now stands, the claim that all the suffering is actually necessary is only a matter of faith based on other considerations alone.

Also note that the theistic explanations for this suffering again make humans the central purpose of the universe—in the end, everything is put here only for our sake. And this once again looks implausible and self-serving when we look at the expanse and age of the universe. Consider Marilyn McCord Adams's explanation.[32] She makes no attempt to explain why God permits evil, and she admits that no amount of worldly goods could outweigh the horrendous evil that humans face. But she argues that the suffering we endure is always outweighed by the pleasure the saved will enjoy in the afterlife. All the agony of animals for millions of years is of no consequence—the universe is all about us and only our fate matters. Theists may also treat animal suffering as justified because the way evolution is set up it was necessary to lead to us. Thus, the millions of years of the suffering and death of those animals do not matter because animals are not ends in themselves but only means to the appearance of intelligent life; the sacrifice of these animals is justified by the glory in paradise that awaits those humans who are saved.[33] But if we start from examining the events in the universe, we see no reason to conclude that we are the central feature of all of cosmic history; instead, we are just part of everything that is beautiful and ugly about the universe. Still, to theists, mankind should say, to quote a T-shirt, "It's all about me." Indeed, once we start from a theistic perspective, it is perfectly obvious that the universe is all about the presence of conscious beings—why else would a conscious transcendental being create?

There is also a defense of natural suffering going back to Darwin: suffering is simply the inevitable result of any natural processes God chose to make humans eventually appear. It is an unavoidable by-product of any way of creating the greater good that results from there being a material universe—it is logically impossible to create a world with free, conscious beings that does not have the possibility of suffering as a side-effect. Moreover, evolution requires death to provide nutrients for other beings and to make room for what comes next, and so God made sublimely ordered killing systems. But this defense too ignores the problem of suffering: God did not have to choose to use evolution—he could have chosen other forms of order. He could have simply started with free humans and the other diversity we see, as Genesis states. Arguably, once God decided to create something physical, he was stuck with creating some physical pain, but if conscious beings are all that is of value in this world and physical existence inherently entails pain, why didn't an omnipotent god create a nonmaterial world that contained nothing but beings of pure mind who had free will but who could not suffer? Angels in heaven, after all, have the free will to choose to follow God or to assert self-will, as Satan and his followers proved. Thus, if creating creatures with self-will was God's only goal and he freely chose to create a material universe that must entail some suffering, he is morally responsible for the consequences.

A counterpart to the *Euthyphro* problem in the "divine command" theory of ethics also pops up: if God could not have chosen other laws or other ways to set up the universe, then there is some independent reality beyond his control constraining God, and thus God is not absolutely supreme. But where would such constraints come from? However, on the other hand, if he was free to create a material world differently, then he is morally responsible for what results from what he chose. Thus, either way, theists have a problem. In addition, according to traditional teachings, after Christ returns in the future there will be a "new earth," presumably free of suffering where the lion will lie down with the lamb; thus, it must be possible for an omnipotent creator to create a material realm with less suffering or totally free of suffering—God could have chosen to create a less painful world but did not. Theologians may reply that God could have done otherwise but he chose to make a material world and to utilize evolution, and we just have to accept it—who are we to question the ways of God? However, that is simply admitting that it is a mystery: we do not understand why there is suffering. Moreover, it in no way lessens a creator's responsibility for that suffering.[34]

That there is a material world at all begins to count against the existence of a theistic god or at least against us having a clue as to why the world is here. Under the "free process" defense, theologians assert that God chose to abide by the laws he made regardless of the consequences, even though it makes him appear not to exist. God set up a universe that runs by natural laws (thus enabling free agents to predict future courses of events and make reasoned choices), and evolution was the way to get to us. Natural forces took billions of years, and God did nothing about it because to intervene would be to reveal himself. Any theistic creator would obviously have the power to intervene and totally control what he created, but God freely chose to limit his omnipotence in order not to tip his hand because he knew from how he set up things that eventually beings with free will would evolve and only they matter. But this does not absolve God of the moral responsibility for all the suffering. If the transcendental reality is a deistic one that merely automatically generates the initial conditions and laws of nature, it could not be held morally responsible for the resulting suffering since it makes no choices. But a theistic god had choices and thus is morally liable for the suffering resulting from his free choices. Humans evolved through a process that involves the suffering of prey torn apart by predators and genetic variations that lead to genetic diseases in humans and animals—did God design that process? And if God could intervene to stop the suffering but elects not to, thereby permitting the suffering to happen, his responsibility is only increased. Theists may reply that the expanse and age of the universe just does not matter to God since he is outside of time and space and has literally all the time in the world to create. But does the suffering endured by sentient beings for those eons just not matter to God either?

Moreover, according to traditional Western religious beliefs, animals have no paradise awaiting them—they suffer only for our benefit. Even if the evolution of conscious beings is their only purpose, is their suffering irrelevant? Why would a moral god permit accidental genetic variations that lead to needless suffering or, worse, guide evolution that way? Or why can't animals evolve by Lamarkian mechanisms so that they could at least control their course somewhat? Or, why is the world designed so that animals have to eat other animals? Killer whales catch sea lions for food, but they sometimes torture them first, tossing them about in the air before finally killing and eating them. Violence, aggression, and warfare are part of the behavior of most primate species, not just humans. (Violence also comes from the inanimate world—from earthquakes and tornadoes that routinely kill to asteroids killing off dinosaurs.) Darwin himself could not persuade himself that a benevolent and omnipotent god would design and create the ichneumon wasps with the specific intention that their young would feed within the still-living bodies of caterpillars that the wasps had paralyzed for that purpose.[35] And it is not only the tools of the predators: the designer also designed the prey with the ability to escape and thus to let the predators starve. Thereby, a horrible dance was created in which each struggles to survive at the expense of the other. All in all, if the animal realm was designed, then the designer has a nasty, vicious streak.[36] Why couldn't an omnipotent creator make animals manufacture their own food from nonsentient sources as plants do? In fact, even some plants are carnivorous. As things stand now, life is one big continuous feast.[37]

Many theologians argue that animals do not really suffer because they do not have a consciousness like ours—since they have no recollection or expectations, their pain is only a temporary discomfort in a joyous life. But based on our shared evolution of mechanisms and the reactions we observe, it is clear that many animals suffer pain the way we do and that some are aware of fear and death. The millions and millions of years of this suffering cannot be downplayed. Biological evolution is based on a struggle (along with cooperation in the struggle against other animals), with animals eating not only plants but other animals—nature "red in tooth and claw" is still an apt description. Can the biological sphere really be seen as an expression of God's "infinite self-giving love"?[38] Theologians who see only joy and beauty in the world are ignoring the part of reality that does not fit their picture.[39] As Darwin said: "What a book a Devil's Chaplain might write on the clumsy, wasteful, blundering low & horridly cruel works of nature!" The philosopher of biology David Hull adds that evolution is cruel, haphazard, "rife with happenstance, contingency, incredible waste, death, pain, and horror"—all evidencing, not a loving god who cares about his productions, but the careless indifference of an almost diabolical god.[40] If nature was anticipating us, as Anthropic Principle advocates believe, it seems indifferent once it got us. Arguably, the amount of natural suffering humans must endure is small compared to that of wild animals, few of which die peacefully of old age rather than being torn apart and eaten or otherwise dying of painful natural causes (disease, starvation, thirst, fire, cold, and so forth). In sum, nature certainly does not suggest that self-giving love is the basis of reality. Some mystics claim to experience an all-encompassing love that makes all life precious. Nevertheless, it still is hard to explain how that reality could possibly be the source of a realm where life is worth so little.

And then there is the natural pain we humans specially must endure. There are the natural disasters. Plagues wiped out at least a quarter of the people in Europe in the Middle Ages. The Lisbon earthquake of 1755 that killed tens of thousands, many of whom were in church at the time, confounded European thinkers—why would a just god do such a thing? Humans are also beset with stillbirths, birth defects, and lethal diseases. Why do children have to die of cancer? How does this develop their souls when they are too young to understand? Why cannot the body produce more pain-killers and at least let them die in peace? If a human tortured children for days and then killed them, we would be appalled—why do we forgive God? (If a child is afflicted just to help the soul-making of *others*, the morality of it all is questionable, and the wanton disregard for life and suffering becomes apparent.) Do miscarriages and stillbirths mean that there is no god who cares or that human life is not valuable? Even ignoring that, if childbirth itself had been designed, why is it so difficult and painful? Saying that it is punishment for the sins of Eve is about as good an explanation as anyone can come up with if they reject evolution as an explanation. (The Indian explanation of the pain that sentient animals and humans suffer—that we get exactly what we deserve because of our own karmic acts in previous lives—is certainly more morally satisfying, although it too does not explain why suffering is there to begin with.) Some natural suffering may be useful for soul-making, but the sheer quantity of it often overwhelms us and seems pointless. In fact, the Buddhist evaluation that everything is ultimately unsatisfying because everything inevitably leads to suffering may be more defensible than the Christian claim in Genesis 1 that the universe is "good" (since it is God's creation) since that creation includes such things as cancer and AIDS.

Alvin Plantinga raises the possibility that nature's evils are attributable to Satan or his minions, thus reducing them to a form of the more easily explained acts of evil resulting from acts of free will.[41] (Why Satan was at work tormenting sentient creatures for at least a hundred million years before we came along or why things are not better designed to be *worse* for us is not explained. Nor does this justify God letting Satan do these things to sentient beings.) Creationists claim that death and the defects in nature are the result of human sin, and so there was no suffering before the appearance of mankind. But if we take the scientific account of the universe seriously, we have to hold any theistic designer morally blameworthy for the natural carnage unless theists can come up with a plausible reason for why it is necessary. In fact, theists admit as much when, despite the essential goodness of creation, they insist that there must be a life after death or eschatological fulfillment to compensate for the suffering and injustices that humans are subject to in this life.[42] For the religious, the pain that preceded us and accompanies us is simply no big deal compared to the eternal bliss awaiting the saved. We may wonder about the morality of all that suffering for the benefit only of the saved—especially since, according to all traditional doctrines of the Western theisms, those actually saved will be relatively few compared to the world's total population. And how does what the saved will experience in their afterlife morally justify the suffering of others, let alone the suffering that preceded us?

Theologians argue that it is better for God to have created the universe than not (since it realizes goods that otherwise would not exist), but they have to address the total price paid for the resulting goods. Their fall-back position on natural suffering is to play the mystery card: "It must all ultimately be for the best; so God must have his reasons even if we do not know what they are." The "why" must have a "because." The suffering of a fawn only *appears* to us to be unjustified: religious reflection on tragic suffering in nature forces theists to postulate hidden outweighing goods.[43] There then must be outweighing goods connected to suffering that are entirely beyond our ken; otherwise the injustice of it all would be unbearable. In this way, to theists, natural suffering is not even prima facie counter-evidence to the existence of a loving god. And no doubt it is possible that there is such a reason—even though we cannot conceive what it is. Indeed, theists may ask why should beings such as we be capable of knowing the reason or have access to what the transcendent knows? Isn't it arrogant to think we should?

Thus, theists can beat back the logical problem of natural evil, if the ploy of invoking mystery is rational. But unless theists have compelling reasons to believe in God on other grounds, most people would agree that invoking mystery and holding on to faith in spite of the empirical counter-evidence is not rational. Faith that there must be some reason would not counter the negative evidence but would be only a matter of relying upon an unfounded guess and merely stating a groundless conviction in the face of what is evident. Calling this ploy "the mystery of faith" does not make it rational.

In sum, if one approaches natural suffering with a prior commitment to theism, one can argue that there must be a god and that everything must be all right—otherwise the suffering is simply too terrible to contemplate. But this ploy is not very satisfying to someone who starts with examining the history of life and the magnitude of suffering.[44] Some natural evil may be necessary to attain some good such as our moral or spiritual development. But much of the suffering appears gratuitous in the sense of serving no purpose for achieving a greater good and being preventable by an omnipotent being, and the theists' ploy of invoking an unknown reason also appears unjustified. What would be the point of a transcendental reality showing us suffering that *appears* gratuitous but really isn't? What good would be accomplished? Merely pointing out the logical possibility that God might have some reason for permitting or even directly causing such suffering is not grounds for rational belief, even if theists have other grounds for believing in God, unless theists can specify what that reason might be. The theists' last move is simply an appeal to ignorance and thus no explanation at all—it is only an admission that suffering is a mystery to us. Similarly, claiming "God's ways are not our ways" or "God does not abide by human morality but has his own" does not get out of the problem but only aggravates it: it affirms that we can never understand why God does what he does.

Thus, the millions of years of pain and death with no humans, the convoluted and violent course of evolution, and the mass extinctions present a major difficulty for seeing the universe as a creation of a moral god. In light of the staggering amount of suffering endured on this one planet alone for millions of years, the theists' attempt to accept both a transcendental cause and the reality of evolution—a God-guided theistic evolution—is not easy. All in all, anyone who takes evolution and natural suffering seriously will be hard pressed to conclude that there is an all-knowing, all-powerful, all-loving god who designed the universe. Evolution does not count against "design" in the sense of fundamental order—it is a natural process that is governed by law-abiding natural forces and random events. But the *particular* order does present a problem for the idea of a *theistic* designer.

Of course, no amount of suffering can count against there being some type of designer—theists claim, after all, that an all-loving and all-merciful god created hell, a place where according to traditional Western religious beliefs the vast majority of mankind will spend eternity being tortured by some of the same god's other creations. (Again, this is hard to reconcile with the justification of natural suffering: humans are supposedly at the center of why there is a universe, and yet the vast majority of mankind still ends up with eternal suffering.) Eternity, it must be remembered, makes the suffering of the last few hundred million years a proverbial drop in the bucket. If a god could create such a hell, no amount of suffering on earth, no matter how hideous and apparently undeserved, can refute the claim that some type of god designed this world. But the suffering inherent in nature does still count against the designer god being the all-loving god of theism.

In short, if a god is revealed through nature, it apparently is not the god of theism. Indeed, naturalists claim suffering counts conclusively against all claims of design by a loving, theistic god.[45] This leads even some contemporary theistic defenders of the design in nature—perhaps aware that the course of nature is convoluted and violent—to insist that we cannot infer anything about the nature of the designer from the design. But if we cannot, we are left with only the abstract claim "there is a design but we don't know the reasons behind it," and that is not much of an explanation (and is a slim basis for religion). If the suffering of evolution and other natural evils serve some overarching purpose, it is a mystery to us.

Is There Fine-Tuning?

The naturalists' third response is that we may see "fine-tuning" for life when in fact there is none. The appearance may only reflect our current, incomplete knowledge of things. Steven Weinberg raises the prospect that many physical constants have reasons to exist unrelated to the fine-tuning necessary for life.[46] It may also be that the apparently independent variables that seem so finely tuned are in fact interconnected and so there is no surprise that they seem so well balanced—the variables in *any* stable universe would appear to be "finely tuned." Some have questioned whether the numbers have been manipulated to make it seem like fine-tuning has occurred.[47] So too, if we vary one variable and hold the others constant, than the fine-tuning making life possible does seem incredible; but if all the variables can vary, then life may well be possible with other combinations. Considering that such a small part of this amazingly vast universe actually permits life and that life only recently appeared and will ultimately disappear as the universe continues on for trillions of years, it may be that life is only a fleeting fluke rather than the goal of some designed fine-tuning. So too, why focus on conscious life? The conditions in our universe are great for making black holes and different forms of, say, carbon. Nonconscious forms of life also are thriving. Only from our perspective does the universe look intricately fine-tuned for our own existence.

It could also be that the universe is simply programmed to get more and more complex, and so conscious life, along with other types of complexity, is inevitable. As discussed in the last chapter, some antireductionists suggest that there may be natural forces (as yet undiscovered) that constrain physical and chemical conditions that would permit the rise of life and then also constrain evolution without there being an established goal by "selecting" among random mutations those needed for more complex forms of life. Such structures for life and consciousness would be as basic to reality as those of physics, but no more encompassing purposes or final goals are present. Life then would not appear by pure chance—random events are part of the process of the evolution of the cosmos, but the dice were always loaded by purely natural biotic mechanisms in favor of life and then complex forms such as consciousness. Other scientists now speak of spontaneous "self-organization" of matter under physical and chemical forces alone.

If there are some biotic structures to the universe, there is no need for any physical "fine-tuning." Life, including conscious beings, may appear under much different conditions than on earth (e.g., non-carbon-based life-forms), and if we knew more about the nature of life than our limited perspective permits, much of what we see as fine-tuning for its appearance may turn out to be an illusion. However, this still does not address the more basic issue of the underlying orderliness for complexity. All the facts highlighted by the Anthropic Principle may be neutralized in the sense that no special principle is needed to explain them, but still where do the laws of self-organization come from? How did the universe come to be self-organizing? Why does nature now have its wondrous creativity? Natural forces may explain the "how" of nature's creativity, but why are these forces here in the first place? Did a transcendental reality set up the natural forces to begin with? Principles for life and consciousness would be no more designed than are gravity and electromagnetism—but then again the latter would still be "designed" in the simpler, goal-free sense. Principles of biological self-organization would be no more (or less) mysterious than magnetism.

If natural structures for life and consciousness do exist, it does mean that these phenomena are integral to the scheme of the universe, but it does not mean the universe is simply "designed for life"—nor does it follow that humans are the *purpose* of all this or that the universe is here just for us. It only means that there are more types of structure to reality and more constants to nature than scientists have yet found. This means that evolution and nature in general are more complex than we now suppose, but that does not affect the fundamental issue of design. The naturalists' response to fine-tuning merely separates simpler "laws" from detailed "designs for a purpose" and explains life by means of the former alone. But this merely pushes the issue of *order* back one step: an ultimate explanation of the underlying orderliness is still missing. We are still left with the basic mysteries of why nature is orderly to begin with and why this particular order.

Multiple Worlds

However, the naturalists' fourth response to the question of design addresses both the particular order present in our universe and perhaps the issue of why there is order at all. It is the "multiple worlds" hypotheses proliferating in science today.[48] Some cosmologists argue for a series of oscillations of the universe—an indefinite cycle of Big Bangs and Big Crunches.[49] Others argue that our world produced by the Big Bang is only one of many different self-contained worlds that bubble up from the timeless "mother" universe in super-space and eventually dissolve back in. Thus, there is a huge field of different disconnected bubble worlds currently existing in unconnected planes of the universe that cannot interact with each other. Each world may have its own natural laws or constants of nature (such as the mass of particles or the strength of gravity), or perhaps all have the same laws and constants but each has its own initial conditions from its own Big Bang. In a kind of cosmic Darwinism, some worlds are mere hiccups that do not produce any stars or life since the conditions do not meld into something that functions; others survive long enough to thrive and produce life. Some cosmologists argue that worlds come and go, but they produce millions of baby worlds through black holes: each black hole is a "white hole"—a Big Bang—that creates another universe. Because the "fire" of origination of the progeny burns all connections to its source there is no connection of laws and so each world may have its own or no laws at all. Other cosmologists argue that there may well be regions of our one universe that are simply beyond our visual horizon (which is set by the age of the Big Bang) in which other laws prevail—we have one infinitely large universe with various pockets of order or disorder. The expansion caused by our Big Bang is one such tiny pocket—its hundreds of billion galaxies are not even a drop in the bucket in the eternal universe.[50] Other regions may have completely different physical laws, or, even if they have the same basic laws we, they may have very different physical conditions or constants. Either way, the result may be very different from what we can see from earth. That is, each pocket may have its own distinct physics, and the laws of our pocket, which we think are universal and absolute, may be only local.

Multiple-world hypotheses are becoming the mainstream position in cosmology. And it is not only cosmologists who have advanced the idea. Some physicists argue that the act of observing certain quantum-level events causes the universe to split into different paths, creating a mind-boggling multiplicity of parallel worlds —i.e., other "universes" existing simultaneously with ours but completely disconnected to each other and thus unobservable even in principle.[51] Each split-off world would have the same laws and constants but different contingent facts. Superstring theory also suggests multiple worlds, each with different particles and different fundamental physical constants. In philosophy, the philosopher David Lewis defends the proposition that all logically possible worlds exist.[52] There may be an *infinite number* of universes, and that idea, like all infinities, is hard to wrap our minds around, and it may also offend those who think that we are unique; but this does not make the idea implausible. In Christianity, the idea that God could not create more worlds was condemned as heresy as early as 1277. And if the idea of a multiplicity of worlds becomes irrevocably embedded in the scientific view of things, theists no doubt will accept that God made more than one "universe" and will claim this is only reasonable—if existing in a material universe is a good thing, why would God create only one universe rather than a series throughout eternity to explore more or maybe all possibilities?

The effect of these multiple-world hypotheses is to defuse the apparent "fine-tuning" in our universe that gives rise to life. Given the unimaginable number of worlds, virtually every combination of laws, physical constants, and initial conditions would occur, and thus in some worlds a combination necessary to produce life would inevitably occur.[53] In sum, multiple-world hypotheses explain the Anthropic Principle "coincidences" as arising perfectly naturally without any such principle. These hypotheses also explain why our world is intelligible: some worlds simply will have beings who can discern some of its structures. Thus, there is no designing going on in our world—we just happen by chance to be in one of the very very small number of universes that produces life.[54] The stronger versions of the Anthropic Principle—that a universe *must* have properties that will allow conscious life to develop at some stage in its history—now sound like no more than misguided boasting.

The idea of sequential or simultaneous multiple worlds does not affect the Cosmological Argument: God can just as easily create or sustain multiple worlds as one. But it wreaks havoc with the Teleological Argument. It counters the claim that there must be a designing agent to produce the features of this particular universe. Indeed, it neutralizes any allegedly designed features in any universe no matter how intricate and impressive. Of course, a god could design multiple worlds just as easily as one, but the apparent design in this universe is no longer *evidence* of a designer—it may be only the result of chance. If we deal the cards out at random enough times, any combination will eventually appear, and the universe has eternity to deal. The probability that one of the worlds will be fine-tuned for life is virtually 100%, despite the fact that there is no overarching plan to the whole thing, just randomness. There is nothing really to explain about our particular pattern because the world had to be one way or another—we just happened as a matter of

not necessarily esp. m endless universes

fact to have won the lottery. Winning the lottery always seems highly improbable to the winner, but someone had to win, and there really is nothing to explain. Moreover, order may be a perfectly natural feature of all universes. Perhaps a TOE would show that there are some cross-world restrictions so that not all possible options can appear; thereby, the role of a designer would again be unnecessary. Or in some universes, there may in fact be no order at all, i.e., total chaos. This possibility would impact the fundamental teleological question of why there is order at all: order would then be only a matter of chance in any given world, and thus no designer even for order *per se* is needed. To naturalists, seeing an intelligent agent at work behind it all, as also happens with the Cosmological Argument, is simply a carryover from the days when humans saw animistic agents at work everywhere in nature behind storms, rain, wind, and so forth. Evolution may have disposes our minds to see agency as the primary type of explanation for unknown phenomena, but we have learned better today.

new Good

But theists argue that naturalists are advancing these ideas of multiple worlds solely on an *ad hoc* basis to counter the idea of a designer: nothing motivates the naturalists but atheism, so they have concocted theories where chance replaces design—as Neil Manson observes, theists allege the multiverse hypothesis "to be the last resort for the desperate atheist."[55] Our best knowledge is that only this one world exists. In addition, naturalists are trying to explain one mystery by introducing another mystery and by multiplying entities. Multiplying worlds also goes against the usual scientific process of explaining things with fewer and fewer postulates. It also introduces vast complexities to explain the phenomena of only one world. If the generation of conscious beings is the purpose of the universes, then the waste of evolution on earth is nothing compared to creating zillions of worlds in which only a handful may generate conscious beings. And naturalists must admit that there is no empirical evidence as yet for any other universes, that the heat of the Big Bang probably would burn any evidence of previous oscillations, and that it is difficult to see how even in principle we could detect a quantum-level split-off since the resulting worlds do not interact.

But naturalists reply that it is the theists who are "desperate" to push the multiverse hypotheses out of science so that the theories can be portrayed as only a naturalistic metaphysical alternative to design. However, these theories have not been advanced merely to counter the Teleological Argument; rather, they are consequences of perfectly good scientific theories arising from the attempt to explain our physics in a universe that began with a Big Bang. To Paul Davies, some form of multiverse is "probably an unavoidable consequence of modern physics and cosmology." In addition, some cosmologists think there may be ways at least to *detect* the presence of other worlds by their effect on our world (e.g., the accelerating rate of the expansion of space may be the result of our space being reshaped by a collision with another space), even if we can never learn their laws and constants or other properties. Thus, these hypotheses may be testable. The theories that gave rise to a multiverse hypothesis also have testable predictions unrelated to the issue of a multiverse and thus may be supported empirically. More-over, not all scientific support is from empirical evidence: theoretical concerns and inferences also enter the picture. And because some scientific theories do suggest an ensemble of universes for theoretical reasons, scientists would in fact have to introduce *ad hoc* assumptions into their equations to *reduce* the plentitude of worlds to just one. The idea of multiple oscillations or bubble universes also fits better with the scientific worldview than one world: there is no reason for just one Big Bang, and multiple worlds would remove one big "singularity" from the naturalistic picture by bringing the idea of the origination of our world within a recurring series of purely natural events. The universe would then be more law-driven. However, the multiple-worlds hypotheses currently is on the most speculative edge of science, and whether they will ever actually be within the purview of empirical research is still an open question.[56]

The Alternatives: Multiple Worlds or a Designer

That there is order in the universe is indisputable, and it can only provoke wonder in anyone who contemplates the subject. The intricacy of the world is so mind-boggling that it feels as if some design is involved, and the need to have an answer to "why?" is hard to get rid of. Since the fact that the variables are exactly fitted to produce life does not seem possible by pure chance if there is only one world—it would be a truly astounding brute fact—we naturally look for some explanation. There are at least three possibilities: multiple worlds, a design by an impersonal deistic reality that sets up the order and lets the world go without any goal, or a design by a theistic god who sets up the universe for a purpose. (This assumes that we now know enough to lay out all the choices—but remember that only a few decades ago we would not have had the multiple-worlds option.) Probability cannot help figure which of these is the most plausible: since we have only one world and one order to consider, there is no data to determine which option is more likely.

Thus, the alternatives we have to choose from seem to be either a multiplicity of worlds or a transcendental designer. Again, a transcendental reality could have designed multiple universes as easily as one, but the point is that we could no longer infer design from the order in our universe: there is no longer a basis to rule out our world's order arising simply by chance from the spectrum of possible universes. (Of course, theists will ask why a combination of laws and constants permitting life is among the logical possibilities to begin with—i.e., why is a life-generating universe among the range of possible universes?) In sum, the multiple-worlds hypotheses do to any cosmic design what Darwin did to the classic design argument: they do not eliminate the possibility of a cosmic design, but they do neutralize any alleged empirical evidence of design. As long as the naturalists' position is plausible, the apparent conclusiveness of the theists' "design" position, even if it initially sounds obvious, is destroyed.

There does not appear to be at present any empirical way to decide between them. We are aware of only one Big Bang, but that does not count one way or the other as to whether there are more. Naturalists have not yet devised tests for other worlds (although such tests may eventually be possible), and theists have not devised a test for whether a transcendental design is at work rather than simple causal order. And theists must admit that whatever scientists find will be taken as "designed." In such circumstances, it is hard to see how theists can make final causation part of science, since they cannot make specific testable predictions or retrodictions. Most theists realize that simply claiming "God just designed it that way" to whatever scientists find is not a scientific explanation. Thus, to find design, advocates of Intelligent Design in Christianity and Islam try to find biological phenomena that evolutionary biologists cannot explain.[57] Currently they are focusing on "irreducible complexity" in organisms. But evolutionary and developmental biologists have not been impressed and have offered alternative explanations within the standard Neo-Darwinian framework. Moreover, design introduces a "black box" that is in principle more opaque than in any naturalistic attempts to explain biological phenomena: Intelligent Design advocates admit that they cannot explain how a designer designs—God's acts of creation are miracles and we cannot expect to study or understand them. William Dembski's test for whether design occurred is merely an argument from ignorance: he eliminates chance and natural forces as explanations of complexity (needless to say, neo-Darwinians disagree with his assessment here), and thus concludes that there is no other option but design—it is not any positive empirical evidence of design.

Intelligent Design advocates are also caught in an additional dilemma: either Intelligent Design is not part of science (because God is a personal agent, his actions of design involve free will and thus are not law-like and so cannot be tested), or the "design" that theists see is purely law-like and thus potentially part of science (but then no god is needed, any more than one is needed to understand gravity or electromagnetism). Either way, God is not part of science. The universe may well be designed, but the argument is relegated to *metaphysics*, since any source of order and laws is not part of the testable body of science itself. In addition, a designer that by definition is transcendental is more "unscientific" than multiple worlds: it is beyond experience in a different sense than other natural worlds—it is not just out of our reach within the natural realm but a different ontological dimension of reality altogether.

Both sides will also claim Occam's Razor. Theists will say postulating a single designer and a single Big Bang is obviously simpler, while naturalists will insist that invoking more natural realms is ontologically simpler than invoking an entirely new type of reality—theists have introduced a metaphysically different reality and three unexplained types of metaphysical causation (for creating, ordering, and sustaining the universe), while naturalists simply have introduced more of the same stuff. To naturalists, the universe is merely much bigger than we once supposed, and drawing a line between the part of the natural universe that we know from what we do not know does not affect the realities postulated—it is the number of *types of realities* in a theory that is the subject of Occam's Razor, not *how much* there is of each reality. And on this point, the naturalists' option is clearly simpler: it reduces the number of metaphysical assumptions we have to make about the universe. Equally important, from the point of view of scientific theorizing, the entire ensemble of worlds is *simpler* than just one: to remove the elegance of the complete spectrum that the multiple-world hypotheses predict and to leave only one world, we would have to add experimentally unsupported processes and *ad hoc* postulates that prevent the creation of other universes.[58] There is simply no scientific reason for only one world. In terms of the issues of this chapter's first section, the only two "natural" states are that nothing exists or everything exists: obviously something exists, and if less than everything exists, we would then need an explanation. Thus, the necessary addition of those hypotheses actually makes the theory of only one world more complex from a scientific point of view than a theory of multiple worlds.

Naturalists' and theists' extrascientific metaphysical commitments may end up determining what theories are eventually accepted in cosmology: multiple natural worlds makes naturalistic sense, while a designer makes the world meaningful in a religious sense. The astronomer Robert Jastrow notes that some scientists are "curiously upset" with the idea that the universe had a beginning because it conflicts with their "articles of faith." It would introduce an event not covered by any laws and so scientists would "prefer" an oscillating universe or multiple worlds free of such a singularity. They "would like to reject" the idea of an origin and believe "it cannot really be true."[59] They are not bothered by the fact that we cannot observe the other alleged worlds. On the other hand, some Catholic theologians have trouble even taking the idea of multiple worlds seriously since the Pope has committed them to the idea that the Big Bang was God's act of creation.[60]

The Problems for Theists

No amount of order and apparent design within this world counts one way or another for whether there is some designer. As Steven Weinberg says, even a completely chaotic universe could have been designed by an idiot.[61] Thus, no scientific discoveries will help with the general issue of whether there is a designer. But even assuming for the moment that there is a transcendental design of this world, what sort of designer is it? What inference does the order in this world warrant? In particular, is there evidence of a loving personal creator of theism? Or, turning the question around, is this the type of world we would expect a loving, all-knowing, all-powerful being to create? Scientific findings impact that religious question. Can we infer anything about a possible creator from the order we observe in nature? As with the Cosmological Argument, all that is warranted is positing what is sufficient to accomplish the task. Theists want to say nature reveals "the glory of a loving

God." Deists may be willing to see a supra-human intelligent reality—a rational mind with the capabilities of a person—behind the universe's order, since the reality obviously is capable of producing intelligent life.[62] (Fred Hoyle, Paul Davies, and now apparently Anthony Flew fall in this group.)

However, must the source of order be a personal agent? Just as the source of three-dimensional space may be spaceless and the source of time timeless, so too the source of life and consciousness in this world may not be living or conscious; perhaps only things within the universe can have those properties. It may be something beyond our categories. In addition, there is a counterpart to the cosmological problem of "who made God?"—who designed the designer? Obviously the designer of the universe cannot be part of the universe, since a living being within the universe could not have designed life and the rest of the universe in the first place. Thus, the designer must transcend the natural world. But, as discussed with regard to the Cosmological Argument, we only get out of an infinite regress in explaining things by establishing something that does not have the property to be explained. But mustn't a designer have the complexity of a person since it must decide how to act? If so, it must have the complexity of what is to be explained, and thus we have not introduced something of a different nature from what has to be explained. Claiming that a transcendental designer thought up the design from scratch and acted on it to carry it out seems plausible, but how did it get the power to think? The idea of "self-designing" is no more plausible than "self-creating." Thus, the problems still remain of where the designer came from (a cosmological problem) and where he got his mental faculties and his ideas (a design problem). In short, theists add the problem of "why this sort of god?" to the basic problem of "why this sort of universe?", thereby complicating things.

In fact, theists have hardest case to win. Theists argue that deists with their impersonal principle of cosmic order cannot provide the ultimate explanation of mind and thus nothing makes sense without a personal god. However, even ignoring the possibility that consciousness may arise from nonconscious transcendental or natural forces, there is still no justification simply in the orderliness of things or even in any programmed goal for an all-knowing, all-powerful, all-loving personal being of traditional theism. Theists may believe it is simpler just to posit one all-powerful creator/designer, but the question is whether the empirical evidence supports such a move—again, only the minimum necessary to explain a phenomenon is warranted, not anything idealized or more speculative or merely simpler for us. However, the problems raised by the inefficiency of evolution and the amount of natural suffering must lead to the conclusion that, even if we argue that the natural order is the rational choice of a free personal agent, nothing warrants claiming that the god is caring or concerned with nature and humans. In sum, the evidence does not suggest the designer is the god of theism.

Natural suffering also permanently limits "natural theology," i.e., inferring the existence and properties of God from his creation.[63] (Finding from fossils that there were extinct species was a major blow to natural theology in the nineteenth century.

It showed that not everything was perfectly adapted, and equally important that the world was not timeless—things have changed over time. Theologians then changed their emphasis: "design" is now seen in the laws and basic orderliness of things and in the overall purpose of making conscious beings with free will.) Theologians may argue that if God exists, then that reality must be reflected in the world he created, and thus identifying the deep structures of nature is identifying a little of the immutable character of God. To John Calvin, nature is "the theater of God's glory." But, as discussed above, theists cannot pick only part of the picture as the basis of an inference. Reality is not all butterflies, sunsets, and flowers.[64] Theists cannot focus on only orderliness, consciousness, joy, and beauty and then say that the evidence of God's nature is there for anyone to see—the rest of reality (natural suffering, the cosmic waste if intelligent beings are the only goal) must be included. Theists cannot have it both ways—claiming the world reflects God's wisdom while screening out the ugliness. They cannot pick and choose what they will accept as "signs" of God's nature without blatantly arguing in circles. God's "perfect morality" and "infinite love for his creation" has to be squared with the agony of cancer, earthquakes, and the ichneumon wasp. The cold beauty of the heavens may reflect the glory of God (Psalms 19:1), but the messiness of evolution, with its accompanying suffering, down here on earth is another matter.

Of course, if the designer is flawed or powerless to help, few would want to postulate him. But if the evidence does not justify more, we would have to postulate an imperfect, bumbling designer who is just doing the best he can (or has a malicious streak) or plead a lack of knowledge on this matter. Moreover, for all we know, the "design" may be, as David Hume argued, a corporate effort by a committee of flawed beings. Nor need the creator or source postulated from the Cosmological Argument be the designer of the Teleological Argument. It is only a theological assumption that they must be the same. The designer may merely be a shaper of a preexisting formless void, as in Genesis 1:1-2. At most, the way things look suggests that a deistic designer could have set things up at the start and just let them go in a mixture of lawful order and chance. Time took care of the rest.

Thus, the order of the world may get us to deism but not to the full-bodied god of theism. Theists also have another problem touched on above. In the Newtonian era, when the order of the world was seen as the product of a deterministic blueprint guiding every detail of history, theists saw it as the work of God. But scientists now have come to accept a randomness to some events, thereby permitting a creativity to nature through evolution rather than the just the unfolding of a pre-established plan, and also allegedly allowing free will in human actions. But again theists see this as the work of God. The newer view is the opposite of a watchmaker designer: the universe is now designed to permit some chance events, unlike the determinism of a clockwork world that would not allow free will. There is an openness to the course of events, even if there is an overall guided plan that guarantees at least some results. "Disorder" becomes part of the designed scheme of things. To theists, "randomness" now was always part of the "design"—it is the creative "freedom" and "spontaneity" through the interaction of law and chance that was God's plan all along. God is just the type of transcendental reality that would let his creation "self-organize." Thus, the determinism of the earlier scientific view—which at the time was also seen as supporting religion—is now seen, upon reflection, as in fact antithetical to religion. But this means that, heads or tails, whatever scientists accept is always seen as proof of a "divine perfect plan," and hence theistic belief is seen as confirmed.

In sum, no matter what, theists see the workings of the universe as the handiwork of God (with the suffering and waste somehow explained away). This suggests that teleology is a metaphysical interpretation of the universe in light of a commitment to belief in God, not any sort of empirical finding. The more concrete teleology of the Anthropic Principle may guide scientific research by suggesting topics to be explored (although the scientific answers will be in terms of efficient, not teleological causes). However, claiming "all is designed" does not lead to any new research topics. Theists can fit any data into a theistic worldview, and thus theists see the data as evidence of a god. They also want to argue that design has empirical consequences: the universe would not be the same if it had not been designed or if the designer had a different character or had different intentions for it. But how can we test transcendental design when whatever is found is considered to be designed that way? The bald claim of "design" cannot lead to new testable scientific predictions. By itself, it simply does not predict any specific instances of previously unknown order or the amount of order to be found.

In fact, if anything—no matter what scientists find—can be seen as evidence of purpose, this raises doubts as to whether teleology actually *explains* anything at all. This also leads to a larger issue: whether invoking God really ever explains nature at all. Theists cannot specify *how* God does any of the ordering. They are only saying "God did it," and that does not help us understand any better than exclaiming "It's a miracle!" It identifies the agent, and to that extent it quiets our urge to know. But we cannot know how a transcendental agent works, and so we are cut off from knowing any mechanism for God's actions. It comes down to a matter of trusting that there is some reason or purpose behind it all. God's will becomes the brute fact that we simply must accept on faith. Thus, we still have to accept a mystery. As discussed in the last chapter, the problem is not that God is supposedly a *transcendental reality* but that he is supposedly *personal in nature* and acting with free will. This precludes the "if *x*, then *y*" conditional of causation as with inanimate objects: God acts with free will, and so no determined results can be forced or even safely predicted. Theists may argue that God will act consistently throughout time or act in accordance with what is revealed in the scriptures of their particular religious tradition, but unless scientists can reproduce the actions this response does not help. (If God freely chooses always to abide by the laws of nature he created, then all we have the natural lawful events and thus God is out of the picture.)

In addition, a god that can do everything does not help us understand why anything in particular is as it is. That is, saying "God did it" could be invoked to explain any event in any situation, and our understanding of why something specific occurred is not advanced at all. Theists will say that God's character is not compatible with any action, but naturalists will reply that no amount of suffering or anything that scientists have found has been taken to be something a god would not do. "Design" is compatible with too many possible states of affairs to say that invoking it increases our understanding at all. In short, design is too flexible: whatever is found is simply labeled "the result of God's will." "Design" is just a blanket explanation for whatever laws or phenomena are found. It does not have the empirical risk of scientific explanations: it cannot be falsified or otherwise empirically tested. If something explains anything regardless of the facts, it is not a scientific explanation of anything. Moreover, design does not work as a meta-physical explanation telling us "why"—the defense of the apparent haphazardness of evolution and natural suffering in terms of "God must have his reasons" is just invoking a bigger mystery to explain a why-mystery. It may be comforting to think that a god is in charge, but the idea does not increase our understanding.

Theists will again claim that God is a simpler brute fact and it explains more and so we should accept it as an explanation. Naturalists will reply that we know the natural universe exists and that theists are arguing from a lack of knowledge and only speculating that there may be a simpler brute fact, and so we should stick with the natural universe as our brute fact. Furthermore, a *person* is not a simple explanation even on the metaphysical level. It is one entity, but one with multiple features and properties. This does not have the elegant simplicity that theologians want—it would take a complex person to do all the astounding ordering of our one universe. Either it must have the abilities at least to set up the forces, laws, constants, and conditions of the universe at the beginning, or it is a incredibly complex Wizard of Oz behind the curtain putting all the levers (if it is active in the universe). Either way, God is "simpler" in name only. Like the casing of a Swiss Army knife holding all the tools, it only adds a new entity encasing a complicated mix of functions. And if an *infinite* person and an *infinite* number of creative acts are required to do all the ordering, that only hurts the claim to simplicity more. An infinitely powerful person might be conceptually simpler than a finite person, but it is not ontologically simple at all. Moreover, given that we cannot know the nature or purposes of an all-knowing god or why he decided to do what he did without revelation or how he did it, it only makes the picture that much more complex, in addition to becoming another attempt to explain a mystery with a further mystery.

Also consider a related problem touched upon above concerning "self-design." Our common sense says that a creator must be at least as complex as whatever he can create—indeed, it is arguable that a creator would have to be an agent of infinite complexity . But theists cannot have it both ways: that God is simpler than the phenomena to be explained and that he has the features such as consciousness that need explaining in the natural realm. Theists believe the transcendent must be conscious to decide to create, to produce conscious beings, to design creation in general, and to monitor the thoughts and deeds of billions of people simultaneously in order to reward or punish. Thus, they also must accept that the transcendent is at least as complex as the beings he produces. But if the transcendent is complex, nothing is gained by invoking God as an explanation: as discussed above, if our complex features require explaining, than his consciousness and other features need explaining at least as much as ours, and more so if he is infinite. Thus, we get no nearer simplicity by introducing a transcendental designer—and being transcendental, his existence is even more mysterious than the existence of his complex creation. (To say that God exists outside the realm of time while all the complex phenomena to be explained are temporal in nature does not get us any closer to understanding how a timeless god could make decisions and create—in fact, it only makes this that much more mysterious.) However, if theists admit that complexity may result from simple rules (as with fractals), then they must admit that the presence of conscious beings does not guarantee that the transcendent is conscious, and this leads to deism or even naturalism.

The traditional theists' answers as to why God ordered reality the way he did always place conscious beings at the center of reality. To them, the universe is worthless in itself; without conscious beings to contemplate it, it would be a meaningless exercise. But to naturalists, making humans the central purpose of reality is highly questionable in light of the age, expanse, and diversity of the universe—over ten billion years and trillions of stars just to bring some conscious beings on this one small planet? Even if conscious life is common in this and other universes, there still is a great variety of other things. The only purpose that the universe evidences is the ever increasing diversity of forms, both organic and inorganic. Seeing reality as a miracle God made just for conscious life is taking a leap. We, of course, like to think of ourselves as important, even necessary or inevitable. We may like to think the entire universe was designed to produce and preserve us, but we cannot cite any science for such an anthropocentric belief.

The reverse danger is seeing the transcendental designer in terms of humans and making what may be impersonal into some mirroring us. If we think of the regularities in nature as being the result of *laws*, then we naturally think in terms of a *law-giver* who made them. But just because we operate with intentionality, must we conclude that the universe too has a purpose? Anthropomorphism when it comes to the transcendent may seem necessary to theists (as discussed in the next chapter), but just because there is order to the universe need not mean *a person* is in charge. Deism is also a transcendental option. Arguably, something capable of making conscious, intelligent minds must have been involved, but whether impersonal forces can produce consciousness in nature is still an open question, and we should certainly restrict any speculation on the matter when it comes to any possible transcendental realities. We normally frame the question as "who made the laws?" but perhaps the correct question is "how did the natural forces arise?" So too, when we look at the world, we can ask "why does anything exist?" without introducing the idea of "made" that changes the question into "who made it?"

Overall, theists have the biggest problems of the three positions here, but their inference to a personal designer still has an intuitive appeal. More philosophers and scientists speak of a theistic god here than with the Cosmological Argument: the Teleological Argument has emotional power because we look with awe upon the wonders of the heavens and the earth. Nevertheless, scientists who are theists tend to be physicists and astronomers who study the more abstract mathematical structures of reality and not biologists who study evolution and are more aware of all the natural suffering. However, we cannot rule out naturalism, deism, or theism. Naturalists may have raised serious problems with the idea of a loving designer, but the question of some designer is not refutable by any empirical evidence, and so deists and theists are not irrational in believing the world can point to some kind of designer. But the problems of natural suffering that humans and sentient animals have endured either refutes the idea of a loving designer or makes it a hidden mystery; add to this the possibility of multiple worlds, and theists and deists must concede that it not irrational to deny the possibility that there is no designer at all.

Is it all then just a matter of "taste" once again as to which option we choose? Does the fact that we can stop the explanatory process where we feel satisfied mean that none of the explanations are more likely than the others? Those with a prior commitment to theism will opt for a religious interpretation of the presence of order, but, as discussed above, it is far from obvious that this introduces any simplicity for explaining the actual order of things rather than merely complications.

The Mystery of the Order of Things

To close the discussion of teleology then, what are the answers to the three questions raised at the beginning? First, why is there any order at all? This we simply do not know nor have any reason to believe we ever will. We may try to remain open on the issue, but our metaphysical predilections may direct us to one of the transcendental answers or the naturalistic one. Second, why does our world have the particular order it has? The multiple-world hypotheses and design remain the two strongest alternatives. Naturalists have problems at present explaining the emergence of life and consciousness on earth. They also have to explain what seems to be an excess of beauty—even naturalists such as Steven Weinberg have to admit that "sometimes nature seems more beautiful than strictly necessary."[65] But theists have more problems. They cannot focus only on what is beautiful to us in nature or on the last few thousand years as all that reveals the purpose of this vast and ancient universe. The god that designed the beauty also designed tsunamis and cancer and millions of years of sentient beings' suffering. If one adopts a transcendentalist position, at best a deistic one seems warranted: natural suffering and the convolutions of evolution are counter-evidence to the idea of guidance by an all-knowing, all-loving god, or a special purpose concerning humans; at most there was an initial setting of natural forces and constants. But no one can answer why there is this particular order to nature, short of establishing a TOE that shows necessity to the natural constants. Third, where did the structure of the universe come from? There is no room for compromise between naturalists who deny a transcendental reality and deists and theists who affirm different forms of such a reality (with all its accompanying how-mysteries). None of the three positions can be ruled out (if theists are permitted to invoke mystery).

Our free rein in the choice of the three explanations points to the mystery of it all. Are we barred from ever knowing the ultimate explanation? We have no reason to believe that we will ever know with certainty which option is correct or whether there may be other options that we cannot dream up. Consider the problems of evolution and natural suffering. With our limited minds, we may fabricate rationalizations for why there is a particular type of suffering, but we cannot be at all sure that a transcendental mind—if there is one—has those same reasons. We obviously are not in a position to declare that there can be no sufficient reason to explain the suffering, but trusting "there must be a reason" may only tell us more about us than reality—suffering is just not acceptable to us, so there must be an explanation. There may be psychological comfort in believing everything is planned or under the control of a divine person, but whether we can be justified with this belief is doubtful. And along with the suffering, why there is order to reality at all remains an intractable mystery. Beings within the natural order, whatever our capabilities may become, cannot know such things.

"The Mind of God"

The transcendental options bring up one further point. Can we know anything of the nature of the designer? In the Jewish and Christian scriptures, there are passages stating that God's thoughts are not our thoughts and his ways are not our ways (Isaiah 55:8) and asking who knows the mind of God (Rom. 11:33). But some theists argue that we can know some aspects of "the mind of God." Moreover, this god is known, not only through special revelations, but through the "book of nature," to use an early Christian metaphor. The structure of the universe arose in a divine mind. Thus, nature reflects the rationality of that mind, and we, being created in the image of God, reflect at least some of that infinite mind. Thus, by grasping some of the structure of the universe, we are grasping some of God's way of thinking. Studying nature through the methods of science will reveal what God had in mind in creating the universe. Science works because our minds participate in the divine reason that structured the universe to begin with. We can reverse-engineer creation and thereby learn something of the mind of God. In short, scientists are making progress in reading God's mind.

For eighteenth century deists, all mystery would eventually be dismissed from reality because we could read the mind of God in nature; our mind would understand and explain it all. Early modern Christian scientists also adopted the image of trying to "think the thoughts of God after him," to quote Johannes Kepler. For Galileo and many scientists of the time, God was the supreme mathematician and the laws of nature are simply ideas in the eternal mind of God. Thus, a mathematical understanding of nature shares God's thoughts. In the twentieth century, many astronomers and physicists still found the idea congenial. The astronomer James Jeans said "the universe appears to have been designed by a pure mathematician" and it "begins to look more like a great thought than a great machine." Albert Einstein too wanted to "know God's thoughts; the rest are details." He said that the most important question he pursued was whether the Old One had a choice in selecting the fundamental laws he chose. More recently, George Smoot when contemplating the measurements of the cosmic microwave background radiation, said "[i]f you are religious, it's like seeing God." Stephen Hawking in a famous remark said that if we find an answer as to why we and the universe exist, "it will be the ultimate triumph of human reason—for then we will would know the mind of God."[66] (He was actually speaking only of a TOE and conceded the mysteries of why the universe exists and is ordered would still not be answered by a TOE, but it is a great line to end a book with.)

But most of the scientists' use of the word "God" today is not meant literally. It typically does not refer to a theistic god but is only a literary device for referring to a rational order underlying the universe. Einstein himself meant at most a deistic reality or a pantheism like Baruch Spinoza's—the idea of a personal god that transcends the universe was only a superstition, he said, that neither science nor religion needed. Hawking is an atheist.[67] Others say that laws are in fact "the laws of God," but they redefine "God" as the fundamental structure of the natural universe, not as a transcendental reality—in short, the phrase "laws of God" becomes a tautology that negates what the word "God" traditionally means. The complexity theorist Stuart Kauffman wants to go further: he wants to "reinvent the sacred" with a "fully natural God" by denying any transcendental creator but using the time-honored name "God" for the creativity inherent in nature itself. This in effect naturalizes God. The same twisting of meaning comes up whenever scientists say they have "discovered God" or have "proved" God exists or that no one can be an atheist—they simple define "God" to mean something in nature that we all know exists and then trivially conclude from this definition that obviously God exists. But in no case is the god of *theism* established so easily.

In sum, these scientists are using personal language for the fundamental, timeless, abstract reality of order and harmony, but they do not mean a transcendental reality that is personal or that intentionally designed our realm. Asking "what was God thinking 14 billion years ago?" is merely a way of asking how-questions about the Big Bang. Contemporary scientists are not engaging in a "search for God" —they deal only with the measurable aspects of the natural order. Theologians too may not literally mean "the fingerprints of God," "God's hand at work in nature," or "glimpsing the face of God," but scientists mean it even less—there is no *transcendental realm* open to study other than perhaps the timeless realm of mathematical truths and laws. Many scientists may accept that science does not rule out a transcendental realm—one can believe that the laws of nature come from God, if one wants to—but the use of the word "God" by scientists does not justify such belief. Indeed, scientists should be cautious about labeling a mystery in their research "God" since theists are quick to jump on the term and load in its full theistic connotations and then treat scientists as theists when they are not. Nevertheless, the language is a bit of irresistible hyperbole used even by the nonreligious to emphasize the importance they attach to their scientific questions and discoveries.

But if we think we really are getting into the mind of a god, problems arise. Again, all we can infer from nature about its source is the minimum necessary to explain what we see. Anything more would be unsupported speculation. And there are severe limits on what we could possibly infer. One must begin by asking how much we really know of the universe, since only with a comprehensive knowledge could we safely infer anything of a creator's mind from nature. There is much to reality we do not yet comprehend and may never comprehend. There may be aspects of reality or regions of reality that are not accessible to our experience and minds. If a multiple-worlds hypothesis is correct, the problem is only increased: we cannot use anything from this one world to infer anything concrete about a possible creator with any confidence—there may not be any basic structures common to the multiverse. There may also be a depth to our universe that we cannot comprehend. Even now we believe that nature is not rational through and through. Scientists have revealed much that is rational (the lawful causal order of things), but they also have revealed parts that appear irrational (the possibility of uncaused events on the quantum level). If this is part of God's overall design, then God permits random events. Indeed, what is "rational" may be only an abstracted portion of the total picture that we creatures can understand. It may be that the way we are built we can see order easily, but there may also be much more disorder to reality that we simply miss. In any case, we cannot focus only on our admittedly incomplete knowledge or on the order we see and argue successfully to the mind of God. At best, the random element forces us to wonder if there is an irrational element to God's mind if reality in fact mirrors God's thought.

Another question is how much we can infer from any product about the mind of its designer. Even if scientists are able to discern fundamental laws and construct a TOE or other comprehensive theory, are they getting into the mind of God? Are the laws of nature somehow mirrored in God's nature, or do they apply only within the world and have no bearing on the nature of the transcendent? How much does the design really reflect the designer, and how do we know? We do not need to know how a builder thinks to understand how a product works, and the same is true the other way around: even if something is the result of intentional, purposeful action, we cannot go from the result that are produced to the mind of its source. We are left with at most an abstract claim of order (if all of the universe and possible other worlds are ordered) and cannot infer with any certainty "what God must be thinking" from the product. So too, there is causality within the universe, but can we infer from that anything about the nature of how the designer exists? More generally, can we infer anything about the realm of the timeless from events within the realm of time? Can a transcendental reality even be like the temporal realm? In what sense would the laws of nature reflect his nature? The difference between radio waves and the music we hear over the radio indicates the problem of the difference between the cause and effect. The problem of the possible difference between the nature of a transcendent cause and the universe is that much greater. If Kant is right, we cannot infer what the nature of the world independent of us (the noumenal realm) is like from our experiences at all—inferring the creator's mind would be at least as difficult. The designer, if there is one, must remain unknown.

This point highlights the limitations we humans are subject to. We are limited both by our physiology and by our limited location in the universe. There are good philosophical arguments for concluding that we cannot even know what it is like to be a bat or a lion—how can we know the mind of a god? Indeed, how could the mind of God be intelligible to us? If God is infinite mind, then we obviously cannot expect any finite creatures to be able to plumb the depths of his mind, but what makes us think that creatures could even be able to share anything of the divine mind? Even if rationality means we are "made in God's image," there is a problem. The problem is not "rationality" in the abstract but that, as postmodernists point out, any concrete thoughts involve more than that and will reflect our point of view. God is sometimes said to be "pure reason," but there cannot be anything like "abstract Reason"—a god might think totally rationally with crystal clarity, complete knowledge, and see all the consequences of any action, but if his mind is at all like ours he would still need beliefs and values to reason from. The context of our reasoning will always be our experiences in this world—it can never be God's context and perspective. Our point of view is always limited and restricted. Moreover, if we share only some of God's rationality, why trust the musings of a "monkey's mind" on matters of the transcendent? Can we really believe a creator would think the way an evolved species at our stage of development—or at any stage—thinks? That sounds disappointing. Yet, even if we were hundred times as bright as we are now or knew hundred times as much about the universe, would we still have any basis for saying we know the infinite mind of God?

One very real danger here is that we infer as "the thoughts of God" whatever we think or like. This occurs in ethics: whatever we value becomes "God's will." This problem is epitomized during wars when all sides shout "God is on our side!" It is the same in matters of beliefs. Revelations today telling us what God thinks usually end up being only what the recipients wanted to hear or already believed rather than challenging them to change. On the other hand, theologians have to explain away what they do not like—in particular, the problem of natural suffering, which, of course, they do not want to ascribe to the mind of God. Some theologians who take the problem of natural suffering seriously think God must suffer along with us because the sufferings of the world would be too horrible if he did not.

They think that must be inferred, although it neither explains the existence of the suffering nor exonerate God of moral responsibility. But what does the pain of evolution tell us about the mind of God or any transcendent reality?

More generally, theologians devise answers as to why God made the universe the way he did in terms of what we humans would want if *we* were creating—even though we obviously are not infinite. We think of what must be the case because we cannot think of alternatives. One is reminded of the early modern scientists who thought all the planets they saw in our solar system must be occupied by humans because no one could think of any other reason why God would have made them. Whenever theologians say "God thinks ...," what they are saying is "What I think is ...." It is a matter of "If I created the world, ..." or "If I were God, ..." This is not as far removed from thinking of God as an old man with a beard sitting in heaven as modern theists like to believe. Indeed, in the end, is the theologians' use of "the mind of God" any less metaphoric than the uses by nonreligious scientists to phrase their own questions?

This discussion also assumes a transcendental reality would have the nature of a person or at least includes the properties of a person. But can we safely assume that? What if the mind is a purely natural product? Is it only something evolved to help our survival? If it is just part of the evolution of nature, we have no more grounds to infer the transcendent has any mind than a body. Then God would be the opposite of an infinite mind—*no mind at all*—just as he is spaceless and timeless. Theologians today do not consider the transcendent to have any form—forms are confined to the natural realm—but may not the mind also be? What does our having bodies tell us about the nature of the transcendent? The transcendent had the power to create beings with muscles, but no one claims God has muscles. How do we know the same is not true for consciousness? Or personhood? Indeed, a reality that causes us to have a body and a brain has no body or brain—how can there be a disembodied mind? What would it be like? We know little of how mind and body are related in the universe—how can we begin to answer whether there can be consciousness without a brain? How even if mind is a level of the natural realm that is not reducible to the brain? How could a mind exist outside the natural realm? How could it think? Or feel? Is the ability to think or feel like the ability to walk—is a body needed for both? We cannot even conceive a god's situation to ask the right questions.

Also consider a simple fact: nothing in our brains is green but we can see green colors in nature. Perhaps too nothing in the ground of reality is personal in nature and yet it supports the personal phenomena of awareness, understanding, knowledge, and will. The transcendent may be a more abstract Platonic structure that generates such personal properties in the natural realm without having those properties itself. Consider knowledge. Computers can record facts without *knowing* anything. Perhaps the transcendental reality too can record deeds (so that we can be rewarded or punished accordingly) without knowing any propositions or having beliefs, much like the Indic idea of *karma* working as a natural law of consequences for our intentional actions. Knowing something is as much an action as riding a bicycle, and God cannot do the latter either. Asking "What does God know?" may make no more sense than asking "How big is God?"

In short, just because there are persons in the universe does not mean the transcendental reality, if it exists, must be personal in nature. To say "God is an infinite mind" is as reductive as any other claim. Nor can deism or an impersonal principle be ruled out. Theists may not be able to abandon all anthropomorphic and analogical reasoning regarding the mind of God in favor of treating God as only a nonpersonal and nonintentional system—they need a reality that involves personal interaction. Nevertheless, there may be no "mind of God" to know. Either way—an infinite mind or no mind—we are engulfed in a great mystery if a transcendental reality in fact exists.

The Problem With Any Ultimate Explanation

A more basic problem with any ultimate solution to the mysteries of why the universe exists and why it has the particular structure it has is that we simply are not in a position to know if we are ever correct. We cannot have confidence in any answer. It is not merely that there are conflicting positions and no compelling arguments for or against one side—we are not in a position to know even how to proceed to resolve the conflict. Nothing in this chapter rules out that there may be a transcendental creator/sustainer or some type of designer, but the naturalists' counter-arguments are strong. No party is being irrational or stubbornly refusing to see the blatantly obvious. Theists can choose a transcendental mystery if they prefer, but nothing in science points to that over the naturalistic alternative, let alone compels it. In short, our empirical evidence and reason leaves us with an open question. Two facts will remain beyond science forever—that the universe exists at all and that it has order. Theists see the cosmological and teleological why-mysteries as tied together: the type of order we have in the universe impinges on why the universe was created in the first place. But no scientific findings concerning the how-mysteries will have an impact on the why-mysteries of being and structure. Even a TOE that reveals why the universe has the order it has will not approach the ultimate issue of why there is causation or order at all.

Some philosophers dismiss these questions as unanswerable and thus a waste of time: we should focus on puzzles we might solve (i.e., science), accept the majesty of reality, and reject all speculation on these why-questions. However, many philosophers do not dismiss metaphysics so quickly. Today we have new ideas in the debate—the Anthropic Principle and multiple-worlds hypotheses. The issues are pushed to more abstract levels than in the past, but the basic problems remain the same as when ancient people first contemplated the night sky with awe and wonder. The vastness and intricacy revealed by science only adds to the wonder of it all. So does what Paul Tillich called the "ontological shock" of the sheer that-ness of reality. A religious response to the mysteries of existence is certainly understandable. But the wonder generated by these questions can also lead to the naturalists' non-transcendental response of awe to the how-ness of the world and to the mystery of its very that-ness.

Our speculation on "why" questions assumes there is a "because," but that is something we cannot be sure of. As discussed, under all the options at some point we are up against a brute fact that cannot be explained—there is no ultimate explanation, no matter how much our minds demand one. Consider the cosmological question. It may be "turtles all the way down" or a "self-caused" reality, but we reach a point where we must give up the quest to find a further cause. Theists press the "why" questions until they get to God. Naturalists do not see that as a legitimate stopping point and think theists are simply not pushing the questioning to the very end and not accepting the mystery of it all. The theists' paradoxical answer of "self-causation" highlights the problem. Self-contradiction is usually grounds for rejecting a claim—a *reductio ad absurdum*—but theists take it as an expression of a profound, if ultimately incomprehensible, truth. But what is the difference between relying on a self-contradiction and simply saying we do not know? The Cosmological Argument only shows that at some point we must give up the idea of causation.

In general, invoking God as the answer to either the being-question or the order-question is not really satisfying since we still have the mysteries of why he exists, why he is what he is, and how he designs, creates, and sustains. It pushes the mysteries of being and order out of the natural universe, but theists cannot argue that at least the natural universe is explained since its explanation is itself a mystery, and we cannot explain one mystery by invoking another and think we really have ultimately explained anything. Mystery is no explanation at all but only an admission that we do not know something. It may be comforting to think there is a big "BECAUSE" to answer all our why-questions, but we have to admit that this may tell us more about our response to these mysteries than anything about reality itself. It may be easier to remain an agnostic on these matters after studying the issues. But it is hard simply to accept the mystery of it all. We want closure and do not want to admit we are only guessing. But mystery is our true lot in the world.

Notes

1.   Munitz 1965.
2.   See van Inwagen 1996 and Parfit 1998 for arguments rejecting the claim that nothingness is prima facie more plausible than somethingness.
3.   Rundle 2004.
4.   See André Linde, "The Self-Reproducing Inflationary Universe," *Scientific American* 281 (November 1994): 48-55.
5.   Paul Davies, "What Caused the Big Bang?" in Leslie 1998, pp. 226-44. Davies, unlike Hawking, does postulate a starter—although it is a deistic reality, not a theistic god.

6. E.g., Stenger 2008, pp. 132-33.

7. Hawking 1988, p. 131. Under his proposal, the universe is finite but unbounded and had no beginning, and the laws of physics do not break down anywhere. From our limited perspective, we may think that there was a beginning at the Big Bang, but space-time is like a two-dimensional surface of a balloon: just as it meaningless to speak of a beginning or end of the balloon's surface, so too the universe has no beginning or end.

8. Ibid., p. 174.

9. Hawking 1985.

10. Hawking 1988, p. 174.

11. See Hawking 1988 and Davies 1992, pp. 39-72, for examples of this.

12. Some philosophers argue that the idea of creation actually *conflicts* with science: any creation of energy-matter would conflict with the law of the conservation of energy which entails that energy-matter can neither be created nor destroyed, and so the universe must be eternal. However, while this law may apply to events within the natural realm, it is not clear why we cannot still legitimately ask where the *supply of energy-matter* came from to begin with. A law conceived for the interactions of events within the natural realm that tells us that we cannot add or subtract energy simply does not speak to that issue. It is very weak grounds for concluding the matter-energy (and hence the universe) must be eternal.

13. See Spencer 1888, pp. 34-46.

14. Moreover, if anything can be its own cause or otherwise explain itself, then a chain of causes or sufficient reasons is not necessary, and thus we cannot guarantee that there must be a *first cause* of everything or indeed any earlier causes at all.

15. In an attempt to defuse this counter-example, defenders of the Cosmological Argument point out that there are *material conditions* out of which the new particles arise and therefore everything that begins to exist must have a precondition. But the fact remains that the *creation* of these new particles is not caused by the conditions but is still uncaused. Thus, the force of the counter-example is unaffected.

16. Treating this "metaphysical power of being" that keeps us from lapsing into nonexistence as a form of "natural energy" only leads to problems. For an example of this error, see Grünbaum 1996. The transcendent is not shooting nuclear energy or whatever into the world from another realm—natural energy needs "sustaining" as much as anything else.

17. One may argue that we cannot expect any experience within the universe to help us explain the universe as a whole, so we must rely upon a priori reasoning. But the only true *a priori* reasons are abstract—the "laws of reason" such as the law of noncontradiction, identity, and excluded middle. As discussed in Chapter 3, actual arguments and substantive reasons are always more concrete and reflect our knowledge in the world. Even our brain processes are a purely natural product; thus, most people would accept that our mind's reasoning is at least affected by nature. Our reasoning is also based on our being in the universe even if we could come up with timeless, culture-free standards—reasoning is as much an experience in the world as any other experience. Thus, it is difficult to see how all of our "*a priori*" reasoning could transcend our situation and explain everything.

18. J. C. C. Smart, "The Existence of God," in Anthony Flew and Alasdair MacIntyre, eds., *New Essays in Philosophical Theology* (New York: Macmillan, 1955), p. 46.

19. See Munitz 1968 and Rescher 1984.

20. Munitz 1965, pp. 12-13.

21. Davies 1992, p. 189.

22. One basic question—must the order be *causal?*—is difficult to address. Even imagining an alternative order for an empirical world is hard, probably because we live in a causal order.

23. Freeman Dyson, *Disturbing the Universe* (New York: Harper & Row, 1979), p. 250.

24. If the laws of nature are in fact *evolving* and not timeless, an order of things is still there. The situation merely becomes more complicated.

25. Weinberg 1999.

26. Bertrand Russell, *Religion and Science* (New York: Oxford University Press, 1997), pp. 216, 222.

27. One theistic response to the why-question is that God took the route of evolution so that it would not be obvious that he created the universe, thereby leaving us the freedom to choose to believe or not to believe in him. This explains why the universe is so old and so big: God hid any trace of his existence by choosing a route that took billions of years to evolve the carbon and other factors necessary to make human life. That, of course, is possible, but the immorality of the millions of years of suffering caused by taking this route still has to be addressed. In addition, angels present a problem: the story of Satan and his followers shows that creatures sometimes still reject God even when he is totally manifested in front of them. That is, they still had the freedom to choose to assert self-will even when God fully revealed himself to them. Another theistic response is that God created the universe in the last few thousand years and only made it look extremely old as a test of faith. Obviously this view cannot be refuted—the universe might have been created ten minutes ago, with everything (including memories of earlier times) also created at that time to suggest that billions of years of time had elapsed. But no one who takes the universe seriously can believe this, nor is it clear why a god would test our faith this way. Nor do theists want to believe God would deceive us.

28. The *logical* problem of evil is simply a matter of consistency: if God is all-loving, he will not want evil to exist; if he is all-knowing, he will know how to avoid it; and if he is all-powerful, he can prevent it—yet evil exists. Does this mean that such a god does not exist? No. The conclusion follows only if an all-loving, all-knowing, and all-powerful god has *no reason* for permitting evil to exist. All that theists have to do is postulate a reason for God permitting natural evil and the logical problem dissolves. Nontheists would require a morally sufficient reason for permitting suffering, but the religious need only suggest a religiously sufficient reason, such as showing how evil is necessary to human "soul-making." If all else fails, they fall back on "God has his reasons even if we do not know what they are." It is certainly not logically impossible that a transcendental reality has such a reason for suffering or sees more to the good in the world than we realize, and only if there can be no reason in principle for the presence of evil in our world does the logical problem remain. So the logical problem of evil is not compelling. Also, one can fall back on faith in a mystery—whether that is *rational* is another issue. In fact, if all we have to do is invoke the mystery that God has some reason unknown to us for allowing evil, then the logical problem of evil is no problem at all. And many theologians such as Alvin Plantinga are indeed content with resolving the logical problem of evil as if that were the end of the matter—if that is resolved, natural evil is no longer a problem for a logically consistent theist. But such theologians ignore the fact that the type and quantity of natural evil bear very negatively on the question of the *likelihood* that an all-loving, all-knowing, and all-powerful god designed this realm, and thus we have significantly less reason to believe in such a god. This impacts the rationality of faith. Indeed, naturalists argue that natural suffering counts conclusively against the rationality of faith in an all-loving, all-knowing, and all-powerful creator.

29. This solution is not as simple as it sounds: it does not take into account the effect on *victims*. When a man is raping and murdering a little girl, why doesn't God intervene to stop the child's suffering? Is protecting the man's free will more important than the girl's

life? Why should the innocent child be allowed to suffer because of the man's choice? Is the only consideration the *freedom* of the *man* and not the *suffering* of the *child*? (And that assumes that a subtle interference from God to give the man a pang of conscience and concern would in fact harm freedom of the will.) If God did set reality up this way, it is blatantly *immoral*: it would be a case of using one person (here, the girl) in a decidedly immoral way and letting her suffer in order to let another person exercise his freedom. It is not moral to cause one person to suffer exclusively for another person's benefit, and yet God stands by in silence while the child is raped and killed. Nor is it obvious how the spiritual and moral development of the *man* is enhanced by letting that act go forward with his act. Wouldn't he have a better chance of becoming a better person if God intervened? How exactly is he helped by this freedom?

30. We tend to see things in terms of what is valuable to us. Calling physical pain and mental suffering *"evil"* is an evaluative term, not a descriptive one. It is morally-loaded, theologically-loaded (what opposes God's plan), and a subjective evaluation reflecting what is good or bad from our point of view. The AIDS virus and poisonous snakes are "evil" to us because they are harmful, but they are just another part of reality and are not "good" or "evil" in themselves, unless nature is merely set up only for our benefit.

31. See Hick 1966.

32. Adams 1999.

33. This is the thrust of Michael A. Corey's *Evolution and the Problem of Natural Evil* (Lantham, Md.: Rowman & Littlefield, 2000). Animals experience pain, but they do not really *suffer* since they do not have self-consciousness (p. 2). Nor can animals conceptualize the future and so cannot anticipate pain. So the temporary "pain" is not really "suffering." In a return to the medieval practice of giving natural phenomena a symbolic interpretation related to their significance for our salvation, Corey sees the ichneumon wasp as really a "symbol" telling us about aspects of the human psychospiritual realm (pp. 157-58). Thus, it is only for our soul-making. In a footnote, Corey finally admits that his denial of the genuineness of evil (because the suffering springs from the mechanisms necessary for the appearance of human existence) is not meant to downplay the very real and hideous nature of the world's evils (p. 166 n. 24). But the afterlife of those humans who are saved justifies the suffering of billions of beings for millions of years. For a more mainstream theologian who also places humans at the center of reality and thus can only see animal suffering in terms of its value for us, see Richard Swinburne, *Providence and the Problem of Evil* (Oxford: Clarendon Press, 1998), pp. 173-75, 189.

34. Why didn't God set up a universe in which at least each moral agent suffers only for what he or she deserves? The Indic idea of *karma* and rebirth is such a natural moral structure: we suffer only because of our own past free actions in this life and in earlier ones. Of course, this presents another moral issue: we should not help others because they are only suffering the effects of their own actions—and if we do prevent the karmic effects from accruing today they will only accrue later. Thus, we ourselves might gain some karmic credit by intervening, but our acts would not ultimately aid the sufferer. Compassion could then be reduced an emotion unaccompanied by action. But at least these ideas do show that a system of natural retributive justice is logically possible.

35. See Stephen Jay Gould, *Rock of Ages: Science and Religion in the Fullness of Life* (New York: Ballantine Books, 1999), pp. 183-87.

36. Christians can reconcile natural suffering with the Bible: God is not all love but has a *wrathful* side that is demonstrated in both the Old and New Testaments and in the animals he has created. See David Snoke, "Why Were Dangerous Animals Created?" *Perspectives on Science and Christian Faith* 56 (June 2004): 117-25. Snoke points out that

we hear about hell more from Jesus than any other speaker in the Bible, and the entire Gospel revolves around the idea of avoiding God's wrath. Thus, the savage side of nature is still a basis for natural theologians to infer something about the nature of God.

37. Naturalists present a different picture: from the point of view of the individual, there is still suffering and death, and thus the question of the meaning of life arises for us; but from the point of view of evolution, violence and death are simply part of the process of how nature generates and maintains more and more complex forms of life. Natural theologians since William Paley can argue that predators cull out the weak and the sick (and thus help the survival of the strong among the prey), but this does not mitigate the suffering—even if they are just being put out of their misery, it does not explain why there was the misery in the first place or why it has to be ended by so much pain.

38. This relates to a practice naturalists find very curious: theists thanking God for the good things that happen to them but not blaming him for the bad—e.g., thanking God for not dying in a tornado that killed others: God is not blamed for the tornado but thanked for sparing them.

39. To quote a recent version of the Roman Catholic Church's catechism: "The beauty of creation reflects the infinite beauty of the Creator. Each creature possesses its own particular goodness and perfection. Each of the various creatures, willed in its own being, reflects in its own way a ray of God's infinite wisdom and goodness." How this can reconciled with the AIDS virus and the ichneumon wasp is not obvious. The counterbalancing theme of Monty Python's Flying Circus's revision to the traditional hymn "All Things Bright and Beautiful"—"All Things Dull and Ugly"—must also be included.

40. David Hull, "The God of the Galapagos," *Nature* 352 (8 August 1992): 486. Also see John Stuart Mill's essay "Nature" in *Three Essays on Religion* (responding to Alexander Pope's claim that "whatever is, is right") for a description of nature as a realm of violence, hideous death, and cruelty, totally devoid of any mercy, justice, peace, or harmony. He notes that one of the things most evidently designed is that a large portion of animals pass their existence tormenting and devouring other animals. Arthur Schopenhauer, for one, thought that the omnipresence of cruelty and suffering in our world renders the idea that a transcendental realm behind this world could be inhabited by a god of love a bad joke.

41. Alvin Plantinga, *God, Freedom, and Evil* (New York: Harper & Row, 1974), pp. 57-59.

42. See Hick 1999, pp. 241-42.

43. Stephen J. Wykstra, "The Humean Objection to Evidential Arguments from Suffering: On Avoiding the Evils of 'Appearance,'" *International Journal for the Philosophy of Religion* 16 (no. 1 1984): 90-91.

44. Some theologians are now taking the problem of natural suffering more seriously. For an overview, see Christopher Southgate, "God and Evolutionary Evil: Theodicy in the Light of Darwinism," *Zygon* 37 (December 2002): 803-24.

45. Theists may argue that their arguments must be considered only in combination with each other—e.g., only in combination with a moral argument and the Teleological Argument does the Cosmological Argument seem compelling and point to more than a deistic source. So too, only then does accepting the mysteries of God and of natural suffering become reasonable. Naturalists, however, see natural suffering as negating any combination of arguments supposedly leading to an all-loving, all-knowing, and all-powerful god.

46. Weinberg 1999.

47. Davies 2003, pp. 257-58.

48. See Leslie 1989; George Gale, "Cosmological Fecundity: Theories of Multiple Universes," in Leslie 1998, pp. 195-212; Tegmark 2003.

49. Any "bouncing universe" may be subject to entropy and so not be eternal. However, we know nothing about the relation of the bounces and little on how the bounces might occur. There may be other forces involved—perhaps an analog to the "dark energy," the repulsive anti-gravity force that scientists have only very recently began to theorize about to explain why the rate of expansion of our universe is increasing. Or the laws of thermodynamics that we have devised for the content of our universe may simply not apply. We know too little to hazard any certainty in this field.

50. A more fanciful speculation is that this multiverse is itself only part of a further mega-multiverse and so on and so on without end. H. G. Wells's suggestion that our universe might be only a molecule in a ring on a gigantic hand has left the realm of pure science fiction.

51. That is, each possibility specified in the Schrödinger wave function is in fact realized but each in a different world. It would appear from our perspective that the number of universes increases exponentially over time, but actually it is not: all physically possible quantum states exist at every instant. Tegmark 2003.

52. See Lewis 1986 on his "modal realism." Also see Nozick 1981, pp. 123-30 on "egalitarianism" and the "fecundity of universes." The cosmologist Max Tegmark argues that there is no reason why only one set of consistent mathematical equations out of the countless possible sets should be endowed with physical existence, and thus there should be worlds in the multiverse corresponding to each such set. Tegmark 2003.

53. There would not need to be an *infinite* number of worlds. To raise the odds that a combination of physical forces that produces life would occur in at least one world and thus to remove the illusion of design, a very large number of worlds would suffice. If there were in fact an infinite number of worlds and pure chance reigned, every physically possible combination would then occur. It would be absolutely certain that a world exactly like ours would occur—it would be inevitable. Indeed, Friedrich Nietzsche would add that everything *recurs* exactly the same way an infinite number of times, just as any string of numbers would recur an infinite number of times in an infinitely long string of numbers. Even if that does not occur, the simplest and most popular cosmological theory today predicts that we each have an exact twin in a galaxy far, far away. Tegmark 2003.

54. Merely surrounding our world with many others does not explain why any given world is conducive to life or up the probability that some world permits life. See Mellor and White in Manson 2003. For the lottery analogy to hold, there must be some mechanism in the universe that permits different possible worlds to be generated—otherwise, the same world could just appear over and over again. Thus, the odds that a world like ours might appear would not increase even if there were an infinite number of worlds. But if there is a *mechanism guaranteeing true randomness*, then an increase in the number of worlds increases the odds that some world will be biofriendly, and an infinite number of worlds guarantees that at least one will be.

55. Manson 2003, pp. 17-21.

56. Naturalists can also advance philosophical arguments for why the idea of multiple worlds fits more easily in a naturalistic picture. For example, just as it took a universe vast both in age and size to generates a few zones (indeed, perhaps only one) where intelligent life could appear and thrive, so too a vast number of worlds must be necessary to generate any where intelligent life could exist. Or, notice that the physical laws of this universe are always "imperfect," i.e., that the numbers involved in physical forces are always slightly off from being simple. This suggests they were not designed but instead are only one possible set out of a larger collection of possible physical constants manifested in various worlds.

57. "Intelligent Design" can refer to anything from a deistic reality that established the general laws of the universe to a theistic designer who established general laws for a specific purpose (making conscious beings) to a theistic designer who micro-manages his creation by micro-engineering every "irreducibly complex" part of creation. Deists can affirm the first option, liberal and moderate theists the second, and conservative theists the third.

58. Tegmark 2003, p. 51.

59. Robert Jastrow, *God and the Astronomers* (New York: W. W. Norton, 1978), pp. 16, 113.

60. E.g., John F. Haught, *Science and Religion: From Conflict to Conversation* (New York: Paulist Press, 1995), p. 101. Some theologians believe that the Big Bang proves *creatio ex nihilo*. That is, scientists have proven there was a creation and that the universe is not eternal and that it is only the desperate move by atheists to advance the idea of an oscillating universe. This will come as a surprise to many cosmologists for whom a multiple-world hypothesis simply follows from other theories. So too, trying to formulate a cosmology without singularities such as a unique Big Bang—one in which natural laws govern everything—is simply science. To naturalists, it is the theologians who are desperate to secure a role for "a god of the gaps" in a world where scientists are rapidly closing gaps.

61. Weinberg 1999, p. 46. Plato's demiurge or an incompetent designer also comes to mind. One is also reminded of David Hume's remark: "many worlds might have been botched and bungled, throughout an eternity, ere this system [of our world] was struck out."

62. Early deists were determinists. Today they may adopt a role for randomness in nature.

63. The multiple-world hypotheses may be used to try to counter the problem of natural suffering—i.e., a moral god is justified in creating all the worlds that on balance have more "good" than "evil" (from God's perspective). But this still does not explain why such suffering exists in any world. In the end, it only increases the amount of suffering over all.

64. For a classic exposition of the position that animals do not suffer but lead joyous lives, see Charles E. Raven, *Natural Religion and Christian Theology* (Cambridge: Cambridge University Press, 1953), vol. 2.

65. Weinberg 1992, p. 250. Of course, beauty may just be a subjective reaction on our part, like thinking that the AIDS virus and earthquakes are "evil" because of their effect on us. Or what we think is beautiful may be the result of our own evolution in particular environments. That is, any being with a sense of beauty will be conditioned to see its own environment as beautiful, no matter how ugly it is to other beings, for evolutionary reasons related to its survival. Nevertheless, theologians do advance an aesthetic argument for the existence of God to explain the beauty of nature and our ability to create beautiful things (classical Western music being the most popular example).

66. Hawking 1988, p. 175.

67. The term "atheist" may become ambiguous when deism enters the picture. It applies to those who deny the existence of a theistic god, either because the concept is held to be logically incoherent or because the weight of evidence is seen as being against the likelihood of such a reality. But the term may be used to deny only a theistic god and not necessarily any transcendental reality (although Hawking apparently would also deny the latter).

— 6 —

Religion

My whole tendency and I believe the tendency of all men who ever
tried to write or talk Ethics or Religion was to run against the boundaries of
language. This running against the walls of our cage is perfectly, absolutely
hopeless . . . [b]ut it is a document of a tendency in the human mind.
— Ludwig Wittgenstein

The meaning of life is the most urgent question.
— Albert Camus

A second way of approaching ultimate explanations is not through philosophical reasoning or the empirical examination of the natural world but through religion. This is not to say that the religious are irrational but only that they approach the issue differently, by asking about the meaning of the world — the "why" of it all. Why are we here, not in the sense of what caused us, but in the sense of what *purpose* explains why we exist? Why is the universe here? Where do we come from? Where are we going? Does our life really matter? Why do we suffer? What happens at death? Why bother to live at all? What gives life meaning? How is any individual a meaningful part of the whole? In short, what's it all about?

From the psychiatrist Viktor Frankl to the physicist John Wheeler, many argue that more important than the deepest scientific question is finding a significance to our life—the most fundamental question we can ask in life is how we fit into the scheme of the things. To the anthropologist Clifford Geertz, our need for meaning is as real and as pressing as our more familiar biological needs.[1] The sociologist Peter Berger adds that we cannot accept meaninglessness, and religion protects us against the terror of meaninglessness by conceiving the entire cosmos as humanly significant.[2] We have a deep need to know that we have significance and value. We want to believe there must be a reason why we are here—"God put me here to do this." To make the contingencies of life understandable and thus more bearable, religions offer explanations that show our place in the scheme of things.

189

Transcendental Realities

For the religious, the search for meaning is set in the context of supernatural realities. In tribal societies, the distinction of "natural" and "transcendental" is not always present, but in the world religions meaning is grounded in realities transcending the natural universe, not in the individual, a social group, or the natural world. Thus, in the traditional understanding of all world religions, there is more than one ontological level to reality: there is the world we inhabit and a reality that encompasses this world (God or a nontheistic counterpart) as its source. There are also supernatural beings—gods, angels, demons, spirits, ancestors— active in the natural order. Humans also participate in this natural/transcendental dualism: part of us is a soul or other reality that survives the death of the body. Thus, the religious sense is that there is something more to reality than both what we normally experience and the order underlying changes that scientists have darkly discerned. We connect to a deeper reality transcending the natural order.[3] This does not mean that the religious must declare the natural world to be ultimately unreal but only that this world is in some way the creation, emanation, or appearance of the transcendental reality.

For all religions, the world has a meaning, and humans are not just an evolutionary accident of a pointless universe. In the words of Alfred North Whitehead, "[t]he final principle of religion is that there is wisdom in the nature of things." All the big questions of life—why misfortunes befall us, why we live, why we die—are answered or explained away in terms of transcendental realities. For the religious, life has a transcendental meaning, and our *summum bonum* is to align our lives with the reality providing that meaning, however salvation is defined in a particular tradition. Thus, those who suffer or feel out of step with reality now have a framework within which to live and to accept the misfortunes that living entails or have a way to end that suffering. Thus, the majority opinion in all the world religions is that the religious are, in John Hick's words, "cosmic optimists."[4]

how/why would a merely physical brain need this?

The Naturalists' Response

Naturalists emphatically disagree with all this. They accept as real only what can be studied by scientists, and since any transcendental realities are by definition not open to empirical study, their alleged existence is rejected. The idea of a god is just an illusion foisted on us by our brain's need for an explanation of everything. Our religious aspirations do not indicate that there is a transcendental source or life after death but are only a product of our hard-wired craving for comfort and reassurance. Even if religion arose and has persisted throughout history because it has evolutionary value in fulfilling important psychological or social purposes related to order and stability (rather than being an accidental, useless by-product), it is groundless: reality has only one dimension—the natural world. We too are only one-dimensional: there are no disembodied souls or other realities that might survive death. Our real home is this world. Life and consciousness are simply natural products of the evolution of the material universe. The universe may or may not end—either way, it is simply the meaningless whirling of mass and energy with no transcendental meaning or purpose, and we are on our own.

This worldview is inspired by the success of science, but it is not entailed by the practice of science. Naturalists still must make one major philosophical assumption: only what scientists can study or infer is in fact real. Philosophers from Socrates to Wittgenstein have realized that some questions about reality are left over even if scientists answer all scientific questions. As Wittgenstein said: "We feel that even when all possible scientific questions have been answered, the problems of life remain completely untouched."[5] Naturalists accept that there may well be more to the natural order than is discoverable by scientists—hence, some how-mysteries may remain—but they deny any transcendental meaning to it all, if they accept that question to be meaningful at all. At best, we should apply scientific methods to all questions and remain agnostic about what cannot be answered. As Bertrand Russell said, "The universe may have a purpose, but nothing we know suggests that, if so, this purpose has any similarity to ours."

And many naturalists certainly do deny there is any meaning to reality: reality just is, and any sense that it has significance for us is only something we impose on it. In an infamous remark, Steven Weinberg said "[t]he more the universe seems comprehensible, the more it also seems pointless."[6] When naturalists look up at the night sky, they do not see the handiwork of God but only darkness occasionally broken by stars that will die. Seeing the events of the world as happening for no reason can be terrifying, but what we want does not change the way things are. To naturalists, the religious fabricate some meaning to comfort themselves out of weakness—e.g., to reassure themselves that everything, no matter how bad it looks, is in the hands of God—because they do not have the courage to face the true senselessness of reality.[7] Science reveals that the universe is unimaginably old, with life a rarity; the future is eventually a universe devoid of life. Is this the sort of universe an omnipotent and omniscient being would create just to generate relatively few conscious beings? This branch of the naturalists would agree with Richard Dawkins when he says "The universe that we observe has precisely the properties we should expect if there is, at bottom, no design, no purpose, no evil and no good, nothing but blind, pitiless indifference."[8] Our insignificant planet is not the center of any vast cosmic purpose but only one planet in a universe with billions of galaxies each with billions of stars. Certainly the universe is not all about us. If we think of the history of the universe as an hour, we have been around for less than a thousandth of a second. The only reason we exist is because of biological evolution, and no purpose is possible because of the random nature of the mutations leading to us.

To naturalists, that we share around 98% of our genes with chimpanzees puts an end to the idea that humans are above animals and not evolved from them.[9] As

Robert Trivers put it, "The chimpanzee and the human share 99.5 percent of their evolutionary history, yet most human thinkers regard the chimp as a malformed, irrelevant oddity, while seeing themselves as stepping stones to the Almighty."[10] Life arose only by chance and evolves only through randomness and natural processes. We are arbitrary piles of the refuse of some star, a fluke that randomly evolved consciousness. The pain and haphazard nature of evolution reveal that there is no plan to any of this. The universe is driven without a purpose and without intelligence. As Stephen Jay Gould said, nature did not know we were coming and does not give a damn about us now. There is no justice, moral order, or any other value embodied in the universe—only nihilism (the denial of any objective basis to values) is justified.[11] Our aging and declining bodies are telling us our fate. When we die, the body simply stops functioning, and our consciousness will fade away just like all of the body's other biological processes. Under this perspective, we scurry around and keep busy, but in the end our actions are meaningless.

However, naturalists can in fact find life within the world meaningful even though there is no transcendental reality endowing the world with a higher purpose.[12] Existentialists may find living in an indifferent world absurd and angst-filled, but naturalists can find it intrinsically meaningful without any answer as to why the universe is here. The universe remains a cold, cruel place where life is only a momentary spark that is soon extinguished, but how we lead our lives can still have value, even if there is no transcendental "purpose" or "meaning."[13] We are all made of the same dust from a dead star, and our biological *connectedness* with every person, plant, and animal is quite enough for many naturalists. Nor need we cast a mark that lasts for eternity to be significant—only our human hubris suggests otherwise. Just being part of the unfolding evolution of the universe and life is enough to give our life purpose and meaning. In fact, some scientists such as Einstein think that it is a sense of an independent self separated from the rest of nature that leads to a sense of alienation. So too, evolution is not merely a matter of selfishness: cooperative and symbiotic relationships are essential to survival. We are also inherently social in nature. A concern for others is basic to our own survival—we know we are not born alone but from another person and cannot survive alone. Cooperation with others is basic to life, and thus there is no need to invoke a transcendental reality to ground a concern for others' welfare.

Under this view, we are part of something truly awe-inspiring and exciting, and we can be truly humble when we consider ourselves and the powers that made us. Committing ourselves to make the universe a better place here and now can make our lives meaningful. As Peter Singer notes, once we accept life as a value in itself we can live a life that is inwardly meaningful by working toward goals that are objectively worthwhile (e.g., trying to reduce avoidable suffering). So why demand that there must be life after death? In fact, we can be more in tune with the wonders and the suffering of this fascinating universe when our focus is solely on the universe itself and not split between this and another realm. And with no life awaiting us after death, each moment becomes more significant and to be savored.

Nor need we see the universe as lifeless—life is rare only because it is so complex and requires certain conditions, not because it is a fluke. Rather, life is as natural and integrated into the whole as anything inanimate. Why is reality organized to produce conscious beings if they are not a natural part of the universe? Nor does evolution reduce each human to a meaningless cog: each individual life can still have significance. There is certainly no reason to believe that merely making more humans is our only significance—if that were so, our offspring would not have more significance than we do. Nor is there any reason to believe that humans are anywhere near being the final product of evolution or that more complex beings will not appear down the road. Nor need we be infinite or eternal or at the center of the universe to be valuable. Theists who believe that, unless we are eternal or our actions have permanent consequences or we are fulfilling some cosmic purpose, our actions must be pointless and there is no reason to get out of bed in the morning simply do not see the true significance of humans.

The mere fact that life and conscious beings exist indicates to many naturalists that we are an integral part of reality—written deep into the universe's structure—and that is enough to indicate that we are significant. Such value is not arbitrary or subjective. Indeed, naturalists can argue that it is the religious who have the problem of finding some meaning to the natural universe (since they do not fully integrate humans into the universe but believe that the most important part of us transcends the universe and that our real home is outside of the natural universe) while naturalists do not. The theists' mistake is to start with humans as central to the scheme of things—instead, we should look for whatever meaning there may be to existence in the cosmos itself in general and in the evolution of life.

The religious may concede that naturalists can lead rich and interesting lives, live fully engaged in their activities, have projects that make their lives worthwhile, have ethical ideals, act morally, contribute to others' welfare, transcend their ego-centered point of view, find life precious, value beauty and knowledge, or greet the universe with awe and reverence—all things that makes life meaningful. But they contend that naturalists cannot *justify* any meaning to the universe. The meaning that naturalists find is only a feeling they impose on the world. Thus, there may be a subjective "personal meaningfulness" that naturalists create, but there is no objective "cosmic meaningfulness" for the whole of existence—this can only come from a transcendental source. Naturalists looking back on their lives must conclude that they have wasted their time on what is ultimately only trivial. But naturalists ask why does meaning have to be connected to fulfilling some transcendental purpose rather than something inherent in the world itself? In fact, naturalists argue that the "meaning" the religious claim is what is actually subjective—the religious merely do not realize that they are imposing a groundless meaning on the world. The religious project a meaning they themselves create onto a nonexistent transcendental reality and then mistakenly think that it is objective.

In addition, naturalists can point out that, even if there is a transcendental source to this universe, this does not guarantee that our lives have meaning. The source may have created us just to watch us suffer or had some frivolous purpose we do not know. Thus, merely invoking a source cannot still our disquiet over the meaning of our lives. There is an *Euthyphro*-type problem here: does God confer an intrinsically valuable meaning upon the universe, or is the meaning valuable merely because God confers it? If the latter, then the meaning may be arbitrary and not really render life meaningful at all; but if the former, no god is needed (since it valuable in itself, independent of God). Indeed, naturalists think that merely fulfilling another being's purpose—simply slavishly doing a task we are assigned by God for his own purposes in order to avoid hell—is not a goal worthy of free beings. We are not the Deltas of Aldous Huxley's *Brave New World*, created to do only tasks assigned to us; the tasks might give the Deltas' lives meaning, but this does not make a universe meaningful. Indeed, simply doing an assigned purpose is *immoral* unless we can know what God's purpose is and can see that it is moral and valuable in itself. Nor is it obvious that any future life necessarily makes this life meaningful. In the end, as Wittgenstein asked, "is some riddle solved by my surviving forever? Is not eternal life itself as much of a riddle as our present life?"[14]

Moreover, naturalists can rightly ask what is the meaning of God? To give a complete answer to the meaning of reality, the religious must go "all the way down" and provide a meaning for the transcendent source. Otherwise, they are merely pushing the problem of meaning back one step: we humans would have a purpose—being the obedient servants fulfilling God's plan, or whatever—but reality as a whole would not. The transcendent may give meaning to the natural world and this may satisfy the religious, but what is the meaning to the totality of reality—the natural realm plus its transcendent source? If by definition what gives meaning to something must transcend it, our picture will always be incomplete. There can be no meaning transcending God giving him meaning, and thus there is no final explanation of our purpose. Does this render the question of the meaning of everything ultimately senseless? The religious may try to finesse the issue as with the Cosmological Argument and argue that the transcendent is the source of its own meaning. But whether this is any more successful than the idea of something being the source of its own reality is doubtful. If God is the ultimate contingent fact and has no further source of meaning, the mystery remains of the meaning of the totality. Thus, according to existentialists, introducing God into the picture does not make life any less "absurd."[15] Moreover, a problem discussed in the last chapter concerning an ontological source for being returns here: if God can be his own source of meaning, why can't the universe be the same? Why can't the universe as a whole give meaning to our lives?

Science and Meaning

Against Steven Weinberg and company, we can ask whether in fact science is capable of destroying the possibility of transcendental meaning. Scientists may find things that conflict with *particular* attempts to articulate a purpose—e.g., by showing that the earth is not the physical center of the universe—but can they find something that would preclude *all* meaning in principle? What sort of experiments or observations could scientists make, even in principle, that would indicate that the universe is without value or purpose or transcendental meaning? Or, conversely, to show that it has such a meaning? There does not appear to be any. The entire issue is screened out of scientific approaches, and thus scientists can succeed completely in what they do—describing and explaining nature's workings—and the question of the meaning of why the universe exists and our place in it will still remain open. Scientists may be able to establish that life and consciousness are integrated into the universe in a way that makes them the product of inanimate, purely natural forces, but a god could still have set up the general order of things for that purpose. On the other hand, no transcendental purpose can be established by any scientific finding—perhaps our universe is only one of a comparatively few universes that produced life and consciousness out of a vast array of universes. In sum, science "makes sense" of the world only in the restricted sense of explaining *how* it works—it says nothing one way or the other about the "meaning" of the world in the sense of *why* the world exists or is as it is.

Scientists, as it were, study the ingredients of a painting's paint and canvas, but they do not deal with whether the result constitutes a picture or not. The question of meaning, like the aesthetic judgment, is "beyond science" in that it is a question that the methods of science cannot address. Or, to use an analogy from Chapter 4: scientists identify the vocabulary (the building blocks of nature) and the grammar (the laws of nature) of a text but do not deal with the issue of whether the text provides a meaningful narrative or not. However, just because meaning does not play a part in the scientific accounts does not mean there cannot be any. Solving all the how-problems of science would not address the mysteries of the ultimate meaning of existence and why we are here. All in all, we cannot be surprised that scientists do not find any transcendental meaning to the universe when science is simply not designed to find meaning in the first place.

What did Weinberg expect scientists to be able to find that would indicate a transcendental purpose? Perhaps he meant evidence of teleology. But modern science has screened out that issue by focusing only on efficient and material causes. Moreover, even if there are no teleological mechanisms at work in reality, this does not rule out a meaning to reality since the meaning may be in terms other than a goal to be achieved. That is, a transcendental reality may have established certain structures and then let things evolve on their own as a valuable end in themselves. Chance may be programmed into the universe as part of that purpose, or what we see as chance may be weighted by as yet unknown natural forces or by transcendental intervention toward the appearance of life, complex life, and conscious beings.

Thus, science *per se* cannot rule out a transcendental meaning to the natural realm. Scientists *qua* scientists ignore the issue of meaning. Many scientists may in fact be indifferent to the issue of why we are here. Others may, as the astronomer Owen Gingerich admits to being, be simply psychologically incapable of believing that the universe is meaningless—and, it should be noted, many scientists are theists or deists. Indeed, concluding from the fact that scientists do not discover meaning through their empirical methods that therefore there can be no meaning is a philosophical mistake that reveals the naturalists' philosophical commitment to science as the only way of insight into the nature of reality.

Religion and Mystery

For many persons, religious concerns surface only during moments of crisis—death, birth, disease, or personal or group misfortune—rather than when life is running smoothly. But the framework of meaning provided by religion shapes the whole life of those who take their religion seriously. Thus, the loss of religious authority in one's life can lead to a sense of meaninglessness. Religious doctrines, stories, and teachings are types of explanations. They are not scientific explanations of *how* things works but of *why* things exist.[16] (However, religious doctrines and myths also encroach on the scientific issues when discussing aspects of creation or transcendental actions in nature leading to potential conflicts of science and religion.) Origins figure prominently in the accounts, although many cultures are indifferent to their high creator god and the origin of the universe itself. Myths are narratives of the events recounting the origin of the world, humans, one's own group, and whatever else is taken to be fundamentally important to a given society. Myths show how we are connected to the world and answer basic questions about the meaning of life, allaying our "why" fears about life. They relieve our anxieties by revealing an underlying order to things. Modern liberals may reduce religion simply to a means of inculcating a moral point of view, but traditional religions encompass all aspects of life, and their ethical codes depend on their worldviews in that the latter are believed to ground the former in the very structure of reality.

Religion may have arisen very early in human history from our awareness of the fact that we cannot control events or the powers at work in nature—no doubt, what was behind death, storms, fires, floods, powerful animals, and so forth seemed terrifying and awesome. But it is the "supernatural" source behind it all that makes mystery an essential part of world religions today. The transcendent is unperceivable and unknowable by empirical means. To theists, even mystical experiences do not plumb the depths of it. This reality is sacred—an otherness "set off" from the everyday. There is a sense of power, ultimacy, absoluteness, and deep reality behind our everyday world. The religious today, living in a world shaped by science, may no longer be able to react to a transcendental reality with awe, fear, or even with the same reverence or adoration our pre-modern ancestors had, but if they are religious they react with a sense of a profound transcendental mystery surrounding their lives that surpasses our understanding and imagination.

The religious emotion is not the awe and wonder at the intricacies of nature or its how-mysteries that drive scientists. It is not a sense of wholeness or beauty to nature that naturalists also can share. Rather, there is a sense of something more to reality than our everyday life suggests—a sense of transcendental realities. Even philosophical amazement at the fact that anything exists or the awe at the very that-ness of the everyday world need not lead to belief in a transcendental reality. But transcendental realities are not postulated as a scientific explanation. Instead, different questions are being answered when the transcendent is invoked: the metaphysical questions of why things exist and what is the meaning of our lives.

Rudolf Otto placed the sense of the "numinous" at the core of religion—the sense of a power that is "wholly other" than the powers of the natural world.[17] The wholly other is beyond our comprehension "not only because our knowledge has certain irremovable limits, but because in it we come upon something inherently 'wholly other,' whose kind and character are incommensurable with our own, and before which we therefore recoil in a wonder that strikes us chill and numb."[18] His focus was on the emotions arising from the encounter with the "wholly other." It is a sense of the *mysterium tremendum et fascinans*—a mystery that is both frightening yet fascinating. That is, the numinous is mysterious (an inexhaustible power defeating any attempt at understanding), yet it draws our attention; its strangeness and unapproachability evoke fear and overwhelming awe (a power producing an inward shudder of the soul), but it also provokes satisfaction, grati-tude, and joy (adoration of the infinitely good and perfect). He considered the numinous state of mind to be *sui generis* and another irreducible Kantian *a priori* mental category for how we experience the world. The sense of the numinous is preconceptual and nonrational. Doctrines remain secondary—"obscure and inadequate symbols" of the encounter. But Otto argued, following Kant, that a conceptual element is always necessary in creating the idea of the "holy."[19] Thus, the "holy" is not "wholly other." But he realized that "rationalizations" in religion from myths to scholasticism can deaden experiences and exclude mystery.[20] Much of his analysis has to be rejected today—most importantly, his slant towards theism (and Christianity in particular) was obvious. He has little direct influence today, although Mircea Eliade carried on the emphasis on the experience of the sacred as *sui generis* and the sacred as a real structure in our consciousness.[21] But Otto's emphasis on a preconceptual, supra-rational sense of a wholly other power does capture something essential to the religious mind.

While the transcendent is the center of answers to why-mysteries for the religious, it generates a new set of how-mysteries in how a god creates and interacts with the world. To naturalists, the religious are mystery-mongers who create mysteries where there are none (since no transcendental realities exist). And it does appear that while philosophers and scientists forestall mysteries as long as possible, the religious readily accept that their deepest beliefs are shrouded in mystery, and thus all religions are rife with both why-mysteries and how-mysteries. In Christianity, there are the mysteries of the creation of the world, people being created in the image of God, incarnation, virgin birth, original sin, redemption and salvation, atonement and the resurrection of Jesus, redemptive grace, the sacraments of baptism and communion, the trinity (how God can be three and yet one), the kingdom of God, and, of course, God himself, to list only the major ones. All are realities that can only be known through faith or revelation, which itself involves a how-mystery. Mystery even has its own branch of theology—mystagogy—with the admission that the mysteries cannot be solved even in principle.[22]

Mystery comes up in religion in another way—secrecy.[23] The term "mystery" entered our vocabulary from the Hellenistic "mystery religions." It referred to keeping the sacred knowledge hidden from the uninitiated because of its power, not to any allegedly "ineffable" reality or experience—one was to remain silent about the mysteries, not because there was nothing to express, but because the knowledge was *dangerous* to the uninitiated. All the world religions have esoteric traditions, although not all forms of religion involve claims to a "forbidden" knowledge that requires special teachers and initiation into the secrets. Christianity entered the Roman empire at the time of the Greek mystery religions and shares some of their features—in particular, a new life after death—if not all of its esotericism. Followers are granted knowledge of the secrets of the kingdom of heaven (Matt. 11:25, 13:11). The New Testament letters of Paul contain many passages alluding to "things that cannot be told" (I Cor. 2:7, 13:2, 15:51; II Cor. 12:1-6; Col. 1:26, 4:3; Eph. 1:9, 3:5, 3:9, 5:32, 6:19), but it is the *disclosure* of things that were once mysterious, not their mysteriousness, that is prominent in the New Testament. God reveals things that otherwise could not be known—things that were kept hidden for ages were now disclosed (Rom. 16:26). And the book of Revelation—the Apocalypse ("lifting the veil")—uncovers the future of God's plan for the world. But in the New Testament, what is revealed retains a mysterious quality: what was previously unknown is now known but still not completely comprehended—in this world, we "see through a glass darkly" (I Cor. 13:12).

The Question of Creation

One difficulty for understanding religious mysteries is that different religious traditions construe the problem of the meaning of life differently. They do not share all the same questions related to why we are here or how we fit into the scheme of things. For example, the prospect of a cycle of rebirth might indicate to Westerners a purpose to life, but to Indians it only defines the problem of living—the objective is to *get out* of the cycle of continuous rebirth. Consider four issues: creation, suffering and death, history, and morality.

First, creation. Not all religious people see a need to answer the why-mysteries of creation: many cultures do not have creation-myths for the universe itself, and in many that have such a myth the accounts of the creation of specific natural and social phenomena predominate. Nor do all religious persons see a need for a creator god to make their lives meaningful. Buddhists argue that our goal is to escape the suffering of the realm of impermanence. Buddhists do not assert that every event in the world is painful—obviously they are not—but rather that even pleasurable experiences are impermanent and we are left with a pervasive dissatisfaction about our life. Hence, the goal is to reach a mental state free of the desires generated by our misreading of the true nature of reality—*nirvana*. Since these desires drive the chain of rebirths, correcting our knowledge ends the chain and thus brings our suffering to an end permanently. Buddhists leave unanswered many questions that theists may think vital—whether the world is eternal or not, whether it is infinite or not, whether an enlightened "person" exists after death or not—because these do not relate to the central concern of ending suffering.[24] Why the universe exists or why suffering occurs is not a concern since, even if there is an answer, we should not waste time dealing with such intellectual matters. When you are shot with a poisonous arrow, you do not inquire about what the arrow is made of or who made it—you just want a cure.[25] So too, what is vital here and now is finding a way to the deathless state, not metaphysical questions about the nature or purpose of the universe. Theists who press on insisting that they need to know why there is a world or why we exist are, from the Buddhist point of view, simply increasing the time they will suffer. Thus, Buddhists have a goal giving purpose to our lives, even though the goal is escaping from this world. Their own metaphysics generates how-mysteries connected to rebirth and the role of *karma* in mental and physical events, but the basic why-mysteries of the meaning and origin of the universe that are such importance to Western theists are dismissed.

Hinduism's doctrine of the world as "play" (*lila*) shows a way that the religious can speak even of creation without needing to see a purpose to the universe. This doctrine does not mean the universe is an illusion in the sense that it does not really exist at all, but rather that a creator god or nonpersonal reality (Brahman) creates for no purpose or reason, no motive or desire—creation is simply an act of its nature. Western theists think that if a transcendental reality created us, then there must be a reason for creation. But under this alternative, the act of creation involves no more intentionality for a creator than breathing does for us. It is done with no purpose, plan, or effort—a creator creates just because that is what creator gods do. The universe is the god's "play" in that sense: the world is an end in itself and achieves no further purpose. There is no "why" to the universe. The universe has always been here and has gone through an eternal series of creations and destructions, of "rolling out" and "rolling back in." Since the universe is eternal, where it came from or where the underlying Brahman or creator god in the series of creations came from is not a pressing issue. We too have no "purpose"—our only goal is escaping from the series of rebirths and returning to the source.

Theists, on the other hand, believe the universe is a creation, and they try to specify a purpose. But surprisingly, this turns out to not be so easy. The Abrahamic scriptures themselves go into details on God's commands concerning such matters as diet, clothing, and haircuts, but they do not give God's grand purpose. Consider some traditional suggestions: God created the world just to be known or loved or glorified—or, in the words of Bernard of Clairvaux, man and the world only exist because of God's need for his love to find both expression and reciprocation. But are we here to just to love or worship or otherwise please God? If the goal of humanity is to know God, that could have been accomplished if we started out in heaven and did not have to go through the suffering of the natural world. It is also a little odd: it would be like dreamers wanting the characters they have created in their dreams to love them. Moreover, any such self-serving purpose also raises a moral issue: if God created us only because he wanted to be loved by us, was creating a universe with so much suffering really a moral act or an act of selfishness, no matter how much good is also created? Some New Age enthusiasts argue God created people merely because he does not want to be alone—but again why didn't God make more angels rather than put us through the suffering of this realm?

These answers, however, reflect just what *we* would think a purpose should be—indeed, it is hard for us to think otherwise, and that is just why all our attempts at setting forth a purpose sound dubious. But theological attempts to articulate the point of the universe, in the words of the theologian John Haught, inevitably sound flat and inconsequential.[26] He also adds that it is not necessarily our business to know the purpose of the universe.

Suffering and Death

The suffering and injustices of life are another problem for the meaning of life that provoke different types of responses in different religions. We all experience problems that alert us to the severe limitations of our existence and that force us to confront the why-mysteries of life. In particular, knowledge of our own death has always been central to religion. Suffering and death become more bearable if we think there is a purpose for it or if there is a life after death that will explain or compensate for the suffering. As long as there is hope, we can struggle and still find life meaningful. Indic religions advance the idea of a cycle of rebirths.[27] And the prospect of an indefinite number of future lives intensifies the sense of suffering inherent in any life, no matter how much pleasure is mixed in. Under Indic beliefs, this beginningless cycle of death and rebirth will continue as long as we have a sense that we are individual selves set off from the rest of reality. *Karma* and rebirth still have their how-mysteries connected to them—e.g., how desires generate a new life and what happens in the intermediate state between death and a new rebirth—but the immediate "why" of our suffering is answered: our present suffering results from our own previous intentional actions.

As discussed in the last chapter, natural suffering ultimately is a mystery for theists—it conflicts with the goodness of a creator and his creation that is central to theistic meaningfulness. Theists do advance theodicies to explain the presence of human evil (e.g., God permits free will and we misuse it), but they do not have a satisfying explanation for why an allegedly all-loving, all-powerful, all-knowing god would permit such suffering. Theists at best have to conclude only that is *not unreasonable* to suppose that God has his reasons for permitting natural evil, even though we cannot fathom those reasons.[28] So too with social evil: we cannot conceive a reason for God letting millions of children die of starvation and disease each year. In the story of Job, when Job asks why the innocent suffer, he is met with only a show of raw power, a voice from a whirlwind asking Job who he was to question God—"Where were you when I laid the foundation of the earth?" (Job 38:4). And that was the last time in the Bible that God spoke to a human being. The mystery remains: we do not learn God's values and purposes.[29] Indeed, the story suggests that our suffering is not that important to God, since he inflicts so much suffering on a person who he acknowledges is innocent and righteous apparently just to test his faith.

John Haught concludes that all theodicies fail because the problem of suffering "is an open sore that theology can never pretend to heal."[30] Even if this is true, it does not cause theists to reject their faith. They believe there must be a reason for God permitting natural evil even if we do not know what it is. The theists' acceptance is based on the revelation that there is an all-loving god that trumps the empirical evidence of unjustified suffering. There is hope, even if we do not understand why we endure suffering. To naturalists, having to believe regardless of appearances begins to look like the claim of some fundamentalists that the universe is really only six thousand years old despite its appearance.

The reality of death also is obviously important here. Arthur Schopenhauer could rightly say that if our life were endless and painless, it would perhaps occur to no one to ask why the world exists and just what kind of world it is. And death itself remains a mystery: we cannot be sure what happens at death until we die. Naturalists see no reason to suppose that there is life after death since nothing in science suggests that anything survives the death of the body. There is no "karmic residue" to pass on or any "soul" that could exist independently of the body in a nonmaterial dimension to reality. However, it is a belief in all the world religions that something in us survives death and that our faith or actions in this life some-how determine the course of events after death. It is essential to the theistic picture of the moral nature of the universe: only with a reward or punishment after death can we explain the injustices and suffering we endure in this life. (However, the need to postulate life after death does show the difficulty the religious have in finding this life and this world meaningful in themselves.) But since we do not know what exactly happens after death, another issue arises: conceptions are radically different in different religious traditions.[31] That heaven is always a vision of what would be an ideal life in this world in a given culture and era gives credence to the naturalists' response that life after death is just something we have made up to make life bearable when things go wrong. The variety of conflicting views also presents a difficulty for the question of religious meaning: it shows that what is actually involved in a life after death is a mystery even to the faithful.

The negative side of a possible eternal life—hell—also presents problems for a meaning based on a moral god. The idea of eternal punishment has developed in the Abrahamic religions, although Judaism has a strong tradition of the wicked simply ceasing to exist at death. However, eternal punishment raises moral issues. Can a finite being can do anything—no matter how horrible—that warrants *eternal* punishment? And there is a larger issue: we need to belong to the "correct" religion to escape eternal punishment, but whatever is the one "true religion" more than three quarters of humanity do not belong to it, and by orthodox doctrines in the West since Augustine they will be consigned to hell for eternity.[32] Why create people when so many will be tortured for eternity for no more fault than that they were born in the wrong part of the world to appreciate (or even know) the "true religion"? Why would a moral creator knowingly create so many beings who would end up suffering through this "vale of tears" and then suffer punishment for eternity? As noted earlier, this greatly damages the claim that the purpose of reality is to produce conscious beings: now the purpose is to produce only a fraction of the humans—the rest are sent to eternal suffering in hell. Moreover, if God is omniscient, the problem is aggravated: God knew in advance that these people would end up being tortured for eternity for their lack of the proper belief or actions. Why create beings if God knew this was what would happen? In addition, what is the point of eternal punishment? What positive end is achieved? Or do we need this threat to be ethical? Even so, wouldn't eternal punishment be pointless suffering?

Some liberal theists cannot accept the moral consequence of the idea of eternal damnation and revise their revelations. Eternal suffering is simply too horrible for a loving god to inflict on any creature, let alone the vast majority of humanity. They postulate types of development in the afterlife—a cycle of rebirths, different worlds, a purgatory—so that eventually everyone is saved. Or they theorize an "inclusivism" in which members of other religions are really "hidden" members of their own religion and so are saved. There is no scriptural justification for these doctrines in the West, but we like better the idea that there must be some mitigation if there is a loving source to the world. Some advocate a "universal salvation" in which everyone, no matter what their beliefs or deeds, eventually end up in heaven after a stay in hell or purgatory. But then what is the point of this world? Why not just put us in the eternal end-state right away? How would anyone be "better" in that state for having suffered through this one? However they deal with this problem, the thrust of this issue is that all answers must be only our speculations based on what we think is best—what happens at death remains a mystery.

History

The idea of history adds a different dimension to the question of meaning. The sense of history that developed in the modern West is that the universe in general and humans in particular have unfolding stories. Prior to this sense, the series of changing events was meaningless. Of course, Westerners were aware that there had been a past, but prior to the modern period all that was of significance to Christians was the creation of the world, God's acts in the history of his chosen people (the Israelites), Jesus, and the destruction of the world—the rest was filler of no consequence. With a sense of history, Christians came to think in terms of ends to be achieved within the world and to see the universe as a venue for a transcendental purpose. The central event is the incarnation of Jesus: it shows God's love for his creation. By acting in the world, God gives history and this world meaning and significance. But not all cultures attach such significance to any historical events. In Indic cosmology, there is a cycle of time in which there is no sense of possible progress or meaning inherent in the world. Instead, the world repeatedly goes through cycles of birth, a pristine state, decay, and collapse. (We are currently in the worst time—the *kali-yuga*. In the current cycle, things will only get worse before a "rolling back" into Brahman and things start out anew in the next cycle.) With this view, history tells us nothing about anything fundamental.

The Western view does come up against one particular problem: the end of the universe. Meaning could be tied to individual lives: every person is valuable, and significance attaches to what we individually accomplish. If so, the ultimate end of the universe is irrelevant. Nor need we value the future over the present. But many people think our lives are meaningless unless our actions "echo through eternity." Our only value is our contribution to some transcendental goal. Eschatology in Abrahamic religions always sees the end as the fulfillment of a purpose and thus is tied to the idea of history. But then doesn't the eventual end of life present a problem? Of course, if the universe has a specific purpose and dissolves once that purpose is accomplished, then there is no problem. For example, if the purpose of the universe is to create beings who can praise God, then that has been accomplished and now the universe can disappear. But as noted in Chapter 4, according to current cosmological theory, the universe will still last trillions of years after no conscious beings are left in it—a blackness filled only by particles. This presents a problem for this scenario of the meaning of the natural universe: the purpose of creating conscious beings will be over and yet the universe will go on existing indefinitely for apparently no reason. How could theists explain that? Theists may dismiss this scientific claim as a groundless projection of things to come and simply say that we do not know what will happen in the far future. *cons's was always a potent*

A world without conscious life is arguably meaningless. It is not just a matter of there being conscious beings to recognize what is going on. To have a genuine purpose or value, something *genuinely new* must appear: if it were simply a matter of billiard-ball-like causation with no conscious agents to introduce change, an omnipotent creator would know all future events. That is, if a form of determinism is correct, why would a creator bother creating since it would know all events and any outcome in advance? A consciousless universe would also be swirling along meaninglessly. Random mechanical events on the quantum level would not change the picture significantly. Thus, without beings who can make conscious and free decisions and thus make unpredictable choices creating innovations in the course of events, there would be only the mechanical changes of causation and either random events or a deterministic course of events. In sum, meaning and value come into existence only with beings who can choose their actions and not simply have things done to them. Thus, it is hard to see how reductive materialists could see any meaning in the universe since there is no real "life" or "consciousness" but only a string of inanimate events. But even with free conscious beings, we cannot tell if history reveals something of the fundamental nature of reality since history remains a matter of our actions; all we know is the future would then not be determined.

Religious traditions such as Buddhism that see meaning in terms of individuals do not emphasize cosmic purposes or end-times. But theists have a basic dilemma: if the end is simply to return to God or to our eternal original state, why were we created in this world at all? Why did God bother going through the whole exercise of creation if the end result is simply the same as the beginning? And if our goal is just to return to the beginning, what is the point of our "soul-making"? Without something genuinely new, reality would be a pointless, painful play. In short, what does creation add? This also suggests that, if there is a creator, the mere fact that we suffer means there must be some purpose to this realm. At least for suffering to be morally acceptable, there must be some purpose other than simply returning to our original state.

Morality

Values and the systems of ethics that encode how we should act are part of every world religion. Giving up our own will to fulfill God's will through service to others may give our life meaning. But are religious ethics necessarily *moral*, i.e, suppressing our self-interest at least somewhat in favor of acting at times for the interests of others? Not obviously.[33] The religious believe that without some appeal to the transcendent all values remain groundless—our values would be purely subjective or just cultural products that can be changed on a whim. In a famous misquote of Fyodor Dostoyevsky, "if God is dead, everything is permitted." (He actually wrote: "if there is no *afterlife*, everything is permitted.") Indeed, if there is no God or afterlife, some Christians say we can do as we please regardless of the consequences for others—in fact, it would be foolish to jeopardize our lives by acting out of anything but pure self-interest.[34] Some theists may need the threat of eternal punishment to keep their urges in check, but people can also be moral without threats. Also note that, even though their acts may benefit others, these theists are acting prudently out of self-interest—avoiding hell and earning heaven—just as they accuse naturalists of doing; only their beliefs about what is real differ. It is the naturalists acting out of concern for others and not themselves even though it involves a sacrifice of their own self-interest who are acting morally.

Also consider the idea that what we should do is commanded by God. What justifies God's commands? Is it only that God commanded them? This leads to a dilemma first identified by Plato in the *Euthyphro* concerning piety: are commands moral simply because *God commands them*, or does God command them because they are *moral*? If there is no reason for God's commands, then they are arbitrary; if there is a reason, then it is this reason and not the fact that God commands them that makes them moral. Theists cannot be happy with either alternative: the first would mean that if God commanded us to torture babies, then it would be moral to torture babies. More generally, this makes his commands arbitrary—simply a matter of might makes right. But the second alternative means that something —morality—exists independently of God and is beyond his control; thus, God has not created everything and is also not omnipotent in the universe but is constrained by something more substantive than only the formalities of logic. Naturalists are perfectly happy with the second alternative. And few people want to say that torturing babies as a rule is immoral only because "God says so" and would be moral if God said so. Morality is not arbitrary, and God could not just as easily have made torture and murder "moral" by a decree—God could, of course, make it a *religious* requirement by decree but not a *moral* one.[35] Hurting others solely for our own benefit is morally wrong by definition and does not depend on any divine decree. Morality involves a genuine concern for others' welfare, and so if we know what the word "moral" means, we do not need God's pronouncements to know that murder is immoral. Thus, God's decrees are superfluous here, and invoking God adds nothing to our sense of morality.

To deal with the *Euthyphro* dilemma, some recent Christians have embraced the first horn of the dilemma and advanced a "modified divine command theory" about the creator's character: a *loving* god could not command an immoral act; and since God by his nature is all-loving, his commands are thus necessarily moral.[36] Morality is simply part of God's character. This may satisfy theists, but it still leaves morality independent of God.[37] God may always act morally—a "perfectly moral being" could, of course, not do otherwise—but this does not explain anything about the claim of morality upon us any more than if I had a morally perfect character and so always acted morally.[38] That is, it does not explain where moral values comes from, why we accept it, or why God has that property. God's character is simply irrelevant to this issue. Morality is still defined independently of God's nature—i.e., why are some deeds such as rape and murder "intrinsically immoral" so that a loving god would never command them? And theists now also have to explain why God's nature must conform to this independent standard. (If theists insist that they would not worship an immoral creator and so would not worship God unless he is moral, then they are admitting that morality is logically prior to determining what is worthy of worship. They would also have to explain why they would not worship an immoral creator. After all, he still created the universe—why is that attribute no longer make him worthy of worship?) In sum, the fact that a loving god would never command an immoral deed simply does not address the *Euthyphro* dilemma.

There are also *practical* problems. The idea of a divine sanction for morality sounds plausible in the abstract until we see the conflicting revelations and inter-pretations of revelations of various religious groups. Which revelations do we accept and whose interpretation of those revelations? More generally, one belie-ver's "will of God" is another's "will of Satan." Remember that the 9/11 hijackers were theists who thought they were doing God's will and expected a reward in heaven. Even within one tradition, there are conflicting interpretations of God's will on most, if not all, social issues. For example, Christians split on capital punishment: Roman Catholics say God generally condemns it, while fundamental-ist Protestants say God heartily supports it. Or consider wars: the religious authorities supporting the government on both sides routinely claim God is on their side—even Hitler had his Christian theologians and repeatedly invoked "divine providence" and the "will of God" in his speeches. How do we know what God's command is in such circumstances? In the West, ethics have evolved over the centuries toward more rights for disenfranchised groups. For instance, slavery is condemned in Christianity today, but it was accepted for 1,800 years—even the Pope had slaves. The change was not based on new revelations from God but on changes in our ethics during the Enlightenment.[39] Some religious factions, how-ever, always oppose these changes as defying the revealed word of God—e.g., Christian ministers in the American South wrote books even after the Civil War justifying slavery on biblical grounds (Lev. 25:44-46; I Tim. 6:1; Gal. 3:28; Eph. 6:5). In short, how do we know who has discerned God's will?[40] Or do we simply give a religious legitimation to whatever social changes we make on other grounds? In any case, how does bringing God into the picture solve our ethical dilemmas?

Also consider the "argument from neglect." Christian theologians have to rationalize why the apparent absence of God's actions in the world in situations where we would expect actions (e.g., preventing natural disasters impacting humans) is not evidence against the alleged existence of a moral god. They counter that absence is just what we should expect of a moral god—e.g., hiddenness preserves our freedom of choice in the area of faith by not overwhelming us with God's appearance (although being in the actual presence of God did not seem to curtail Satan's freedom to assert self-will and rebel). Moreover, if God is not capricious, then if he intervened even in *one* case to prevent suffering, then logically he would have to intervene in *all* relevantly similar cases, and this would curtail our freedom of choice (since he would be constantly correcting us). God cannot prevent undeserved suffering by miracles or even warn us about upcoming dangers—all he can do is tell us in prayers that he loves us and that we should love one another. Thus, not only do theists not take the absence of God's intervention to help the suffering as counter-evidence, they now see such absence as actual *evidence for* the existence of a *loving* god.[41] But one does have to wonder on what grounds can theists now claim to *know God is moral*, no matter what he does. What is the difference between a "hidden loving god" who does not act where we think a moral agent should and an impersonal deistic source or no god at all?

Some theists and nontheists think of ethics as a "natural law," a property of the universe like the properties of physics but social in nature rather than physical: the basic laws necessary for a society to survive have their foundation in human nature, and so we can discern a core of ethical beliefs common to all societies by our reason. But considering the variations in cultural standards for many issues (e.g., what is inappropriate sexual conduct), this leaves the natural ethical laws very abstract at best—one should not murder (i.e., not kill a human without one of the justifications accepted in a particular society), not steal (if a society recognizes private property), and so forth. There may be more agreement across cultures on the most basic features than postmodernists are willing to admit, but there is disagreement over universal features (e.g., over what constitutes justified killing or whether not killing extends to animals as in Buddhism and Jainism), let alone over the less basic points of different ethical codes.[42]

On the other hand, naturalists too have a problem in trying to ground ethics. Materialists in particular have the problem of how harming conscious beings is any different from merely kicking a stone. Other naturalists, however, can value consciousness as a different and more valuable feature of nature than inert objects and thus justify treating humans and other sentient beings differently. But the basic problem still remains: if the universe just is and serves no transcendental purpose, why should I ever sacrifice my own interests for someone else's? Bertrand Russell did not believe morality is a matter of "personal taste" and yet he did not know how to justify it and was aware of this inconsistency. Morality has the "feel" of an objective claim upon us, not something arbitrary or subjective. But according to naturalists, this problem is not sufficient to believe there must be a transcendental law-giver telling us what our values are since, as just discussed, theists have their own problems, and so how does invoking God help us out at all?

Arguably, naturalists in fact have an *easier* time justifying morality since under their metaphysics we are all interconnected entities within one naturalistic whole and all life-forms are related by descent, with persons being dependent upon other persons for their existence—thus we should naturally be concerned with others' welfare too and not just our own. Theists, on the other hand, believe that God is somehow responsible for the creation of each being individually and that we have to be *commanded* to love one another.[43] Nothing in naturalism entails rejecting morality—only if one comes to the issue with the mind-set that morality must be underwritten by an authority transcending the world or it is groundless can one conclude that. Certainly nothing in the metaphysics of naturalism compels that looking out for oneself is the only relevant consideration in deciding how to act. Indeed, exclusive self-interest is arguably inconsistent with naturalism's metaphysics: realizing that nothing centers around oneself and that we are part of one integrated whole is no basis for selfishness—one cannot see oneself as a separate entity to be valued over everything else if we are part of an interconnect whole.

Even moral realists, who believe morality is an objective feature of reality like the laws of physics, confess that since the *Euthyphro* dilemma eliminates God as a possible source, the source of morality remains a problem.[44] This leads many naturalists, despite the feel of morality, to reject all forms of moral realism. But many still reject any subjectivism or nihilism. To them, morality is an institution we humans created to aid in our evolution. Because the vast majority of people are religious, societies may ultimately need a religious legitimation to justify following an ethical code, but this is just a "big lie" to satisfy our misguided psychological need. But while ethics may only be our "conventions," this does not mean its values are *arbitrary* in any sense: only values that permit societies to survive can end up being adopted. Thus, ethics are justified on pragmatic grounds. Still, morality's only objective claim is that a society could not survive long without the adherence to certain basic ethical tenets—without laws against such conduct as murder and theft, societies would self-destruct. Thus, these laws seem to us to be objectively real but are not—they are just social norms we make up because we need them.

In sum, trying to answer "why should I keep a promise when it will benefit me not to?" or in general "why should I be moral?" is not easy.[45] All in all, morality—its source, its claim on us, its requirements, how it and any particular ethical values are justified—remains more mysterious than we like to admit.

Is There a "Meaning of Life"?

We must also conclude that the meaning of life, if there is one, will remain an ultimate why-mystery. The "true meaning of life" would reveal our true relation to the rest of the world, and to do this we must step outside the world. Any "sense of the world" lies outside the world, as Wittgenstein said,[46] and we would need a god's point of view to see it. The universe is inscrutable between naturalistic and the various religious hypotheses on meaning. Theists believe that the true meaning of life is at least partially revealed by a transcendental reality, but two problems persist. First, the religious must rely upon such revelations: no empirical study reveals any purpose beyond the proliferation of diverse organic and inorganic forms and a trend toward complexity. Relying upon revelation, however, does not help: the foundational revelations of the different traditions conflict on basic points—as the religious scholar Huston Smith has said, "people have never agreed on the world's meaning, and never will." Moreover, there is no apparent means to adjudicate between the competing answers. The conflict of revelations may disappear if one culture ends up dominating the entire world through social, political, or economic forces; the winners, of course, will see this as the will of their god triumphing.

Second, salvation in all traditional religions involves *getting out* of this world. The goal is salvation in some other-worldly terms. However, how can the only point of this world be to get out of it? Nothing we can contribute to the world matters or is of worth. In the end, if the meaning of life requires a transcendental goal, isn't the universe itself ultimately meaningless? The religious may respond

"So what?" but anyone who takes science seriously may still wonder why such a vast and old universe exists. If God went through the bother of creating a universe billions of years old and of such vast dimensions (assuming that is what happened), mustn't there be some reason for this universe itself and for life other than only the salvation of a number of beings that is very small compared to the effort?

In the end, we are left with choices. The religious are "cosmic optimists" looking for something positive in the suffering of our lives. To the religious, it is impossible that all this is the senseless spinning of atoms—there must be some reason and meaning to it all. In classic "philosopher's disease" type reasoning, they argue "We're here, so there must be a reason." The alternative is emotionally unacceptable. The mere fact that anything exists is taken as evidence that there must be some point to it all. Many religious people may be able to rest comfortably, agreeing with Arthur Koestler that merely knowing life must be meaningful—that there *must be* a reason for all this—is enough, even if we cannot know what that meaning is. However, this faith is not actually finding a meaning to life but simply giving up the search for one. Moreover, accepting "purpose" in the abstract means it is still a mystery to us, and no religious life could be built around only abstractions. Thus, whether the religious must specify some concrete purpose or meaning in order to have a full religious way of life is an issue. So too, we want to believe that each individual human life, along with our species as whole, has some special value. In addition, many demand a heaven and hell because we are convinced there must be justice in the universe and it is not evidenced in this world. The belief in such a transcendental meaning can produce a very positive life of hope and trust, making suffering and death easier to take. With our fears allayed, the religious then do not have to face the mystery of it all.

The religious may be correct that there is some cosmic meaning, but just because meaninglessness is unacceptable to us does not mean the universe must have a meaning—reality does not have to conform to our desires. In addition, the various competing religious traditions' answers to the question of the meaning of life begin to look more and more like only our own attempts to find meaning: if a transcendental reality exists, we seem to be imposing meaning onto it, not receiving revelations from it. It may be that even if there is a transcendental purpose to the world, it is nothing like any of our purposes. The danger that our brain will persist in finding a "because" even when there is none also cannot be downplayed. (And much of the reasoning on the meaning of life does seem to boil down to "I like the idea of x, and therefore it is reasonable to believe in x.") Do we have any reason to believe that the religious scenarios are anything but wishful thinking—just "warm and fuzzy stories" to comfort us, as Stephen Jay Gould said? Do the religious have reasons for hope, or is hope just groundless optimism since the history of the earth so obviously conflicts with the idea of a loving god? Do theists focus on the future since the past and present are obviously not evidence of a god? It is hard to accept that there is no reason for all this universe being here and for us in particular. But even if we must believe there is a reason, that alone does not give us the right to spin out just any meaning that makes sense to us. It may be that since we are beings who act intentionally that we most naturally feel there is an intentional being behind the whole shebang. Naturalists by their lives also show that humans do not have an innate "thirst for transcendence"—or at least some people can satisfy that thirst with phenomena in the natural realm.

Four options remain: one of the many revealed meanings is correct, if there is a meaning we are not in a position to know what it is, there is no transcendental meaning, or we cannot know if there is a transcendental meaning or not. No side appears more "reasonable" than the others in this regard. We do not seem to be in a position to ever know.[47] How can beings within the universe know the meaning of it all? To a naturalist like Robert Solomon, realizing that "why?" has no answer is "the singular fact that now defines our existence." But perhaps as Viktor Frankl said, our concern about a meaning of life is "the truest expression of the state of being human." We are doomed to live with this deepest personal mystery of all.

Indeed, the problem of meaning is felt more forcefully today, even though the physical suffering millions upon millions of people endured before the modern age was much greater than it is for many now and thus the need for meaning was greater then to make the suffering bearable. The loss of an overwhelming sense of a transcendental reality today may account for the difference: we no longer are certain the gods exist, and so the counter-evidence seems more compelling. Existentially, the question of the meaning of life hits home only in moments of personal crisis when the old answers one has grown up with no longer seem to work. Most people do not even ask the question of the meaning of life, being content with busy lives and too engaged even to be conscious of a problem.[48] But when "the experience of nothingness" arises and traditional answers fail, the universe appears indifferent, disillusioning, and terrifying. God is not seen to intervene in natural and social disasters when we think a loving creator would. Ironically, it is easier today for the well-off and comfortable to see life as pointless.

The Transcendent and Language

Problems that the religious have with language highlight the mystery of transcendental realities. Central to all traditional forms of religion are ideas of a transcendental source, either a personal god or gods or a nonpersonal counterpart—a "power" "behind" our world, the "ground of being." It created or at least continually sustains this realm, even if how it brings the world's existence out of nothing is a mystery. How to specify the transcendent's nature is difficult because in all religious traditions it is "wholly other" than what is in this world. This only intensifies the problem of specifying its nature. The transcendent is not "infinite" in the sense that it is merely more of the same finite stuff as this world (unless one endorses a form of pantheism or panentheism). It is not a "being" or like any other type of object within the world, and it does even not "exist" since only things in this world exist. It is not even "ultimate reality" in the sense of being on the top of the same scale on which we could place anything from this realm. All the words in quotation marks indicate the problem that any language will have in designating the transcendent. The transcendent is beyond our worldly categories and yet we humans must use them. What can we say of its nature if it is truly transcendent? But the religious also want to affirm features and so resort to worldly categories: the transcendent is the cause or ground of our being, it is loving (or is neutral), it is personal (or is nonpersonal), and so forth.

Theists in particular routinely affirm positive characterizations of the transcendent. They also want to deny the opposites—e.g, that God hates his creatures. In addition, the religious also immediately qualify any positive descriptions because the transcendent is wholly other and hence not conceivable by our finite minds. Theists say that God is moral and just, but they qualify this because of the suffering in nature and evolution by saying "God's morality and goodness is not like ours." But how are we to know what these words mean when referring to God if they do not mean something like our morality and goodness? Why are we using those words and not their opposites? As John Stuart Mill rightly asked, if in ascribing goodness to God we do not mean what we normally mean by "goodness," it is then incomprehensible what we do mean, and thus how do we know God's goodness might not be totally different from what we love and venerate? The same problem arises with "being" or "creator"—theists want to affirm both but also to deny that God's "mode of being" and how he "causes" things are like anything natural. Nor is God a "person" like us—there is no brain, body, or gender—and yet is still "personal" in nature (an individual who thinks and wills). Arthur Peacocke wants to say that God is "transpersonal," "more than personal," or "at least personal" and yet still has the attributes of a person.[49]

All this introduces the problem of trying to depict the transcendent in terms of metaphors (e.g., "the hand of God" to depict God's actions in the world) and analogies to things within our realm (e.g., God's goodness is to the transcendent what our goodness is to us). A metaphor involves seeing one thing in terms of something else. However, metaphors are limited: what is referred to both is and is not like the subject of the metaphor, or else no metaphor would be needed. Part of how we learn in general is by extending known terms to new applications. But in the case of the transcendent, there is a problem: our terms are developed for use in the natural realm—how do we extend them to a reality utterly unlike that? In science, theoretical models are helpful in leading to new avenues of research. But nothing like this works for a reality utterly unlike ours. Even the word "transcendent" is metaphoric: "*trans*" means "beyond," and the transcendent is not literally on the other side of natural phenomena. No spatial relation taken from within the world literally apply: the transcendent is not "above," "beyond," "behind," "below," or "within" the world. And if we introduce terms developed just for the transcendent—such as "God" or "Brahman"—still we will think in terms of worldly objects and entities. Thus, use of even those terms will have a metaphoric element.

Analogies are no better since they are based on our ideas for the world. All our concepts are about the finite, but we cannot apply them by attribution or proportionality to the infinite—in mathematical terms, no ratios work when one of the terms is infinite. But the bottom line is that the role of the familiar is firmly implanted in how the religious see the transcendent through language.

But if the transcendent is utterly different from anything in this realm, how can any metaphor even begin to work? Metaphoric language can be referential—we can refer to a person as a lion and there is a referent (the person) to ground the metaphor even if we cannot translate the metaphor into literal language without some loss of the cognitive content. But how can we ground reference in a total mystery? Metaphors involve seeing one thing in terms of something else, but at some point some term will have to be meant literally if we are to identify what is being referred to.[50] Even the theologian Paul Tillich, who thought religious language is symbolic, asserted that "God is being-itself" is the one literal, nonsymbolic statement that can be made.[51] Critical realists can distinguish "sense" and "reference" and argue that the transcendent can be referred to even if we cannot specify any of its properties (the "sense" of the thing) except through changing, inexact, and limited metaphors. In science, realists have the problem of a common referent surviving changes in theories. In religion, the problem is whether terms from different traditions referring to transcendental realities can actually have the same referent. In addition, critical realists are left with the problem that every metaphor has both a positive and negative component—how do we tell which aspects are similar and which are dissimilar?

Most importantly, if no part of the metaphor can be specified literally—if, for example, God's "goodness" is so unlike ours that even the word "good" cannot apply—we are left with a complete mystery. We literally do not know what we are talking about. The religious could not specify why one metaphor is more useful or appropriate than another. Why is referring to the transcendent as a person "less misleading" than referring to it as an inanimate object? Why is God like a shepherd and not a sheep? Why aren't all metaphors equal? There must be grounds to distinguish appropriate from inappropriate metaphors, i.e., some thread of commonality to the appropriate metaphors, and this common thread can be specified literally. Saying that the metaphor of "the hand of God" depicts the action of God in history does not mean that theists know how God acts, but they still are asserting that God literally acts in history. Metaphors may be required, but they do entail simpler statements—i.e., the grounds for why a particular metaphor is appropriate for envisioning the transcendent and another is not. And if the simpler statements are not literal but are metaphors all the way down, the initial metaphor cannot be said to be stating anything about the transcendent. The only alternative is that every metaphor is appropriate to the transcendent—anything goes.

In fact, can metaphors ever give us insights into a transcendental reality? In science, the worked-out metaphors of models can lead to new experiments and new discoveries. Models are explanations insofar as they give a sense that we now understand, but they only suggest new ways of looking at that domain. Experiments still must be conducted. The metaphors from the everyday world are never "true" pictures of the domain being studied since there are always part of the model that does not reflect that other domain. This highlights the problem of metaphors as a map of a new realm. By organizing how we see a new realm, they can easily mislead us: we do not know what aspects can be carried over into the new domain until experiments are made. In addition, the religious may get caught up in the metaphor itself and have it end up restricting their vision.

So how do we know a given metaphor actually illuminates the nature of the transcendent or merely quiets our questioning by creating an image that satisfies us? Attention may become directed to the cultural element used as the symbol, not the domain being studied. In science, new experiments can show how metaphors help and when they stop helping, but there are no experiments or other empirical examinations in religion, only conflicting revelations. Without experiential controls, how can we tell if the metaphors tell us anything about the transcendent or its relation to the world? In this regard, the transcendent is more like Kant's noumenal realm than simply another realm of the natural world to be explored by scientists. Metaphors depict the transcendent, but they are drawn from this realm and reflect our current ideas of what the transcendent must be like—in the case of Abrahamic theisms, the conceptions end up reflecting an idealized, powerful male. Today some theologians suggest "mother" may be a more appropriate metaphor for a creator. But the important point is that metaphors change because cultures evolve, not because we are receiving new revelations providing us with more information about the transcendent. The Bible will have a say in what metaphors the Christian community ends up adopting, but in the end there is no testing except to see if a metaphor reflects a culture's current doctrines about God. The same religion may have multiple metaphors in words and visual symbols for the same alleged property of the transcendent, and different religions may have conflicting doctrines and thus different sets of metaphors. Is there any way to determine whether any metaphors are better than others except by what a religious community thinks illuminates its particular basic doctrines? Of course, cultural changes that instigate new metaphors may end up modifying the basic doctrines, but the religious are not receiving new experiential in-put from transcendental realities.

The problems with positive characterizations of the transcendent in itself or in its relation to the world has led some theologians and mystics over the centuries to espouse an "apophatic" or negative theology denying any positive characterizations are possible. As Thomas Aquinas put it, we cannot know what God is but only what he is not; nor should we pry into the secret of where God came from but simply accept that there are things we cannot grasp. Nicholas of Cusa's doctrine of "learned ignorance" is the recognition of the limits of theological thinking: the highest learning that we can attain is to know that we are ignorant of God. According to him, God as God really is cannot be known by us (even after death), and all attempts at positive characterizations must all be rejected—God is simply beyond the reach of the human intellect. Anything we conceive even by analogy will be infinitely short of the reality because there is no proper relationship between the infinite and finite.[52] The prohibition in the Abrahamic religions against creating "images" of God is thereby extended to word-images and not just visual portrayals. The transcendent is a nameless, unfathomable, incomprehensible, ineffable mystery. Words used for the transcendent only point in the direction of being beyond all we can comprehend. This leads to silence as the only appropriate response.

Silence, however, is hard for any religious community to maintain. All the classical saints who speak of God being beyond our knowledge—Augustine, Aquinas, Anselm—nevertheless continue to make positive characterizations of him. Indeed, how could the object of worship just exist in the abstract without any known features? How does anyone love or pray to a total mystery? How can anyone have faith or trust in a blank? And, philosophers will ask, how can you know something exists and yet not know *any* of its properties whatsoever? And what value would a totally featureless transcendent have in our lives? How can mystery be the answer to the meaning of life? For any religious life to be possible at all, some positive characterizations—God as a loving creator, or whatever—are needed. The religious need to believe they have comprehended at least something of the transcendent for their own understanding and for distinguishing appropriate from inappropriate religious responses to the transcendent. Thus, despite the philosophical problems with depicting what is supposed to be "wholly other," the religious typically proceed with positive characterizations while at the same time admitting the limitations of any such utterances and the tension between "is" and "is not" in any metaphor.

Is Knowledge of the Transcendent Possible?

Of course, many religious believers believe we do know something about the transcendent. Revelations recorded in sacred texts such as the Bible and the Qur'an are believed to reveal something of the nature of God and what he wants us to do. As discussed in the last chapter, natural suffering presents problems for any "natural theology" based on the idea that the world is God's self-revelation. Many theologians, both liberal and conservative, want nothing to do with natural theology—they view the design argument as an idolatrous attempt on the part of finite humans to grasp God in rational or scientific terms, thereby diminishing the mystery of God by seeking to bring it under the control of the limited human mind.[53] These concerns have led many theists to argue that we can know anything of God only through his direct revelations. Thus, God is not in fact "wholly other" even though his mode of existence and properties are in some ways utterly unlike ours. The transcendent has features of which we have clues through revelations or mystical experiences even if we only see through the glass darkly and cannot fully understand. Theists argue that we cannot *fully* comprehend God, but this does not mean that we cannot comprehend *anything* of God or that conceptions of God are meaningless simply because aspects of God exceed our ability to know. This affirms that the transcendent is a mystery but also affirms that the transcendent is not utterly beyond us and thus not a total mystery. Even Rudolf Otto made positive statements about the "wholly other."

In particular, theists want to affirm a god who is personal in nature and a caring creator with whom we can have a personal relationship. Theologians go further and debate mysteries over God's alleged properties (e.g., how a timeless, incorporeal being could create or act in nature), but the personhood of God is always the starting point. So too, revelation according to theists requires an act of will on the part of God and hence a reality with the properties of a person. Talk of a "person" will remain metaphoric in that the transcendent is obviously not a human being, but theists assert that it has at least some personal features. Thus, theists argue that the much maligned idea of anthropomorphism in the end has to be affirmed.[54] Through evolution, our minds may be disposed to see agency as an explanation for anything we do not understand, and so anthropomorphizing *nature* is a mistake, but theists argue that we must see the transcendent that way since any creation of persons and a complex world involves intentional agency, and hence the creator's personhood is required. However, God's nature is always more than we can understand, and so a critical realist stance towards the metaphor is needed. Personhood will always be based on our experiences of living, limited, embodied humans in this world and thus can only be applied with limitations. Figurative language (e.g., God as a father or mother or king) will also be needed in religious ways of life to keep the transcendent from becoming too abstract. But the point is that even abstract language will be metaphoric since the transcendent's mode of being and acting is not like ours. The metaphor may hit part of the transcendent, but to lose the "is not" part of a metaphor would produce an idol by reducing the transcendent simply to what we can comprehend. There is also the danger of seeing the transcendent as too much like a human—of making God in our image. The religious may make the transcendent simply a mirror of themselves, e.g., whatever we think is good is what God thinks is good.

In the end, the religious want it both ways: that the transcendent is knowable and is more than we can ever know, and that our language both applies and does not apply to it. Metaphoric speech is necessary to conceptualize the transcendent, but it also shows the limitations of any characterization of this mystery. "Is not" will always apply to all concepts from our realm. Our conceptions of the transcendent will remain precisely that—our conceptions with all their limitations.

Theology and Mystery

More elaborate religious doctrines are a natural result of the human tendency to ask and answer "why" questions. In this way, theology plays a vital role in the demystification of the transcendent. It is the religious theorist's central task of explicating and defending the belief- and value-claims entailed by doctrines and practices of their tradition that ends up defusing mysteries. This role of theology is not to defend or enhance mystery—instead, any sense of an underlying mystery is harmed by looking for explanations and reasons. But this need not replace in the minds of believers the experienced "God of Abraham, Isaac, and Jacob" with an inferred "God of the philosophers and scholars," to use Blaise Pascal's phrase. The mystery of the transcendent can be affirmed and the full religious life can remain intact even when asking about aspects of God. But as Pascal also warned, "if we submit everything to reason, our religion will contain nothing mysterious."

The problem is that theology directs attention away from any mystery and toward what we can comprehend. Here we run the risk of making human reasoning the standard for ultimate reality or believing we can fathom the ground of reality. In effect, theologians risk taking *human* ways of thinking and planning and then seeing them as God's. As noted in the last chapter, in their attempts to understand, many theologians believe they can get into the mind of God—their books constantly tell us what God "thinks," "feels," "wills," "desires," "intends," "values," and "loves." Even if these terms are only meant metaphorically, they still direct attention to what *we* think, feel, will, desire, intend, value, and love. It is always a matter of "putting yourself in God's shoes." can we believe any answer? Our everyday reasoning is usually permeated with wishful thinking, and the danger here is that even our most refined religious thinking about the transcendent will be the same, especially when there are no empirical ways of checking our claims.

The Ontological Argument is the classic case of taking the concept of a reality and filling it with whatever we value or whatever seems "reasonable" to us and then claiming it is real—in effect, we create God in our own image.[55] It is an attempt to prove from our reasoning alone that a reality exists—a reality greater than which nothing can be conceived, no less. Most philosophers dismiss it as fallacious since we cannot argue from a concept to a reality. We cannot go from the logic of our concepts to what must be in the world—the argument at best shows only that the *concept* of "a reality greater than which nothing can be conceived" entails the *concept* of "existing," not that the concept is instantiated in reality. That is, if God exists, he is by definition "a reality greater than which nothing can be conceived," but that God actually exists does not follow from our definition. Others find it unconvincing since the argument would work only if we first assume that the reality is "necessary" in the sense of being "true in all possible worlds" when it does not appear to be a necessity of thought that all worlds must have such a reality or a source (i.e., a world may itself be "self-existent," just as theists claim God is).

But the important point here is that at best the Ontological Argument proves only an empty shell—we fill it with what we consider "perfections," i.e., what we value from our own point of view and not necessarily what is actually real. (Of course, the Ontological Argument works equally well for attributes we do not value, e.g., the argument just as easily establishes a "maximally evil" reality.

Theists may consider an immoral or morally neutral reality to be "morally intolerable" and thus not worthy of worship, but this is just an indirect admission that theists demand that reality must reflect what *they* consider worthy of worship.) Consider one alleged property—being "all-powerful." We think a creator cannot be anything less than omnipotent, but we also like to think that we have free will and thus can exercise some independent power in the world. However, if we have some power, then God does not have all the power—God cannot control the free choices we make. Theologians want to keep both ideas so they hypothesize that God voluntarily withdraws some of his power to allow our freedom. We prefer to think that God chose to give us free will and the accompanying power, but what are the grounds for claiming this, other than that is what we like? In the end, what we think is best decides the issue since we have no empirical data to work with.

If the *a priori* approach does not work for studying the natural world, why would we think it would work for an alleged transcendental reality? Just because we have nothing else to work with does not mean our *a priori* reasoning is reliable. We have no idea what a "perfect reality" is except as a projection of what we want—the set of properties we think are best (e.g., power, morality, knowledge). The early Christian theologian Origen argued that the body of the resurrected must be spherical because the resurrection body will be perfect and the only perfectly symmetrical solid is a sphere—are contemporary theologians' thoughts about God any different in the type of reasoning? Even if we could somehow prove through the manipulation of our concepts that a reality must exist, we still are left to our own devices to determine what its properties are.

The idea that God's existence or his particular attributes can be *proved* is giving way under the force of criticism to the easier idea of showing that it is *not impossible* that a theistic god exists and that belief in such a god is *not irrational*.[56] Under the newer approach, the Cosmological and Teleological Arguments become, not arguments for the existence of God, but ways of showing how the world makes sense in light of the assumed existence of God, and thus indirectly showing why it is reasonable to believe in God. This approach is not a "natural theology" inferring the existence of God from natural phenomena but a "theology of nature" that simply *presupposes* the existence of God and uses him to explain nature by interpreting the existence and phenomena of the world to fit a theistic picture. That is, if God exists, then *x* "makes sense" (where *x* is the world, order, consciousness, morality, or whatever). The arrow does not point from natural phenomena to the existence of God but from the existence of God to natural phenomena; major problems since Hume and Darwin have been presented for the former, but the latter is immune to science since theologians can always say God created whatever laws of nature scientists find. In this way, knowledge of God allegedly contributes to our understanding of nature, but God does not play a role in science. Looked at that way, the ineffectiveness of the Cosmological and Teleological arguments to make the nonreligious religious is easily understood. But even under this approach, the doctrines standing between us and the mystery become the center of attention. The theologians' task still directs attention to seeking answers to religious how-mysteries. The focus is on explanation and not on the underlying mystery.

In short, theology becomes the enemy of mystery. Since Tertullian, Christian theologians have acknowledged that there is more to God than can be understood by our reason, and hence there is an abiding core of unfathomable and incomprehensible mystery. But theology's anti-mystery thrust has always countered that. By the fourth century, the Christian saint Athanasius could remark that it is the superstitious portion of mankind that is fond of mysteries in matters of religion and thus likes best what they understand least. Indeed, to all Christians, in the beginning was *reason* (*logos*), not *mystery* (Jn. 1:1). "Mystery" with a capital "M" becomes just another name for God, with attention now focused on what is knowable about God. This obviously stifles any sense of a mystery. Religion cannot stop with awe and mystery; some conceptualization is needed, as Otto noted. But theology makes the conceptualizations central. Indeed, religion in general shares the problem of substituting concepts for mystery, even when a focus on conceptualizations may be misleading. At best, religion is like an inoculation against mystery: it gives the faithful a small dose of the mystery of the transcendent in order to prevent them from being overwhelmed by it. Most of the world religions can be seen as being founded by people who saw the prevailing religion of their era as deadening any sense of transcendental mystery. But in the end, all religions, like philosophy and science, fight mystery in favor of explanation. A culture of control overrides a sense of mystery even when the mystery is the source of our world.

The Demise of Mystery Today

It is often noted that today there has been an "eclipse of mystery" in Christianity as well as the other two Abrahamic religions.[57] There is an erosion of concern for mystery in Christian liturgy and theology. The intellectual leaders of the liberal branches of religions today seem more comfortable with secular thinkers who have no interest in any alleged transcendental realities and no interest in mysteries in general. The sense of awe and power that Otto described in the encounter with the holy is absent from mainstream religious rituals even if reference to awe and power remains. Many people wonder whether we will ever be able to have the sense of a sacred reality surrounding us that our pre-modern ancestors had. In the words of the theologian Sally McFague, even the most religious are secular in ways our ancestors were not—"We do not live in a sacramental universe in which things of this world, its joys and catastrophes, harvests and famines, births and deaths, are understood as connected to and permeated by divine power and love." We can still contemplate the majesty of nature—the mountains, the sea, the night sky—but do we sense a transcendental power behind it all? To those most influenced by the modern consciousness, we live in a secularized world that has no "depth" dimension to reality (and hence no transcendental mysteries) but is only the one-dimensional world studied by scientists. Lives no longer have a transcendental context.

Some Christian theologians in the twentieth century did emphasize the importance of mystery to religion, notably Gabriel Marcel and Karl Rahner, for whom mystery is not a lack of knowledge but an ontological reality: God is the "holy mystery" and paradoxically is both self-evident—indeed, the only self-evident reality in human life since God is closer to his creatures than we are to ourselves—and also the absolute, incomprehensible mystery totally beyond our reach.[58] But few theologians today even refer to it.[59] God is no longer awesome and terrifying but a something we can understand and deal with as a creator, friend, and source of salvation. Any reference to a mystery of God is dismissed as just mystification or New Age flakiness. Postmodern theologians see no transcendental mystery at all; at most, they may note purely natural how-mysteries. In fact, some actually celebrate the loss of mystery, arguing that mystery is no longer needed in religion: religion is reduced to a matter of ethics alone, and the loss of a sense of mystery—indeed, the loss of any sense of transcendental realities—is only an indication of the budding maturity of our species, revealing our autonomy and freedom. (Religion's anti-mystery thrust is also carried on by New Age practitioners. Karla McLaren notes that they "can't handle mystery—not even a tiny bit." They have "no tolerance whatsoever for mystery" but give "an answer, reason, and a source" for every aspect of life and the universe.[60])

As in philosophy and science today, mystery is not welcome. We may think that the religious would accept and even celebrate an underlying mystery to reality as central, but the practical nature of religion pulls in the opposite direction. It is hard to lead a religious life with nothing but mystery. Thus, there is the natural tendency to replace mystery with something we can get our mind around and thus a tendency to confine the transcendent to something manageable. Elaborations of the concept of "God" provide the specificity needed for devotion. Once we have something to focus on, some or all of the mystery disappears and we turn to alleged attributes and powers. Thereby, the religious do not have to face the mystery. Images and concepts replace the transcendental mystery. It is not merely that there is tension between the otherness of God and humans' need to have something we can grasp—rather, the latter tends to push the former completely out of the picture. We naturally side with something we can handle over mystery.

Christian churches acknowledge many mysteries in principle, but more typically what is known about God through the revelations recorded in the Bible (or personally received) becomes central. God is a mystery only if we do not know him, and through revelation we know him. Revelations defuse the why-mysteries of the transcendent somewhat by revealing some of God's nature and his agenda for us, although they leave intact the how-mysteries. Revelations supply basic answers to our questions of life, and any residual mystery in the background can be ignored. Mystery may remain at the heart of a revelation, but any terrifying mystery of the unknown is also lessened or dissipated totally.

Nor are all religious believers concerned with issues surrounding the beliefs entailed by their basic faith-commitments. For example, the mystery of suffering and evil does not disturb many. Some theologians have simply given up trying to explain God's possible reasons for permitting evil or even to show only that it is logically possible that a perfectly good god could permit natural suffering for the millions of years and the human evil exemplified most concretely in the suffering of the children who died in the death camps during the Holocaust. Why didn't God intervene to help?[61] Even if theologians can counter the "argument from neglect," still we apparently are on our own. Rather than trying to explain God's ways, these theologians simply *assume* that God is good (or holy, just, merciful, or benevolent) and that his goodness is somehow consistent with the evils of life. They then move on to practical issues of faith, such as how we can be moral in this world. Indeed, a theodicy may lead to accepting suffering as part of God's plan and hence may impede our resisting evil or promoting social justice; thus, this theological project is rejected on moral grounds.[62] Such thinkers have given up wondering why and just accept such suffering as a grave mystery we are not meant to understand. They stop our questioning by putting the question aside unanswered. It is an open admission of not knowing by those who remain faithful for other reasons and an affirmation that there is an answer regardless of what we experience.

Problems of a Transcendental Mystery

In sum, a sense of mystery surrounding even the transcendent—and any accompanying sense of awe and wonder concerning any alleged transcendent power behind nature's phenomena—no longer figures centrally in mainstream Western forms of theism. Even in Pentecostalism, the central experience is of being saved from hell and thanking Jesus for the act of redemption, not any mystery. On the other hand, naturalists can have a sense of great mystery and awe about reality—both for the many how-ness mysteries of science concerning the natural order and for the that-ness mystery of the fact that anything exists at all. Scientific how-ness mysteries do point to deeper "how" mysteries within the natural realm, but only for the religiously-minded do they also point to a possible transcendental source that may explain why there is order, why conscious life evolved, and so forth.

Today there are "religious naturalists" among liberals in Christianity and Judaism who try to make the scientific and philosophical senses of awe and mystery into something religious, but it is mysteries of the *transcendent* that are distinctly religious, and for naturalists those are nonexistent. These naturalists deny any transcendental realities that can design or act within the natural realm, or any life after death for humans. Thus, they deny the traditional realism of theistic traditions and naturalize all ontological claims about a transcendent reality—e.g., to Stuart Kauffman, "God" becomes the name for the "ceaseless creativity in the natural universe."[63] Some religious naturalists may accept a form of a deistic source, but they still deny theism in any of its traditional forms of a god active in this world. They may still enjoy participating in the social institutions of the theistic religion of their youth, but they are not theists in any literal sense. They exploit scientists' fascination with the how-ness of nature or philosophers' bewilderment with the that-ness of being to try to import a sense of mystery back into a religion with no sense of transcendence. Mystical mindfulness of the natural world becomes an "experience of God." In short, under religious naturalism purely natural awe and humility become "religious" simply by calling some aspect of nature "God." But religious naturalists cannot make everyone religious or tell avowed atheists and agnostics that they are actually have "religious experiences" simply by claiming scientific and philosophical wonder is really religious.

We live in what John Hick calls a "religiously ambiguous" universe,[64] and we are left with uncertainty concerning whether there are any transcendental realities. Is the mystery of the transcendent generated simply by our *reflection*—thinking about the problems of existence—or is it based on some *awareness* that there is something beyond the natural world? Are we aware of a limit to our thought about the transcendent because we are aware of something more, or is it only that we feel there must be something "more"? From Hegel to Wittgenstein, philosophers have argued that to know a limit is already to be beyond that limit. Are we aware of a genuine mystery at the edge of the natural world, or does merely the use of our concepts generate it? Is the transcendent a "known unknown," i.e., something actually experienced even though we cannot know anything about it? Are we aware only of a question or of something more? Just because we can formulate an idea of a transcendental reality and can formulate a question does not mean that there is anything real there. Obviously, this is were naturalists and the religious part ways, with the religious claiming we have some awareness of a transcendental reality and are not merely formulating metaphysical explanations.

But another problem arises here: the concept "God" is not merely shorthand for "the mystery of the transcendent." It is a theistically-loaded concept, not a neutral label. A paradigm of faulty reasoning is simply to label an underlying mystery "God" and then to feel justified in reading into it all the current ideas of one's theistic tradition. Theists traditionally believe specific things about a god—the transcendent is "all-mighty" and holy and thus must be an all-powerful, all-knowing, and all-good person. Are these attributes based on some real knowledge, or do theists simply posit them based on the idea that the transcendent must be the most extreme in every value we value? As with the Ontological Argument, is it just a case of us thinking "Well, God must be the most of every-thing we value and more"? Isn't postulating the simplest conceivable person (i.e., one with infinite powers and knowledge) the way to go? But just because such a reality seems simpler for us does not mean that it is real. We can use such words as "all-knowing" or "eternal" to identify by extrapolation the most extreme case of something we do know, but that does not mean that there is any corresponding reality.

Moreover, theists' intuitions conflict and are hard to harmonize. That theists want both an all-powerful god and that our actions matter was noted above: God must be "self-limiting" to explain why we have free will or any power at all. The medieval Kabbalist theory of God's voluntary "withdrawal" (*tzimtzum*) of his power in order to let something exist "outside" of God is one attempt at reconciling these intuitions. But this sounds only like our own *ad hoc* idea: we like the power, tidiness, and comfort of an omnipotent creator, so we posit one and then adjust it to fit our sense of free will.[65] The possibility that there may not be any god or that we do not understand is not admitted. Similarly, "omniscience" leads to a problem with free will. One can know only what is true. So, does the fact that God knows what I am going to do next mean that I now have to do it and so I have no free will in the matter? How can God know the future unless it is already fixed, or is he "omniscient but not in the usual sense"? Theists do not like the idea that the future is determined, but they cannot imagine God is not all-knowing. So they restrict the idea of "all-knowing" from knowing all past, present, and future facts to knowing only what has actually already occurred or only the laws of nature or only our experiences, although any such restriction would limit an omnipotent being's ability to plan and provide for us. Or theologians argue that God has decided not to exercise his omniscience when it comes to humans. Once again he "self-limits." Similarly, with creation: how can an immutable, changeless reality create a world at all? Doesn't that act indicate a change? How can God be changeless and simple and yet choose and act? But if God changes and evolves, then under the reasoning of the Cosmological Argument doesn't he need a changeless source too? How can we think of "creation" that does not involve an act of effort and time? In the end, any answer to any of these questions will just be speculation on what makes sense to us and not necessarily reflect reality since we cannot check them. All of these efforts at reconciling our conflicting intuitions begin to sound suspect and futile.

Or consist suffering. Because Jesus suffered on a cross, many Christians find it easy to see God as suffering along with his creation. Indeed, some theologians have recently decided that the unexplainable suffering of millions of years of evolution is simply too horrible without God also suffering. Others are disquieted by the idea of the creator suffering. Theologians also have trouble with the conflicting ideas of God being timeless and immutable but also compassionate. Anselm and Aquinas argued that God "acts as if he felt compassion although he does not actually do so" (since an all-knowing reality would already know the eventual outcome of things and not react emotionally to the day-to-day course of events) or "has something akin to joy and delight in creation" but does not "feel" the way creatures do (since an immutable reality has no feelings) or "is touched by our suffering" but not in the "usual sense" or "experiences a torturer's joy at torturing but not in the way the torturer experiences it." But all of this begins to sound forced and very strange. How a god can have compassion or suffer and not feel them the way we do is a meaningless use of the terms. Theistic intuitions are simply conflicting: they want a god who is unchanging but also touched by love and suffering. Theists cannot have it both ways, but they do not want to give up either.

The concepts applied to the transcendent do point away from the worldly, but can they do more? Can we say anything positive about the transcendent with confidence? Well, it is transcendent—but it is also the immanent ground sustaining the universe. God is timeless and spaceless since he exists "outside" of the natural world—but this does not tell us anything about God's nature but only what it is not. Theists need a god who can act within the natural realm to answer intercessory prayers and to perform other miracles. Otherwise, all that is left is a deistic reality that created and sustains the natural world and at most set up laws and forces so that conscious life could evolve. Arguably, this source is even more awesome than a theistic god since it made a self-running universe that it does not have to intervene in to change things, but it is also remote, uninvolved, and unknowable. It can be admired and trusted, but since it cannot be interacted with, it cannot be loved or called upon for help; intercessory prayer would be useless; it simply leaves us on our own.

And then there are the how-mysteries. How does God create and hold the world in being? If God is the source of order, how does he regulate the universe? Or how could a transcendental reality act within the natural realm to perform miracles? How can a disembodied mind act to create, sustain, and intervene in this realm? Some theologians argue that there is a "top-down" type of causation—modeled on one interpretation of mental causation—that requires no gaps in the natural causal order: God can intervene without interfering with the laws of nature just as humans can enact our intentions without violating those laws.[66] But the use of mental causation is only a case of using one mystery to explain another, even greater mystery, and so it is not very helpful. Other theologians try at least to locate the "causal joint" within the natural realm where divine action performing miracles takes place. For example, today some theologians think God acts on the apparent quantum-level indeterminacies in order to bring about phenomena on the everyday-level, e.g., directing the course of evolution to lead to humans. According to the widely-held interpretation of quantum physics, there is a permanent ontological gap in the causal order at that level, and so there is a permanent home for a god to act undetectably by science—it will never be pushed out by further scientific findings (if the current theory remains standing). But major problems with this idea have been demonstrated.[67] The next generation of theologians may use an "unmanifested" dimension of the strings from superstring theory or something else from the then-current theory as the causal joint. But none of these attempts address the basic mystery of *how* a transcendental reality can act anywhere within the natural realm in the first place. Indeed, transcendental interaction with nature would be unique, if it occurs, and so it is hard to see how we within the world can have any idea about how it works. Another point noted in the last chapter expands the problem: the relation of the universe to the transcendent is unique, whether it is creation, emanation, or something else. Creation is not comparable to any relations within the world, and so all metaphors or analogies mislead.

In sum, the transcendent may be invoked to explain the "why" of things in the world, but we are still left with a host of how-mysteries. A tradition's revelation will explain why a miracle occurred even if we still have no idea how it occurred. Moreover, we are still left with major why-mysteries—in particular, why the transcendent exists at all. Advocates of the Cosmological and Ontological Arguments believe God is "its own cause," but few other theists are convinced. As discussed, the idea of how anything can be its own explanation is hard to make intelligible. And without an answer as to why God exists, a complete answer to the ultimate mystery of mysteries—why anything exists—remains as unanswered under theism as under naturalism.

Another basic why-mystery theists have is why God created at all. Did he have the freedom to choose, or was it done from necessity (e.g., divine goodness or joy automatically flows into a creation by emanation)? If the latter, then this act is not "personal" in nature—the reality could not have done otherwise. If the former, aren't his actions in the final analysis, if not arbitrary, at least in need of some explanation? Theists cannot feel comfortable with either horn of this dilemma. A "perfect" and self-sufficient reality would lack nothing and thus would have no need to create or for anything at all. Nothing finite, no matter how many worlds God creates and how valuable they are, can increase what is already infinitely valuable. In addition, why would a perfect being bother creating a world and then leave it alone? But then again, why would he have to get involved with it? As noted above, theologians have supplied possible reasons for creating—e.g., God wants to be known or loved, or God wants to experiment through evolution. But the problem is that we can only give reasons why *we* would create a world, and we have no reason to believe a transcendental reality would conform to our ideas and wishes. How could we know a creator's reasons? No creature can see "from God's point of view." We have to impute some human motive that sounds good to us. Any answer we supply would reduce God to a being with a limited imagination. We may placate our need to know by making up some reason that satisfies us—"Why else would a god create the universe?" Theologians may ask what else can we do but use our reason. However, it would be more honest simply to accept that we have no idea why a creation has occurred (if one did).

The uniqueness of creation also applies more generally to God's nature: it will always remain unique from our point of view. This means there is nothing for our reasoning to work with, since our reasoning (including what we think is *a priori* true, logically impossible, "self-evident," or a deductive certainty) depends upon how we think about the world. What makes sense to us should not be the criterion of what is real. As noted in the Chapter 4, nature is constantly surprising scientists and proving itself to be counterintuitive—why should we think we have a clue about the fundamental ground of reality, especially when we cannot examine it at all? We are finite beings with limited brains formed through evolution, with limited mental abilities and perspectives, and so what makes us think our intuitions must be reliable in these matters? Is it no more than our need to think we know?

This means we face the prospect that there is no way to anchor even metaphors for the transcendent in worldly experiences. The transcendent is unique, and thus there are no grounds for comparison between the finite and the transcendent—even calling it "unique" is misleading since it makes it one thing set apart from other things.[68] Using the human mind as a model for God is little more than a sophisticated version of seeing God as an old man with a beard in the sky. We are left with Xenophanes's famous remark that if oxen and horses had hands and could draw, oxen would draw the shapes of their gods like oxen and horses like horses. It is us "writ large," to quote the atheist Ludwig Feuerbach. Our conceptions of God must still remain only projections of human nature. As the theologian John Haught asks, is the propensity to think of God as personal perhaps more a manifestation of our immaturity than a realistic appreciation of the inexhaustible mystery of reality?[69] Any metaphors end up being qualified into oblivion—e.g., God is good but its not our type of goodness, God is a person but not like any natural being, God has a mind but not like ours, or God caused the world but it is not like any natural cause. What is the difference between saying a term is being used in "a nonordinary sense" or in "a unique sense unlike any other" and saying the term is meaningless when applied to the transcendent? We are left with the *via negativa* as a way to point to away from things of this world, and nothing else.

Between Mystery and Religion

The conclusion of this discussion must be that even with religion we do not have the answers to the big "why" questions, and in fact we end up with the why-mysteries intact, only now with more how-mysteries. But mystery is permanent in religion since we cannot conduct experiments on the transcendent to answers our questions. Mystery immediately surrounds all the conceptions and attempts at comprehension, and thus mystery is central to religion in a way it is not to philosophy or science. Reinjecting more of transcendental awe and mystery into religion may be needed to keep religion vital. Placing more emphasis on the human inability to comprehend the transcendent would keep the "living" theistic "god of Abraham, Isaac, and Jacob" from becoming only the "dead" deistic "god of the philosophers and scholars." (Connected to this is also the need for religious experiences: once God becomes only an explanatory construct rather than a presence, any religious sense of awe is gone, even if the construct is a transcendental one.)

Granted, religions always need conceptions to direct people's attention to the mystery behind the world, but metaphors can trap as well as lead. The question is what is the center of attention. The religious are stuck with using concepts from the realm of the familiar, and thus the familiar may control a mystery they confront. In the end, our conceptualizations close us off to the mystery. Even if the religious want to argue that in the final analysis the transcendent is not "wholly other" and in fact we know something of it by revelation or other experience, there is still a problem: the religious will assert that the transcendent has a depth that is never revealed and thus that humans can never fathom completely. In short, in Augustine's phrase, "if we could comprehend it, it would not be God."

Thus, in religion there is always tension between the pull of the mystery of the transcendent and the control of conceptions. Religion manifests both the presence of mystery and our need to explain—in effect, theists want both "transcendence" and "closure" by affirming both a transcendental mystery and the applicability of certain concepts. It is the one enterprise where mystery is absolutely central and where one would think that mystery and humility would be emphasized—that there is more to the transcendent than we could possibly understand—but religion also ends up countering mystery. The tension leads to paradoxes that both affirm and deny the conceptualizations. Theists cannot worship without some conception of the object of worship. And the religious naturally focus on what they know for worship, prayer, and the rest of a religious life, while acknowledging that there is more. But, as Augustine asked, how can the faithful worship what they understand? Mustn't the object of worship always be more than that, an ineffable mystery? Do theists in effect worship only the rational image or the "something" behind it?

Indeed, the religious may downplay mystery altogether in light of their need to make concrete commitments. If the transcendent is literally "wholly other," then there is only mystery, and an utterly mysterious reality could play no role in a religious life. Hence, it would be irrelevant—a *deus otiosus*. Salvation in any religion also does not depend on comprehending or investigating the mystery of the transcendent or the theistic how-mysteries. Few Christians would disagree with Paul when he said that if they understand all the mysteries but have not love, they are nothing (I Cor. 13:2). A religiously adequate image of the transcendent becomes more important, and the religiously committed may not remain open to the full mystery of a transcendental reality since they may deal with only a conception of it. The question becomes: can the religiously committed remain open to the mysteries of a transcendental reality, or do they end up substituting our concepts for the reality? Thus, the religious may end up only paying lip-service to the core of transcendental mystery at the heart of things.

Within Christianity, Eastern Orthodox theologians emphasize mystery more and "the foolishness of pretending to have comprehended God."[70] But Western Christian theologians seem to be on the side of closure, the enemy of mystery. Even if they acknowledge that God is a mystery, they plow ahead anyway. It is rare that they even mention mystery at the conclusion of their works after chapters of detailed analysis and attempts at explanation. They put explanations and conceptions in the place of mystery. The "death of God" theology, i.e., the end of the image of the transcendent in terms of a personal being among beings, may have removed a conception out of step with today's beliefs, but it did not lead to a rebirth of mystery. It reflected more the liberal theologians' adopting the loss of transcendence—and hence religious mystery—that is prevalent in academic culture.

Thus, unlike in the past, today mystery must actually be constantly reaffirmed in the religious life. If the religious let symbols predominate in their consciousness, the mystery is dispelled, and a spark vital to religion is lost. The mystery of the transcendent is domesticated, and the belief-component of faith becomes a matter of intellectual assent to propositions. With a loss of a sense of transcendence, liberal churches in Christianity, as Huston Smith notes, dig their own graves: without a robust, emphatically theistic worldview to work within, these churches have nothing to offer except rallying cries of "be good!"[71] To counter this, theologians today could become defenders of transcendental mystery, even if speaking of God's majesty seems totally out of place in liberal circles in American society today. They would have to concede that we humans can know little, if anything, of the transcendent's nature. Theologians work at showing that the beliefs of a religious way of life are not inconsistent or that it is not irrational to hold them, but they should concede that they are not providing new insights into any transcendental reality. Less energy would need to be devoted to making a mystery palatable to the faithful—a more religious response is simply to accept it and to remain humble before the unknown sources of existence. The role of "true theology," Ian Ramsey reminds us, is "to stir vision, and to remind us of the mysterious."[72]

A more minimal theology is possible. For example, science can never in principle give a complete picture of action in the world (since it deals only with abstract causal conditionals), and thus there is room for God's actions in the world. Naturalistic metaphysics, not science, conflict with religious commitment to the transcendent, but one can rationally reject naturalistic metaphysics, and thus theists can admit their ignorance on theistic how-mysteries without a crisis of faith. The how-mysteries could then be left as just that—mysteries. So too, theists could see "creation" in terms of a sustaining source rather than as an origination-event, thereby at least removing the focus on the how-mysteries connected to questions of a first event and the origination of the universe. They can "leave the details to our creator." However, a theology may become too thin: the loss of content can lead to a loss of commitment. Can theologians come up with an emotionally satisfying story when the view of the transcendent is stripped of all of our attributes? That may be too abstract for any actual religious way of life. Moreover, there are limits to how much accommodation be can made to science. To theists, neuroscience cannot even in principle provide a complete picture of a human since something in us survives death and is not purely natural. So too, humans cannot be seen as mere accidents of evolution; rather, we are somehow planned into the process from the start. Thus, neo-Darwinian theory must be wrong in at least one important respect—the role of random mutation.

Two further problems also must be noted. First, one cannot simply yell "Mystery!" any time someone raises a problem with a religious doctrine without hurting a religion's credibility. The religious cannot advance religious doctrines and then when asked to explain them glibly say "Oh, we don't pretend to know what that means. It's a mystery! Just accept it. You'll understand it after you die." If the religious themselves cannot understand their own doctrines, they are admitting they are not being rational in the way of life they follow. (Nor is clear what it even means to say we believe something we admit is ultimately unintelligible.[73]) Claiming a belief is a mystery might make it impregnable to criticism, but it does not salvage doctrines from the problems both believers and nonbelievers have with their intelligibility and believability. If a religious doctrine cannot ultimately be understood, the religious may have to admit they literally do not know what they are talking about and are just following a tradition.

Second, religious mysteries of the transcendent must be distinguished from the faithful jumping to the conclusion "God did it!" for every unsolved *scientific* mystery. Responding to every how-mystery with "God works in mysterious ways" not only sounds flippant but does not explain anything. Falling back on a claim of "God did it" to explain the emergence of life or consciousness or whatever is the current gap in our scientific picture of the world is not the same as accepting the underlying mystery of the presence of the entire world. The former can rightly be called "mystery-mongering," and it stifles empirical investigation. It also does not relate to the religious sense of the mysteries of the transcendent. Scientific how-mysteries must be distinguished from the religious mystery of the transcendent and the meaning of things, just as natural and religious senses of mystery and awe must be. (There are two possible qualifications. First, if the transcendent guides the course of events as part of a purpose or design for the world, there will be a religious how-ness mystery to the course of natural events, even if the guiding action is not detectable by scientific analysis. Second, miracles, if they occur, would fall through the scientific nets since science only deals with repeatable acts, not unreproducible acts which the acts of a free transcendental agent would be.[74])

Any closure of mystery in religion is weaker than in science since the transcendent cannot be investigated to answer questions. We are not learning more about the transcendent, if it exists, although theologians do adjust their view of the transcendent as our knowledge of the world expands and our values change. So too, in any era and culture theologians can always find metaphors and analogies to use for the transcendent that quiet believers' need to think that they understand. Theists can utilize what is available to give expression to the transcendental mystery, but the limitation of all metaphors must be remembered. Metaphors may tell us more about the culture from which they arise than any transcendent reality. And again, any conceptualization may also end up standing in the way of experiencing a reality and accepting a mystery. All religions want to remain connected to images from the past, but the faithful may be trapped by what was once wisdom.

The beginning of wisdom in religion according to Hywel Lewis is to appreciate that we may talk of the transcendent in terms of "explanation," "source," or "ultimate perfection" and "know that these are the appropriate terms to use, and yet know that in using them we are speaking of 'something we know not what.'"[75] He also argued that the religious life requires a fuller god than one that results from accentuating mystery. Indeed, those who rely on revelation see the stress on mystery as one-sided, if not totally inappropriate. Still, the depth of the mystery of the transcendent cannot be denied if a religious life is to remain vital: reducing the mystery to a comforting old man in the sky or some other familiar, manageable image saps it of its power. The religious thus will have to balance accentuating both conceptualizations and mystery. Whether this dance of openness and closure can be resolved is doubtful—as soon as one is accented, the other demands attention.

Nevertheless, the bottom line is that the religious will have to recognize that we know very little, if anything, about any transcendental reality—if it exists at all. It may not be spiritually satisfying, but we are doomed to live with it. Theologians will continue to pontificate on "the mind of God," but we should see their efforts as the very human speculations they are. This should lead to more tentativeness and humility about one's own religious pronouncements and more tolerance toward other religions—indeed, the openness resulting from the lack of certainty could be liberating. But this, of course, is easier said than done. (The same can be said for confirmed atheists; agnostics can be more open to the big questions, while atheists too think they have closure.) If one thinks that God has revealed himself to one's own tradition through its scriptures and leaders, one naturally concludes that one's tradition has the answers to the religious mysteries—in fact, it would obviously be wrong to compromise or even respect other religious claims if one is convinced that one has the word of God oneself. This sense of certainty unfortunately has been accompanied throughout history, not only by a sense of confidence and serenity, but by authoritarianism and fanaticism. It can polarize a society and the world, and it has led to religious wars.

Notes

1. Geertz 1973, pp. 140-41.
2. Peter L. Berger, *The Sacred Canopy: Elements of a Sociological Theory of Religion* (Garden City: Doubleday & Co., 1967), pp. 56, 100.
3. The Buddha opted not to discuss transcendental realities for soteriological reasons—to avoid creating images in our minds that may block our becoming selfless. This leads many scholars to declare that early Buddhists denied such realities. However, Buddhists quickly adopted a less negative approach toward transcendental realities.
4. Hick 1989, pp. 556-69; Hick 1999, pp. 47-73. Even if the world is destroyed or a natural disaster wipes out most of mankind, members of the Abrahamic traditions have a ready-made explanation: we are being punished for not being faithful, as happened once before with Noah's flood. This would make even the destruction of mankind seem justified in their eyes, and hence it is bearable no matter how horrible it is.
5. Wittgenstein 1961, Prop. 6.52. He adds there is no legitimate *question* left after science and that is the answer. The "yearning for existence" lies outside what can be said.
6. Steven Weinberg, *The First Three Minutes* (New York: Basic Books, 1977), p. 154. He later qualified his statement by saying "I did not mean that science teaches us that the universe is pointless, but rather that the universe itself suggests no point." Weinberg 1992, p. 255.

7.   Walter Stace argued that we do not typically live by truths. Rather, nature deliberately puts illusions into our mind to induce us to go on living, and the Great Illusion is "the religious illusion that the universe is moral and good, that it follows a wise and noble plan, that it is gradually generating some supreme value, that goodness is bound to triumph in it." "Man Against Darkness," in Klemke 2000, p. 92.

8.   Richard Dawkins, *A River Out of Eden* (New York: Basic Books, 1995), p. 133.

9.   The data from the comparative anatomy of animals is conclusive evidence to neo-Darwinians of our descent from common ancestors. Non-Darwinians can argue that this only indicates biotic laws at work in nature and not a common descent or that a god chose to design with a limited range of options. But theists have a problem: why would a god, who had no constraints on his choices, just happen to choose to create in a manner that looks exactly like evolution? Unless one is willing to accept this amazing coincidence, the data does suggest natural forces are responsible for the history of life.

10.   In Richard Dawkins, *The Selfish Gene* (New York: Oxford University Press, 1976), p. v.

11.   Can *beingness itself* be a source of *value*? It obviously has value to us since we would not exist otherwise. But, as Milton Munitz says, beingness ("boundless existence") has no features that can make it a center of infinite value in itself or a source of meaning or purpose to finite entities. Being has no meaning or purpose. Munitz 1986, pp. 279-81; Munitz 1993. Theists would agree, but they see a purpose to the universe endowed by a god.

12.   Ford 2007, pp. 225-26.

13.   A story from India's great classic the *Mahabharata* can be seen as a similar response to our situation (although its actual context involves the acceptance of life after death and transcendental realities): a man is chased through the jungle by wild elephants and jumps into an abandoned well to escape; he clings to a vine, but the herd is still outside so he cannot leave; he looks down and sees a snake at the bottom waiting for him to drop; he looks up and sees mice chewing away that the vine he is holding; over the well, there is a beehive and angry bees are now swarming about him, stinging him; through all this, a drop of honey falls from the beehive and the man catches it on his tongue—the secret of life is to be able to enjoy that drop of honey in those circumstances.

14.   Wittgenstein 1961, Prop. 6.4312.

15.   Not just existentialists share this view. See Thomas Nagel, "The Meaning of Life" in Klemke 2000, pp. 6-7. That life is deemed "absurd" in this sense only means that it has no ultimate explanation, not that it is "irrational" or "not worth living."

16.   The theologian Gabriel Marcel (1950-1951) drew the distinction between a problem and a mystery: a problem can be resolved, but a mystery envelops us and thus we cannot get any distance from it in order to make it into an objective problem; hence, a mystery cannot be resolved but is something that can only be lived. Mysteries involve us so personally that we cannot extricate ourselves from them and confront the issues as objects. Such mysteries include God, death, evil, suffering, and love. However, he saw the basic mystery as an *ontological reality*—the transcendental god of theism. John Haught, another Catholic theologian, follows Marcel on this point (Haught 2007: 23, 25-26).

17.   Otto 1950.

18.   Ibid., p. 28.

19.   Kant in *Religion Within the Limits of Religion Alone* does mention the "holy mystery (*mysterium*)" of religion as something that cannot be made known publically "and it may well be expedient for us merely to know and understand that there is such a mystery, not to comprehend it."

20.   Otto 1950, pp. 26-27.

21. Eliade 1959. As a phenomenologist, Eliade tried to remain neutral on whether the sacred is in fact real or not. The sense that the sacred is real is, he argued, an innate structure of our mind, present since the earliest cultures, and thus humans are *Homo religiosus* even if we have lost that sense in modern society. But he remained neutral on whether this structure of consciousness reflects something real outside our consciousness, while Otto thought the idea of the holy was divinely implanted in us by an objective reality.

22. Rahner 1978; Bacik 1980.

23. See Bolle 1987.

24. *Majjhima Nikaya* II.427. These questions may be unanswerable in principle or they may in fact be answerable but are left unanswered solely for soteriological reasons. Either way, Buddhists contend that we are wasting our time dwelling on such metaphysical matters.

25. *Majjhima Nikaya* I.63.

26. Haught 1995, pp. 181-82.

27. Some early Christians accepted the idea of rebirth (e.g., the theologian Origen), arguing that we have to go through every type of experience before we can be saved.

28. Theologians draw a distinction between a *theodicy* (which advances God's *reasons* for permitting evil and hence explains it) and a simpler *defense* (showing that God *may have* morally sufficient reasons to permit evil; evil is then compatible with the existence of God, and so there is no logical contradiction between faith and evil). But with only the latter, theists would be admitting they do not know if there are reasons—they would be banking on an impenetrable mystery and alleged moral reasons we have no access to. An explanation that quiets our questioning would be missing, and the basic mystery would remain intact.

29. There are passages in the Old Testament that state Yahweh is the cause of both good and evil (Isaiah 45:7; Lamentations 3:38; Amos 3:6). Some Biblical scholars suggest that God has a dark side and has to be held accountable for his part in the "slaughter-bench" of human history. See, e.g., John K. Roth, "A Theodicy of Protest," in Stephen T. Davis, ed., *Encountering Evil: Live Options in Theodicy* (Atlanta: John Knox Press, 1981), pp. 7-37. The standard position in Judaism is not that God is "all-loving" but that he is *just* toward all—he does not love all humans equally, but he does dispense his justice equally to all.

30. Haught 2000, p. 55. John Hick also accepts that suffering ultimately is unjust and inexplicable, but he believes there is positive value to this mystery and thus that Christians should not give up their faith. Hick 1966, pp. 371-72.

31. See Hick 1999, pp. 240-52 for the problems connected to death.

32. Christianity and Islam are the two largest world religions, each with under 25% of the world's population as nominal members. Thus, whatever is the "true religion," more than 75% of the billions of people alive today do not belong to it and are thereby, under traditional doctrines, condemned to suffer eternally in hell. Add to this those who are nominal members of the true religion but who still die unsaved. Islam is more generous to other theists, but many treat Christians as polytheists worshiping three gods rather than one three-part god. Christianity has more exclusivists than Islam, i.e., those who affirm that there is no salvation outside of accepting Christ as our savior. Indeed, there is exclusivism even within Christianity: Catholics question whether non-Catholics will be saved, and fundamentalist Protestants question whether Catholics and liberal Protestants will be saved. All in all, under the traditional doctrines only a small percentage of humanity appears headed to heaven. God will let the rest be tortured for eternity by some of his creatures.

33. For more on religion, morality, and ethics, see Jones 2004.

34. William Lane Craig, "The Absurdity of Life Without God," in Klemke 2000, p. 43. He argues that there are no *objective* moral values if there is no god. "The Indispensabil- ity of Theological Meta-Ethical Foundations for Morality," *Foundations* 5 (1997): 9-12. Even ignoring the *Euthyphro* problem, it does not take a god commanding us to know that gratuitously torturing a baby is morally wrong: it is as objective a claim, and as true by definition, as "2 + 2 = 4"—we do not need a god to tell us the former any more than we need a god to tell us the latter. Not everyone may adhere to moral injunctions, but this does not change the fact that such torture is not other-regarding and hence by definition not moral. Craig also argues that we need belief in life after death and postmortem rewards and punishments to be moral. But, again, if we act solely out of self-interest (here, fear of hell or hope for heaven), we are not being moral. Lastly, Craig argues that there is no sense of moral duty or obligation or accountability without belief in God. This may be true, but this then would only mean then that morality is a basic value we either choose or do not choose—morality may not have a command over us, but this does not change its objectivity. The idea that morality has a demand on us may come only from a theistic position. Craig believes that anyone who does not believe in a postmortem reward or punishment is being "just stupid" to act any way but selfishly—for them, there is "no reason to be human." But does he really believe that (to use one of his examples) a mother who risks her life to save her children is being "just stupid" unless she believes in life after death?

35. The story of Abraham's willingness to sacrifice his innocent son Isaac (Gen. 22) is a case of *religious* requirements overriding basic *moral* requirements. Some dismiss the moral problem by saying that human sacrifice was common in that part of the world at that time. But then why did the angel praise Abraham for his willingness to go through with it?

36. See, e.g., Philip L. Quinn, *Divine Commands and Moral Requirements* (Oxford: Clarendon Press, 1978). William Alston argues that a more plausible form of the divine command theory is one in which our *obligation* to be moral but not moral *goodness* is dependent on God.

37. Theists cannot get around the issue by definitions. They cannot say that the source of the world by definition must be moral. True, a theistic god by definition is indeed worthy of worship and hence moral, but the question then becomes whether such a god actually exists—e.g., how do theists know the source of the universe is not a nonmoral deistic reality or an evil creator?

38. Saying "God is goodness itself" introduces the problem of whether "goodness itself" is personal in nature or not an active agent. Either way, it becomes simply another way of saying God is purely moral in character.

39. See Stenger 2008, pp. 202-3. There were both religious and secular sources.

40. "God's will" can also lead to a fatalism or to being an entirely meaningless claim: whatever happens could not happen without it being God's will, and so whatever happens is by definition God's will.

41. See Daniel Howard-Snyder and Paul K. Moser, eds., *Divine Hiddenness: New Essays* (Cambridge: Cambridge University Press, 2002); for the other side, see J. L. Schel- lenberg, *Divine Hiddenness and Human Reason* (Ithaca: Cornell University Press, 1993).

42. The meaning of life cannot be merely to help one another with our suffering. This would leave the questions of why there is suffering and why anyone is here in the first place totally unanswered and thus not give the complete picture of the meaning of life. See Baggini 2005, pp. 58-71.

43. That is, Christianity and other theisms are inherently individualistic on this point: each person is in some way created by God, and everything is a matter between each person and God. We have to be *commanded* to love our neighbor because this is not a natural

response under such a worldview. Under naturalism, we are all part of the same naturalistic whole (including animals and all of nature). We realize we are not on our own—we are all connected, and so cooperation is necessary and natural. A pure selfishness does not follow from a naturalistic metaphysics. We cannot think only "What's in it for me?" without being out of step with the way nature really is. There may be things worth dying for. On the other hand, since we do exist, we must assert some self-interest—being totally selfless is also out of step with being a part of our natural universe. Of course, we can go beyond self-interest into being competitive and selfish, but we cannot deny our need for others. In short, we cannot consistently look at things only for what is in it for ourselves. Thus, under naturalism concern for others is a natural extension of concern for oneself.

44. If altruistic behavior is established in the animal kingdom, does that prove that morality is grounded in nature and does not need a transcendental source, or does it only prove that God designed nature too and animal altruism is only a manifestation of his loving divine will? How could we determine which option is correct?

45. The question for us then becomes why should we be moral? Why should we adopt a moral value-system in circumstances where acting selfishly would be to our benefit? Why should we sacrifice our self-interest for the welfare of anyone other than loved ones when it costs us? Theists again want "Because God says so, and you'll go to hell if you are not moral" to be the answer. But this answer cannot in principle be grounds for being moral: it only reflects self-interest, and this conflicts with the underlying concern for others that is the basis of morality. The naturalists' attempts to ground moral concern in nature usually also reflect self-interest—e.g., evolution built in a tendency into beings to fairness because cooperation and honesty are better in the long run for both the individual and the species. Even the "enlightened self-interest" of "ethical egoism" is still a form of self-interest: it may involve a concern for justice and the welfare of others but only as a means to one's own good, and by definition morality involves sometimes giving up some self-interest in favor of others' interest. (The question is why we act: if we act out of concern for others and also gain a benefit as a side-effect, we are still being moral.) In the end, the question "Why should I be moral?" will always be difficult to answer if one is asking for a self-interested reason to give up some self-interest. But obviously, trying to answer it in terms of "What's in it for me?" will be self-defeating: we are looking for slef-interested reasons to give up some self-interest. Ultimately, the only answer can be grounded in the fact that the other people (and animals) are capable of suffering and thus warrant a moral person's concern, not in any self-regarding reason or in God's commands. In short, the answer to "why help others?" is simply "Because they need it." Being moral becomes a basic value we either accept or reject. It becomes a fundamental decision on what sort of person we want to be.

46. Wittgenstein 1961, Prop. 6.41.

47. See Cooper 2003, pp. 136-42 for a defense of the idea that life must be mysterious.

48. The question of the meaning of life does not arise as long as we are unself-consciously and fully engaged in the act of living, taking the world for granted; our natural state is one of faith, and most of us never reach the point of self-consciously asking "what does life mean?"—the question only arises when we stop, look around, step back, and say to ourselves "Something is wrong here!" (Ford 2007, pp. 251-52).

49. See Arthur Peacocke, *Theology for a Scientific Age* (Minneapolis: Fortress Press, 1993), p. 360 n. 13.

50. The natural vagueness of language has led some philosophers and linguists to claim that *all* language is metaphoric. However, the fact that even a simple term like "cat" may not have exact boundaries in its literal usage does not change the fact that there is a literal sense to the term and a symbolic one—there is a difference between extending the

term "cat" to a new species and calling a human being a "lion." Analogies and metaphors are parasitic on literal uses and thus remain different than literal uses.

51. Paul Tillich, *Systematic Theology*, vol. 1 (Chicago: University of Chicago Press, 1951), pp. 238-39.

52. Hopkins 1990. Nicholas of Cusa did say that we know God is one and "being itself" in which all opposites coincide—God is the undifferentiated, absolute maximum that precedes all differentiations and unites all opposites as their cause. But he ended up affirming positive characterizations of God as loving and so forth, based on Jesus's teachings.

53. Haught 2000, p. 34. As a logical matter, to avoid circularity natural theologians would also need independent evidence of a designer's powers and plans; without an explanation of why God would create the universe we see, postulating a god is only speculation and that limits its power to explain anything.

54. See Ferré 1984.

55. See Jones 1993, pp. 149-66. The father of the Ontological Argument, Anselm, believed that "that than which nothing greater can be conceived" is a reality that cannot itself be conceived. He did not want to make God into "the greatest conceivable reality" or to tie him to any conception. Thus, God still ends up greater than can be understood or conceived and thus is beyond conception and hence ineffable. Nevertheless, the Ontological Argument is still based on comparing the *greatness* of conceptions and thereby ties the being of the argument to what we humans can imagine or conceive. We must have some conception of this reality to have a way of comparing its greatness to that of other beings. The argument is centered on our conceptions—that reality than which "*we cannot conceive* anything greater"—and not on an inconceivable transcendence. If God is in fact beyond our conceptions, we are putting a conceivable being in his place for the argument to proceed, and thus the argument is not about the true reality at all. In any case, the argument once again makes our conceptions the focus of attention.

56. If one adopts a postmodern stance, the entire enterprise of proving one's beliefs is rendered pointless and groundless. Even if we reject postmodernism, showing merely that participation in a given religion is not irrational or that a belief is reasonable to hold significantly lowers the bar from having to prove that the religion's doctrines are true —indeed, it is hard to show that the general idea of a creator god is irrational, especially when one can always modify a god's properties if a contradiction between, say, omnipotence and omniscience is suggested, or one can adjust one's religious beliefs to be compatible with the scientific and other beliefs of the day. At least liberals have no problem adjusting their beliefs to fit objections and thus can claim that the resulting beliefs are "rational." That is, a belief may not be confirmable, but if one side cannot show that their opponents' position involves a logical contradiction or is clearly contrary to the best evidence, then it is at least *not irrational* to hold their belief. We do not need compelling evidence to be rational. An intermediate position is to argue that theism is "more reasonable" or "more probable" than naturalism or vice versa. But arguing the claim that a theistic god is a better explanation of the existence of the world or the source of order is difficult in light of the problems discussed in the last chapter. And one's metaphysics may determine what counts as "reasonable." In the end, one can believe in a transcendental reality if one is so disposed or reject it if one is so disposed—everyone who at least addresses the issues is rational. Of course, not all theists or naturalists have given up the fight to either prove or disprove the existence of God. Naturalists point to logical problems in the concept of God (to show that ultimately the concept is incoherent, no matter what believers may think) or to empirical evidence that make the idea seem less than probable. To them, the preponderance of available evidence supports atheism and not either a transcendental position or the openness of agnosticism.

See, e.g., Martin and Monnier 2003, 2006. But theists can always make adjustments to beliefs to get around the logical problems. Indeed, *ad hoc* adjustments are fairly easy to think up. To give two examples: a perfect creator cannot exist because whatever he creates would be perfect and the universe is not perfect (but a perfect creator could have reasons to create something less than perfect; or maybe the universe is perfect for some hidden reason); a transcendental being cannot be omnipresent since anything that transcends space cannot exist anywhere in space (but a ground of being is both transcendent and present in all it sustains; being present everywhere, it can act in specific situations thereby causing miracles). In general, it is hard to prove a logical contradiction. Consider the popular argument against omnipotence: can God create a rock that is so big that he cannot move it? If he can make such a rock, then there is something he cannot do (he cannot move it); or if he cannot make it, then again there is something he cannot do (he cannot make it)—either way, there is something he cannot do, and so he is not omnipotent. The simplest reply is that God could make a rock that fills all of space and so there would be no place to move the rock, thus rendering moving physically impossible and making the whole concept of "moving" logically problematic in this context. That God cannot do the logically impossible (e.g., if we define 2 plus 2 to equal 4, then even God cannot make 2 plus 2 equal 3) has not been taken to be a limitation on omnipotence ever since William of Occam rejected it, and God not being able to move something when moving it is impossible is not far removed.

57. See Bacik 1980; Haught 1986.

58. Marcel 1950-51; Rahner 1978.

59. See Macquarrie 1973. Gordon Kaufman is one contemporary theologian who makes much of a sense of mystery. But this is misleading. Although he attempts to disown "religious naturalism" (Kaufman 1993, pp. 254-63), he reduces "God" to merely a symbol for the forces controlling nature and human history. He then treats scientific wonder about the how-mysteries of nature as if it is religious awe and mystery when no transcendental mystery is involved. The "ultimate mystery" becomes a matter for science. Thus, he can retain the symbol "God" and the concept of "religious mystery," but he empties both of anything like their usual content. An unfathomable depth to nature is not the same as the transcendental mysteries. Science remains a matter of the how-ness of nature, not of the why-ness or that-ness. Science may well increase our sense of the mystery of reality "horizon-tally" (within the natural realm), but this cannot be converted into the "vertical" mysteries of religion (the relation of the natural realm to a purported transcendent source).

60. For her insider's account, see Karla McLaren,"Bridging the Chasm Between Two Cultures," *Skeptical Inquirer* (May/June 2004): 52.

61. Some conservative Jewish rabbis see the Holocaust as a punishment from God for the apostasy of European Jews. This sense of punishment can be seen without a theistic god: some Japanese think their defeat in World War II has the result of Japan's "group *karma*."

62. See Terrence W. Tilley, *The Evils of Theodicy* (Washington: Georgetown University Press, 1991); Sarah Pinnock, *Beyond Theodicy: Jewish and Christian Continental Thinkers Respond to the Holocaust* (Albany: State University of New York Press, 2002).

63. Some religious naturalists are antirealists, but they differ from antirealists in science: they take religious language that appears to describe a transcendental reality as only about an *attitude* or *orientation toward the natural world*.

64. Hick 1999, pp. 13-17. He suggests that there will be an *eschatological* resolution of the mystery in an afterlife.

65. Also note that such a voluntary self-limitation of omnipotent power would not mitigate a god's responsibility for natural suffering since this self-limitation is not truly binding but could be withdrawn at will.

66. See Arthur Peacocke, *Theology for a Scientific Age* (Minneapolis: Fortress Press, 1993); John Polkinghorne, *Science and Creation* (Boston: Shambhala, 1989).

67. Peter Hodgson, "God's Action in the World: The Relevance of Quantum Mechanics," *Zygon* 35 (September 2000): 505-16; Nicholas T. Saunders, "Does God Cheat at Dice? Divine Action and Quantum Possibilities," ibid., pp. 517-44.

68. See Munitz 1986 for a statement of this position.

69. Haught 1986, p. 128.

70. To the Eastern Orthodox, God's essence (Gk., *ousia*) is transcendent, hidden, and utterly beyond all understanding. God is known through the uncreated operations (*energiai*) in his work, like gravity is known through its effects but not in itself.

71. Huston Smith, *The Soul of Christianity: Restoring the Great Tradition* (San Francisco: HarperCollins, 2005), pp. xx-xxi.

72. Ramsey 1964, p. 69. Robert Fuller (2006) wants to make wonder a central emotion of religion today. He sees all senses of wonder as ultimately pointing to a transcendental reality—the existence of an unseen, more general order of existence from which this world derives its meaning and purpose. Wonder potentially wards off literalism and authoritarianism in religion and leads to a life attuned to the widest possible world of personal fulfillment. Naturalists, however, can see wonder as leading us to consider the possibility of causal powers behind things (and thus energizing the scientific quest) but not leading us to powers of a transcendental sort. Both groups can see wonder leading to contemplating an unseen order of things, but only the religious see this as naturally leading to something transcending the natural order. (Also see Miller 1992 for a postmodern treatment of wonder as necessary for an openness to reality that is necessary to gain understanding. But he also thinks that philosophical wonder about being leads to religious awe.)

73. To say "I don't understand it, but I believe it" is irrational. It is rational to admit there may be more to the transcendental than we can understand. And it rational for scientists to observe subatomic phenomena through experiments and thus believe that such phenomena exist even if they do not understand how the phenomena could exist. But it is not rational to admit that some claim cannot be *understand* (i.e., is unintelligible) and still to claim to believe it.

74. Miracles, i.e., the transcendent's intervention in the workings of the natural order, are a necessary part of theism but not part of deism. As discussed in Chapter 4, science is about discerning recurring causal relations—if conditions x exist, then phenomenon y will occur. Miracles do not contradict this since the natural conditions that scientists study for their conditionals would not be the set of conditions present in a miracle—God's action would constitute part or all of the causal conditions. In short, miracles would fall outside scientific study. A miraculous event is not uncaused, but its cause is not controllable and thus is not repeatable by humans—we cannot force God to repeat his acts—and therefore is simply not open to scientific study. And if the consequences following from these new conditions obey the causal laws, then no causal laws would be broken by God's actions. Scientists *qua* scientists thus can make no pronouncements whether miracles do or do not occur. In his recent study of "special divine action," Nicholas Saunders concludes that Christian theology is in crisis on this point but that theists should continue to assert that God is active in the physical world despite not knowing how it is possible. Saunders 2002, pp. 215-16.

75. H. Lewis 1961, p. 237.

— 7 —

Religious Experiences

Fire! God of Abraham, God of Isaac, God of Jacob. Not of the philosophers
and the wise. Certainty, certainty, heartfelt, joy, peace. Joy, joy, joy, tears of joy.
— Blaise Pascal

All that I have written seems to me like so much straw compared with what
I have seen and with what has been revealed to me today.
— Thomas Aquinas

In effect, Chapter 5 was about the god of the philosophers and Chapter 6 the god
of the theologians. This chapter is about the god of the prophets and mystics. The
switch is away from an explanatory creator or designer god to the "living" god
experienced by prophets and mystics. Without some claim to have actually *experienced* the transcendent, religion's factual claims are no more than speculative
metaphysics. But these experiences raise important questions for the issues at hand.
Do religious experiences, if genuine, answer any fundamental why-mysteries of
reality? And do they remove how-mysteries or actually only add more?

Types of Religious Experiences

The notion of "religious experiences" is notoriously vague. Some people use it
describe all of their religious way of life or the cumulative religious history of
mankind. Theologians often describe faith as a "religious experience" in order to
make religion seem more empirically-based or even scientific. Prayer and worship
entail a belief-commitment that a god exists, but that does not make them *experiences* of the transcendent. A sense of overwhelming awe or astonishment at the
wonders of nature or at the mysterious depth to nature revealed by science or at the
mere existence of the world are other types of experience that are often labeled

237

"religious." However, these senses are only philosophical or scientific wonder. Pondering the question "why is there something rather than nothing?" may evoke the response "this all must be the creation of a god!"—this would indeed be a "religious experience," but this reaction is not empirical support for what is ultimately a claim about a "god of the philosophers" since it is simply a feeling of what we think must be the case. "Seeing" the world as a visible sign of an invisible reality also is not an experience of the latter but a result of reflecting on these matters. What many see as "signals of transcendence"—order, beauty, truth, morality, perfection, self-consciousness, a sense of meaning, a sense of community, a sense of absolutes—may reveal more about ourselves than reality and cannot be counted as evidence of any transcendental realities.

Theologians also exploit the notion of "intuition" to make our thought into a "religious experience."[1] For example, they speak of the "deep-seated human intuition" that death is not the final word. But this is not any sort of experiential claim but only an expression of belief and hope based at most on some deep thinking about the issue. It is merely a product of our thought and is not an experience of a transcendent presence, power, or reality. It is a matter of suspecting there is some reality out of a sense of not knowing, not an awareness of a reality. It is based only on our own reasoning, and calling it an "intuition" or "insight" cannot change that. An intuition may merely embody our certainty about our own speculative ideas or hopes that something exists—it is not a "disclosure" of a reality. It is not any new experiential in-put or a new cognition or a new "mode of knowing." In short, it is a product of reflection, not the impingement of the transcendent on our consciousness. Some thinkers speak of our "thirst for transcendence," a genetic "religious impulse," a desire to find purpose or to escape our finitude, the conviction that our questions must have answers, or a longing for contact with "something more" as built into our very consciousness. But this would only place the sense of transcendence squarely in our desires or our natural evolution and not in any impinging reality. If anything, this would lead to the conclusion that our sense of mystery is *not* evidence of a transcendental reality since we would search for an answer even if there is nothing there no matter what.

Being confronted with a mystery is a result of inquiry. "Mystery" is a concept arising from our questioning, not an experiential one. The phrase "the experience of mystery" is ambiguous—being aware that we do not know something versus sensing a reality or presence that, upon reflection, we do not comprehend. But "seeing" the limits of our understanding is not an experience of any further reality. It only indicates that we are aware of a question that we have no answer for or may not even know how to begin to answer. Even being able to conceive that there may be "something more" beyond what we think is not an experience or evidence that something more exists. Nor is this being "pursued" or "summoned" by a Mystery, as some theologians say, as if mystery were itself an entity impinging upon us.[2]

It is only those religious experiences in which a transcendental reality would have to be present for the experience to be genuine that add to the discussion of genuine mystery, and only this category will be discussed as "religious experi-ences" from here on out. Surveys today show that a majority or significant minority of all classes of people do believe that they have in fact had some type of numinous or mystical experiences, e.g., awareness of a presence or power other than those felt in the natural world. Alistair Hardy in his study of the British found an aware-ness of a benevolent non-physical power that appears to be partly or wholly beyond the natural order and far greater than the individual self to be widespread; this leads him to argue that such experiences are deeply-rooted in human nature.[3]

If these experiences are genuine, they add to our stock of cognitive experi-ences. In the experience of seeing a tree, the tree is among its causal roots. Scientists may be able to identify the neurological basis for the sense-experience and how the tree affects it. But, since we are part of nature, how a transcendental reality can impinge on our consciousness introduces a new issue. It is a special class of experiences and perhaps involves a special mental faculty not active in other experiences. But the impinging of the transcendent would alter our consciousness in the way that a new idea or intuition does not. It also means that a genuine transcendental experience cannot be explained away solely in naturalistic terms of an experiencer's social, psychological, or physiological conditions.

Two types of such experiences are prominent in the traditional forms of the world religions: numinous and mystical experiences. With the numinous experi-ence, we sense the presence of a transcendental reality and power. (In addition, whether or not they are transcendental experiences, the visions in "near-death experiences" may be the source of our idea of an afterlife and of both heaven and hell.) There are different forms of awareness of the transcendent, from a vague sense of presence to being touched by it to being engulfed, as Blaise Pascal felt for two hours one night in 1654. But not all numinous experiences are theistic, nor are all theistic experiences a "feeling of absolute dependence" as Friedrich Schleier-macher described.[4] But note that joy is not the only element of such experiences: along with fascination, Rudolf Otto rightly included that the element of *terror and dread* in religious awe that was also present in early religions, although it appears absent in the world religions today (but some experiences on the mystical path may be described so). However, that the transcendent seems to initiate the experience is common to all numinous experiences.

Revelations

One way to end a mystery is for God simply to tell us the answer in a revelation. Revelations are a primary type of the numinous category of religious experiences. They are the transcendent's self-disclosures that "unveil" (*re-velum*) a mystery —pulling back the curtain that conceals the transcendent from the natural world and telling us something about the nature of the transcendent or how to live that we would not otherwise know. It is natural to think in terms of a personal being communicating to us, e.g, the angel Gabriel revealing the Qur'an to Muhammad. Many Mahayana Buddhists also take the Lotus Sutra to be revealed. They involve communications—specific messages, not a vague presence of the divine. They are what a transcendental being chooses to tell us about his nature or his purpose for us. They are communications initiated by the transcendent, and it is the communicator's claim to authority that makes the messages binding for followers. To the faithful, when a prophet speaks, it is God who is speaking.[5]

Some theists speak of a "universal" or "general" revelation in the sense that the world itself is a general revelation to all of humanity. So too, we are all allegedly aware of an all-powerful, all-knowing, all-loving ground of reality with which we can have a personal relationship. In addition, theists assert that God gives "special" revelations to members of select groups. The three Abrahamic traditions accept ancient Israel and its founders as the only such group in the ancient world. Christians see Jesus as a unique revelation of God himself in the incarnated form of a human—"the image of the invisible God" (Col. 1:15)—that is the decisive self-disclosure of God. Christians today still claim messages from God, but no new revealed doctrines have been accepted by all Christian traditions. Muslims accept Mohammed as the "seal" of the prophets, i.e., the final prophet whose revelations correct and complete all earlier revelations. Baha'is claim newer revelations, with their founder Baha'u'llah as the last of the great prophets. Mormons see Joseph Smith as restarting the tradition of revelations in the nineteenth century in America, with revelations continuing today. Specific messages imparting religious truths are communicated in special revelations—there is no such thing as an ineffable revelation. Revelation has to be propositional, unlike with mystical experiences: some concrete information about God or his purposes that the prophets could comprehend and express in their culture's language was conveyed to them.

Such revelations do address basic why-questions. For example, only through revelation would we know that the world is good (since it is the creation of an all-loving God). The messages communicated, along with mystical experiences, are the only experiential source for our knowledge of the nature of the transcendent, the meaning of the world, and what we should do. Theists believe that not all mysteries of the transcendent are revealed—only what God wants us to know is of himself and his purposes. Nevertheless, God is no longer a totally unfathomable mystery. The mere fact of revelation entails that the transcendent is not a deistic source but has the nature of a person—a being that thinks and freely chooses to communicate—and thus not "wholly other" even if ontologically he is totally unlike his creation. Thus, with revelation some why-mysteries are banished for the faithful but some remain. In addition, a major new how-mystery is introduced: how can God communicate with mortals? How can something utterly not of this world—nontemporal and nonspatial, or at least bodiless—*act*? How can it make a vision, dream, or audition or otherwise implant ideas into our heads?

Critics point to the multiple conflicting revelations throughout history (including those that continue to arise around the world today) as evidence that people are deluded in believing that any of them are from a transcendental source. The evil acts done throughout history by people who claim access to God also have not inspired confidence in the whole idea of "revelations." There is also the suspicious fact that revelations today usually tell the recipients what they want to hear rather than challenge them.[6] When Utah was denied consideration for statehood because of polygamy, Mormons received a revelation banning it. It must also be admitted that the revelations specific to one religious tradition look dubious to members of the other traditions. And it must be asked why all classical revelations are so ethnocentric. Why is there a "chosen" people? Why would the truly transcendent be concerned with only one fraction of the people of the world? (Revelations in regions of religious conflict, such as with the Baha'i, often are more universal.)

If this issue then turns on who has the *correct* revelation, the problem is how we can determine this when revelations are beyond our reasoning and other experiences. The problem is not the diversity of competing revelations *per se* (since one of them might be "correct" or "legitimate" or "authoritative") but the lack of any neutral way of resolving the conflicts or determining which revelation is best. Each religious person in the end must decide what he or she accepts as revelation (if anything) or defer to ecclesiastic authorities. This involves the risk of faith. But it also means that in determining what is revelation, we bring our prior beliefs to the table. For example, do we let our sense of moral rightness determine what we accept as a true revelation? Were a modern Abraham to claim that God commanded him to kill his son, how many people today would accept him as a prophet of God rather than a madman? In the end, do *we* decide what is the word of God?

Mysticism

Mysticism is a form of religiosity that has shaped all religious traditions.[7] At the center of mystical ways of life is an internal quest to still the conceptual and emotional apparatuses of the mind; in one track, the mind retains sensory or non-sensory differentiations, and in another, the quest culminates in a consciousness void of all sense-experiences and mental images.[8] The objective is to correct the way we live by overcoming our basic misconception of reality—in particular, a sense of a distinct self set off from the rest of the world—and thus to experience reality as it really is in the way open to humans. Through the mystical quest, we come to realize the reality present when all our conceptual and emotional content and structuring is removed from our mind. No new messages from the transcendent are revealed, but we allegedly have an inner contact with a transcendental reality and gain knowledge of reality. (Of course, a mystic may also be a prophet, but no new revelations are received in the mystical experiences themselves.) In enlightenment, both our knowledge and our will are corrected; and, free of self-will, we can align our life with the way reality truly is and thereby end the mental suffering that results from trying to manipulate reality to fit our own desires and images.

Mystics claim to experience a power and reality, not merely give a new interpretation to the world. Other types of religious experiences and other altered mental states—such as alleged parapsychological powers, auditions, and visions—are often associated with mysticism. However, Zen is not the only tradition to dismiss visions, sounds, and sensations occurring during meditation as hallucinatory "demon states" (*makyo*). Mystical experiences are sometimes called "visions" or "seeing," but they should not be taken to be sensory; the same holds for tactile terminology—"touching," "penetrating," "embracing," "grasping." The inner quest central to mysticism involves a process of "forgetting" or "fasting of the mind" of all cognitive and emotional content. It involves a calming or stilling of the mind—a "withdrawal" of all mental powers from all objects. It is a process of "unknowing" all prior knowledge.[9] Thereby, silence and an inner stillness arise. And yet throughout the process one remains awake—in fact, mystics assert that only then are we conscious to the fullest extent possible for a human.

Mindfulness and the Depth-Mystical Experience

There is not one "mystical experience" but two significantly different types: the mindfulness state and the depth-mystical experience. Both types are allegedly cognitive. The quest may lead to sensing an undifferentiated unity to the flux of experienced phenomena free of conceptual distinctions (the "mindfulness" state) or to an awareness of a reality underlying all of the experiencer's subjective phenomena and, under one interpretation, all objective phenomena (the "depth" experience). Thus, both are about the fundamental reality of things, but the natural world and its source must be distinguished: mindfulness is about the dynamic realm of "becoming," while the depth-mystical experience is about the changeless "beingness" underlying subjectivity or the objective realm. Mindfulness thus is not a low-level, failed, or partial experience of the depth-mystical reality but a distinct type of experience. In this state, there is a sense of the beingness of the everyday world while the mind still has sensory or other mental content and thus differentiations. The meditative techniques and the resulting physiological effects also differ for the two types: mindfulness involves a passive receptivity and attentiveness free of conceptualization in observing the sensory or other events in our mind (as with Buddhist *vipassana* techniques), while the depth-mystical meditation involves focusing attention and concentration (as with Buddhist *samadhi* techniques). But meditators cannot force the mind to become still by following any series of steps. As long as we are still trying to "get" enlightened, we are still in an acquisitive state of mind and thus cannot succeed. No act of self-will nor any preparatory activity can force enlightenment—one must let go, surrender to reality. But once meditators stop trying to force it, the mind stills itself and the mystical experiences occur automatically. In the depth-experience, the transcendent ground appears active and the meditator passive; in mindfulness, the same holds for natural phenomena.

Mindfulness loosens the grip that our mental concepts have on our sense-experiences and inner experiences.[10] Mindfulness is not a discrete experience, as with the depth-mystical experience, but an enduring state of consciousness. Mindfulness exercises destructure the conceptual frameworks structuring our perceptions. Normally, we abstract apparently permanent objects from experience and react to our creations. But mindfulness dissolves the conceptual boundaries we have imposed on reality. We usually see rugs and hear trucks, but with pure mindfulness we would see patches of color free of rugness and hear objectless noises. Our awareness becomes focused on the beingness of the world, not the uniqueness of any object or the structures operating in nature. In the resulting state, an experience of a fundamental uniformity of being and connectedness of all we experience both internally and externally comes through—the simplicity of the "such-ness" or "that-ness" of all reality—along with more intensity to sensory input. This sense of beingness can lead to a sense of being a part of a unified whole or to the glow to perceptions of "nature mysticism."

With mindfulness, we see what is presented to the senses as it is, free of our purposes, feelings, desires, and attempts at control. We are living fully in the present, witnessing whatever arises in our consciousness, free of expectations and fears, without judging and without a sense of possession. The mind then mirrors only what is really there, without adding or distorting whatever is presented. Mental categories no longer fix our mind, and our attention shifts to the underlying beingness of things, although mental structuring remains present in all but a pure mindful state in which we have sensations but no structured perceptions.

With the depth-experience, attention shifts from the realm of change to the still center of being. Depth-mystical practices involve focusing attention and emptying the mind of all conceptual and sensory content. This concentration leads to an inner stillness in which one is free of all differentiated mental content and yet still awake. When the mind is stilled completely, an awareness of a dimension of reality not known through ordinary experience and self-awareness allegedly bursts through—a direct implosion of a transcendental reality, unmediated by theories or concepts, with a resulting sense of reality, certitude, and typically finality. What occurs is not "objective" in the sense of an awareness of something existing independent of us, nor "subjective" in the sense of experiencing something we create. There is a sense of being rather than experiencing. It is a state of consciousness free of any intentional object. It is like a light that is turned on but not illuminating any object. With the absence of differentiations in the mind there is a sense of unity to our being: all of reality shares one simple and partless source, or our true self is isolated. But with a mystical experience, we are not "united to God" in any ontological way that was not previously occurring. Only our misguided sense of individual existence within the natural realm is now abolished, and we now conform with the transcendent's "will." The experiencer does not "obtain" or "become" anything new—all that changes are our knowledge, will, and emotions.

Mystical enlightenment can be defined as ending a sense of an individual self in an enduring state of consciousness. It reveals an inner silence at our core. It involves knowledge of the fundamental nature of reality (as defined by each mystic's religious tradition) and subsequently living a life in accordance with it (typically by following the ethics of the mystic's tradition). The enlightened state in depth-mysticism is a continual state of mindful awareness while in contact with the transcendental reality. The sense of a separate self in the everyday world is replaced by a continuing inflowing of the ground of the true self or of all of reality. The enlightened now live in the world in a state of freedom from the attachments and concerns generated by a false sense of an individual self—they act literally selflessly, i.e., free of a sense of self.

Reality

Mystical experiences shake up our habitual ideas about what is real, but do they in fact tell us anything new about what is real? That is, are there claims about the nature of reality that are justified only by mystical experiences? Mysticism shifts attention to the sheer that-ness of reality rather than its what-ness (what things are individually) or how-ness (how things work or came to be). There are two issues here: for mindfulness, the being of the experienced world apart from concepts; and for the depth-experience, the relation of what is experienced to the being of the natural world.

With mindfulness, attention is shifted to the being of the world independent of the conceptualizations we impose. To convey the sense of what is real and what is "illusory" (i.e., temporary, dependent, and changing), the *Chandogya Upanishad* gives the analogy of the clay being real while the pot (i.e., the temporary form the clay is in) is not. This does not mean that mindfulness mystics dismiss the world as "unreal" in any strong sense (depth-mystics are another matter). The world's being is affirmed. We "create" distinct objects by imposing our ideas on the world, but the beingness underlying these "objects" is real. The idea of ontologically distinct entities—each self-contained, self-existing, and independent—is uniformly rejected. Most importantly, this includes the idea of a distinct self. But the "objects" of sense-experience are "unreal" only in that sense. We misread sensory experience and construct an "illusion" of distinct, permanent entities. But what we conceptually separate as "entities" are still real eddies in a flowing and integrated field of events. For example, a tree is not identical to the earth, water, and sun light that gives its life even though it depends on those things, but neither is it totally distinct; it remains a unique and identifiable part contributing to the whole of reality. In Buddhism, the interconnectedness of everything becomes prominent, but this does not deny the reality of what is interconnected. Only the permanence and independence of entities we conceptualize out of the flow of events are denied.

Does mindfulness reveal something about the world's beingness that the unen-lightened miss? After all, nonmystics have argued that things are impermanent since Heraclitus first noted that we cannot step in the same river twice: part of what constitutes the river will have changed by the time we try stepping into it a second time, and this is true of all things—"everything flows and nothing abides." So too, that all things share the same "being" and that the universe is one interconnected whole are points that nonmystical naturalists can easily accept. Neuroscientists claim there is no "self" in our mental makeup. And the problem of how language operates if there are no permanent entities to refer to is now prominent in post-modern philosophy. No special experiences are needed to understand any of these points. But with mindfulness, the beingness of things is brought into awareness. Sense-experience still occurs and can be structured or unstructured, but now the beingness of everything is the center of awareness. In short, mystics realize something experientially, not merely see a logical point that nonmystics can also assert. In the Buddhist analogy, it is the difference between an intellectual accep-tance of the idea that water will relieve thirst and actually drinking water. Mystical selflessness expands one's sense of connectedness and thus the scope and signifi-cance of one's beliefs and values. Thereby, it alters one's views—in fact, the "something more" of the depth-mystical experience may fly in the face of what mystics normally expect and may drastically alter their understanding of the basic doctrines of their tradition. But mystics may have no new propositions to expound: the beliefs and values of the enlightened state are typically those of the tradition the enlightened were trained in, although mystics now know them to be true in a way they did not before. And, as discussed below, nothing about the *nature* of "being-as-such" is revealed—the mystery of being remains as full as before.

Depth-mystical experiences present a different type of problem: mystics apparently experience an immutable and fundamental reality transcending the natural world, but this reality is open to radically different interpretations. The popular view today of "mystical union" is that all mystics treat what they experi-ence as the power underlying being and thus as the source of the natural world. But one common interpretation in theistic traditions is that what is experienced is only the depths of the self—we can never experience God. Thus, the distinction between creator and creation can remain intact for mystics. Samkhya-Yogins affirm a pluralism of selves and a distinct matter with no underlying creator or ground common to both substances. The popular view is not even an accurate interpreta-tion of Advaita Vedanta for which there is only one reality and thus no "degrees" of reality or "dependence" of one reality on another. Nor has any classical mystical tradition adopted a pantheism equating the transcendental realm with the natural world, creator with creation. Neoplationism is often considered "pantheistic," but the emanated material universe is not the One. Theravada Buddhists bypass the issue of the relation of a transcendent source (if there is one) to the world and deal only with the components of our experienced world. For theists, there is an underlying, self-emptying source of the world's being—a personal god or a nonpersonal, inactive ground like the Godhead of the medieval Christian Meister Eckhart's Neoplatonist-influenced system. It is not a being among natural objects nor a being "outside" the universe. Nor is it the "beingness" of the natural world or even "nonbeing" (since the latter suggests nonreality) but beyond all categories. As Plotinus put it, the source of the world's being must be "beyond being."

Mystics emphasize that the reality experienced in the depth-msytical experience is the opposite of anything in this world. It is timeless, spaceless, changeless, and so forth. Some mystics (including Plotinus and Eckhart) go further: the reality is not alive, not intelligent, has no knowledge, and is not conscious—it is the source of life, intelligence, and consciousness but has none of these features.[11] (This is an issue for the Cosmological and Teleological Arguments.) It is ineffable: we can name it, but our concepts cannot "capture" it the way they do objects. Is it totally featureless except for some type of oneness and reality? Is it an experience of naked beingness?[12] And how does this uncreated "power to be" (Advaita's *sat*) underlie the natural world? Water in its gaseous, liquid, and solid states always has the same underlying molecules—by analogy do all things in the world in conscious, animate, and inanimate states have the same underlying "being"?

The overwhelming power of the depth-mystical experience can lead to the belief that what is experienced alone is real. Thus, there can be nothing else of value or significance. Advaita, with Brahman as the only reality, is the paradigm of a system that dismisses the natural world as "unreal" since it is not changeless while reality is permanent and changeless. (Advaitins, like Parmenides, do not deny that obviously there is the *appearance* of change but only that appearances can be deceiving, as with optical illusions. Similarly, mystical experiences expose the world of change to be only one giant deception.) Thus, under this system there is no creation since Brahman is eternal and nothing real is created—the world is just a meaningless, unreal play. However, Advaitins have the mystery of why there is any manifest world at all—Shankara admitted that even in the enlightened state the world does not disappear, comparing the situation to a person with an eye disease seeing two moons even though he knows there is really only one.[13] Calling the illusion (*maya*) simply a misreading of what is there does not answer why the illusion is there to begin with. Shankara said the world has an "undefined" (*anirvacaniya*) status between being real (which Brahman alone is) and being unreal (as "a married bachelor" is). But the theologian Ramanuja can rightly ask what is the difference between saying that and simply admitting that the natural world is *real*? Even saying the world is like a dream is not to dismiss its reality: the dream itself is real— it occurs even though it is dependent upon the dreamer and its content is not part of the objective realm. So too with the world.

Other traditions take the manifest world to be real. Emanationism is common. The Upanisads set forth a cosmogony in which the underlying source (Brahman) emanates forth a real world like a spider spins out a web. Daoism also presents an emanationist cosmogony without a personal creator.[14] In the West, Neoplatonists have taken an emanationist view of creation: the world and selves emanate out of an eternal, uncreated godhead in different stages—God's being or love automatically "overflows" and makes the manifest world. (The how-mysteries connected to emanation—the solidification of divine energy, as it were—are no less than for any other method of creation by the transcendent.) The natural world is treated either as fully real or at least as having a "degree" of reality. The simple event of being transcends both creator and creation. Even the creator god is an emanated reality, not the uncreated source of all reality—the godhead is the only power producing the natural realm. (A story from Hinduism might give Abrahamic theologians pause: the first being who emanated out mistakenly concluded that he was the creator of all that flowed out after him.)

Theists in general affirm that our world is a reality totally dependent upon God. They reject the automatic and impersonal nature of the emanation process—they want an intentional role for the personal transcendent in creating the world. But to deistic emanationists, our being is God's being (and thus all being is of one nature), although this is not a pantheism: God's creatorship and our creaturehood are never lost. Today panentheistic theologians take the commonality of God's and creation's being as their starting point. For them, everything is "in" God in some sense. However, the stronger idea of the world as God's "body"—an idea around since the Hindu Ramanuja—has two fundamental problems: how God's transcendental mind can be embodied in the material world, and whether God then determines all actions that occurs in his "body," thus negating free will. If the universe is created and not eternal, it would also mean that God was born.

Not all mystics follow Plotinus in equating matter with evil.[15] Many emanationists see the world as essentially good since it is the overflowing of a good source. But theists dislike emanationist metaphysics since the problem of natural suffering would be aggravated if the world is the being of God—the evil would then be in the very *substance* of God. Theists do not want AIDS and cancer to be in the being of God or to be an aspect of the divine. Also an immanent god makes the suffering of nature seem more a part of his plan—whatever exists and whatever occurs is divine as such. Thus, theists prefer a creation from nothing: God creates something that is distinct from himself out of nothing, and thus evil does not touch the underlying source. But whether this is a distinction without a moral difference is an issue: God would be responsible for permitting natural suffering in either case—being a distinct creator would not absolve him of responsibility for what he created if he could have created differently or if he could intervene in creation to change things. Any overwhelming sense of love and joy in a depth-mystical experience also aggravates the problem: how can a joyful, loving source produce a world that has so much suffering and where life is so cheap? Eckhart wanted to claim that God suffers along with us, but in the end he had to admit that suffering for God is so joyful that it is not really "suffering" at all.[16]

Mystics often have a sense that there is a fundamental rightness to things at the deepest level. But any overwhelming sense of a loving source that some mystics have may lead to denying that evil is real—everything is actually perfect as is. Seeing God in everything can thus lead mystics to moral indifference or antinomianism: everything shares the same being or presence of God; so everything is fine the way things are, and there is no need or right to change anything; or everything is innately good, and so we can do whatever our body desires. Suffering and death do not matter, if they are considered real at all, since they do not affect what is real. The billions of years of evolution are irrelevant because events in time are irrelevant to the timeless. Does this mean that any sense of love that accompanies some instances of selflessness is no more than simply feeling connected to all that exists rather than any indication of the nature of the source of things? In sum, any sense of an underlying love only deepens the mystery of natural evil.

Consciousness figures prominently in mysticism. For Advaitins, it is not an activity of the brain but an inactive reality constituting the reality of all subjective and objective phenomena. To classical mystics, the idea of individual selves in the natural world is just another illusion: there are no self-contained, independently existing realities, and this includes "selves." There are streams of consciousness but no entities in the natural realm underlying them. There is no separate center—no "I"—in the world that thinks, feels, and acts, but there is the thinking, feeling, and acting. We create the everyday self—the ego—by identifying with our awareness of our mental activity. We think there is an independent, self-existent entity, but in fact "self-consciousness" is just another analytical mental function—one that simply observes the rest of our mental life. By identifying with this one analytical function we reify a separate entity and set ourselves off against what is real. Indeed, thinking that there is a self to be defended and enhanced is the central illusion keeping us from seeing the interconnectedness of all of the natural world and experiencing the underlying source. But theists and Samkhya-Yogins once again must be mentioned. Theists are adamant that individuality is an irreducible part of reality, keeping the ontological separation of self and God intact, even if we "forget" it during some experiences. We can even replace our will with God's, but we are still there. Samkhya-Yogins believe there are no egos (*jivas*) in the material world, but there are separate uncreated and deathless conscious entities (*purushas*): there is the entity of pure consciousness, and the objective is to isolate our own entity from matter. (All *mental activity*, including the ability to will and know, is considered evolutes of *matter* in this system.)

All depth-mystics think they have experienced something absolutely real within themselves, but this variation in interpretation leads to the issue of whether even they know what they have actually experienced—is it the depth of the self, God, a nonpersonal Brahman, or only ordinary awareness free of content? Because of this openness in interpretation, the depth-mystical experience itself cannot answer any of the mysteries of the nature and the status of the natural world and consciousness. Having a depth-mystical experience may permit the experience of a previously unrealized dimension of reality, but it does not answer the questions of whether what is experienced is alone real or whether the world is equally real. Nor does it provide any information about the relation of changeless "being" is to the realm of dynamic "becoming." Nor, no matter how powerful the experience is, can it disprove the reality of what is experienced outside that experience—a sense of a timeless reality or being unaware of the passage of time in a mindful state does not negate the presence of the realm of time. Thus, mystics, while outside the depth-mystical experience, must still decide what they take to be real—the experience does not resolve the metaphysical problems.

Depth-Mystical Knowledge

This leads to another issue: does the depth-mystical experience reveal something of the fundamental mysteries of any transcendent realities that revelations and reason do not? In particular, does the depth mystical-experience give us an insight into the source and nature of the world's being? The depth-mystical experience involves a different state of consciousness than occurs in the other types of knowing. Mystical consciousness is a mode of being that one becomes rather than merely some newly acquired knowledge. Theistic mystics claim their depth-experiences thus give a "participatory" knowledge of God, not knowledge of an object by "acquaintance" or reason. Theistic and nontheistic mystics alike emphasize the direct awareness involved in mystical knowledge. It is the distinction in Buddhism between representational knowledge-that (*jñana*) and insight (*prajña*) in which the knower, the act of knowing, and the known are not distinguishable entities. Self-knowledge is the closest analog in ordinary experience.

One problem unique to mystical knowledge is that analyzing or even remembering the depth-mystical experience will always be done in a different state of consciousness than having it, and thus mystics cannot remember it in the way they actually experience it. That is, unlike with ordinary experiences, thinking about the experience takes us out of the state of the experience in question. Conceptualizing alienates us from the depth-mystical experience, but to communicate what is experienced mystics must leave that state of consciousness. Once out of that experience, what is experienced is reduced to an object of thought, even though mystics assure us that transcendental realities are not accessible through thought—we can know God (i.e., experience him), but we cannot know him (i.e., any attempt at intellectually understanding him transforms him into a graspable object). Thus, any conception of the transcendent—including calling it "the transcendent"—makes it an object among objects even though its reality is wholly different from anything within the natural world. Describing a banana as "a fruit" does capture something of its nature even if it does not "exhaust" it. However, mystics contend that no concept "captures" anything about the transcendent. Thus, the Ontological and Cosmological Arguments make the transcendent into a thing in some way comparable to other things—not what is experienced in mystical experiences. Thus, to mystics, those arguments miss the mystical reality. Mystics may have experienced some fundamental reality, but, if the depth-experience is free of all structuring and hence is not conceptual, what do they *know*? Any answer will involve a conception, and if all conceptions are rejected, where does that leave the mystics? Even knowledge by "participation" entails statable propositions. If we cannot formulate ideas about a reality, can we truly be said to *know* something about it rather than just be *aware* of it?

The change in states of consciousness introduces another major problem. Even if we grant that mystics are aware of a transcendent reality, nevertheless what significance the mystics see the experience as having is given *outside* the depth-mystical experience itself once a mystic has returned to a "dualistic" state of consciousness. All evaluations of the status of these experiences are made outside the depth-mystical experience. In this regard, even enlightened mystics are in the same boat as the rest of us, even though they have a larger experiential base from which to make their decisions concerning what is real. After they return to an ordinary or mindful state of consciousness, what they have experienced will be an object of intentional consciousness for them too. But only outside the depth-mystical state will mystics be able to decide what sort of insight the experience is.

And what sort of insight the experience is taken to be will depend on factors outside the experience—the beliefs and values of their religious way of life—and not just what is given in the experience itself. The resulting sense of oneness tends naturally toward a monistic ontology. But not all mystical systems involve an all-encompassing nonduality in which the apparent diversity in the world is in the final analysis unreal. The depth-mystical experiences in fact has been fitted into radically different ontological schemes. As noted above, Samkhya-Yogins interpret the depth-experience as the *isolation* of a self from all matter, and within this fundamental dualism of matter and consciousness, they accept a plurality of selves. Theists have also incorporated the depth-experience in two different ways—unison with God's will or a sense of experiencing the ground of the self—while retaining the reality of persons. Of the classical traditions, Buddhists weight the insights of mindfulness over those of the depth-mystical experience and give a pluralistic interpretation of the nature of our experienced reality.

Theists can weight the depth-mystical experience as the most important ontological insight, or they can see the depth-mystical experience as only a partial insight into a theistic transcendental reality and decide that numinous experiences offer deeper insights into that reality. Some theists contend that numinous religious experiences are a deeper form of the depth-mystical experience than that experienced by nontheists.[17] However, if the depth-mystical experience is indeed devoid of all differentiated content, the numinous experience of a personal divine reality cannot be a form of the depth-mystical experience. Rather, both theistic and nontheistic mystics must in fact have the same experience but only interpret it differently after the experience is over. Still, some ranking of the different types of experiences is needed in any religious tradition.

Thus, the depth-mystical experience itself does not determine the decision of how it is interpreted. Indeed, that the depth-mystical experience is taken to be an insight at all—rather than just an interesting and powerful mental state—depends on factors outside the experience. Even though mystical experiences affect the mystic's worldview or ethos, this process nevertheless occurs only outside of the depth-mystical experiences in "dualistic" consciousness. Some interpretation must always be present. Of course, the interpretation given the experience may be ready-made by a mystic's tradition prior to the experiences, and this raises the question of what is more fundamental in the claim to mystical knowledge—the mystical experiences or the beliefs and values of the mystic's tradition?

Constructivism and the Claim to Mystical Knowledge

Postmodern constructivists believe that *all* experiences are intentional and have some content. Thus, the depth-mystical experiences, contrary to the mystics' own claims, must have some conceptual content and structure—there is no such thing as a "structureless experience."[18] Beliefs and concepts penetrate the experiences themselves; they are not applied after the fact in an interpretation. Moreover, constructivists add a second claim: all the belief-content and structure of the experiences is supplied solely from the mystic's existing tradition—the trans-cendent, even if it exists and is in fact experienced, contributes nothing to a mystic's actual beliefs. Thus, depth-mystical experiences cannot add anything at any point to the beliefs or values that a mystic adopts. The religious beliefs that the mystics bring to their experiences are not merely one of the components contri-buting to their experiences but are the *only* cognitive component and constitute its total cognitive content. Meditation does not involve a deconditioning of a tradition's framework permitting new cognitive experiences but simply helps the meditator to internalize fully the tradition's beliefs and values. Enlightenment is the final internalization of a religion's beliefs and values—the culmination of long periods of intense study, practice, and commitment to those specific beliefs and values. The experiences themselves are not potential sources of any fresh cognitive in-put for a system of belief but are merely an intense feeling of previous beliefs. The second part of constructivism, in short, rules out the possibility that mystics have any knowledge of a transcendental reality, even if mystical experiences are experiences of such a reality—the nonexperiential cultural belief-systems control everything. Thus, there is no such thing as "mystical knowledge" in any real sense.

Constructivists offer no reason for this claim other than the fact that ordinary experiences are structured. That we may have a cognitive state of consciousness without an object is simply impossible for philosophers who have the preconceived idea that all mental states are intentional. In ordinary perception, we do not experi-ence a patch of colors and interpret it as a rug. There is only one act: seeing a rug. This may apply to mindful states of consciousness and revelations, but in the case of the depth-mystical experience, the mind is allegedly empty of all differentiated content, and only after the event do mystics interpret its significance. Construc-tivists simply apply their theory without examining the possibility that the depth-mystical experience is unique. Nothing mystics say could in principle provide counter-evidence—it is all automatically interpreted to fit the preconceived theory.

Nonconstructivists defer to the mystics here: only mystics are aware of both differentiated experiences and the depth-mystical experience, and they claim the latter is radically different in type; mystics sense distinctions after the experience and are aware that they are not in that state during the experience. Thus, to nonconstructivists the depth-mystical experience transcends all historical and cultural contexts and is same for all mystics since there is nothing in them to differentiate them. Nonconstructivists can readily agree that the images and interpretations of the experience that mystics form in their dualistic consciousness *after* their depth-mystical experience are shaped by the beliefs and values of a particular mystic's tradition, but they ask if in fact this experience is free of all differentiations—as the writings of even many theistic mystics suggest—then what is present to structure the experience itself? They also point out that the content of mystical experiences often comes as a surprise to mystics (which indicates their doctrines are not controlling their experiences) and that it seems more likely that mystics such as Shankara and Eckhart interpreted their tradition's revealed texts to fit what they have experienced rather than vice versa.

The important point concerning the roles of beliefs and values can be maintained without endorsing even the first part of constructivism. Constructivists are correct in pointing out that there is no one abstract "mystical worldview" or "mystical value-system" but instead a variety of more specific mystical systems, often more than one within the same religious tradition. Mystics do bring their cultural beliefs and values to their experiences, and these influence their later understanding of their own experiences. But the knowledge-claims come only after the experience, whether or not the depth-mystical experience contributes to the claims or are conceptually structured. Nevertheless, nonconstructivists, like the mystics themselves, want to affirm both that the depth-mystical experience has some cognitive content and that the experiences themselves are free of all interpretative and conceptual content.

Is There Mystical Knowledge?

Should we accept depth-mystics' claim to knowledge? Unfortunately, discussions of whether mystical experiences are an insight into reality quickly degenerate into a war between religious and naturalistic metaphysics. Mystics contend that the depth-mystical experience is a direct experience of a reality in a way that no other experience is. After the experience, it "feels real" to the mystics themselves. In fact, it seems vividly real—an insight, not an hallucination. Thus, mystics argue that it is evidence of being in contact with a transcendental reality rather than a mental aberration. Naturalists, of course, disagree. They need not deny that such an experience occurs, but they do deny that a transcendental reality is among its causal roots and thus they reduce the experience to an experience involving only elements of the natural order. For them, the brain only evolved to help us survive in the natural world, and when we succeed in removing all sensory and other content while remaining awake the brain simply malfunctions badly—the depth-mystical experience is nothing but our mental gears spinning without engagement.

There seems to be no way to resolve this dispute. For example, the bliss or joy of mysticism does not necessarily mean that mystics are connected to a fundamental reality rather than simply "blissed out" when the brain is malfunctioning. In general, any scientific account of the brain-events occurring during meditation or the depth-mystical experience will not help: since the experience occurs, obviously there must be a physiological basis permitting it, and there is no reason to suspect that scientists cannot identify the workings someday (e.g., what parts of the brain are active during meditation). But merely identifying the physiological bases will not determine whether the experience is an insight, any more than identifying the apparatus of perception could disprove the claims of sense-experience.[19]

But even if we give mystics the benefit of the doubt and simply *assume* that the depth-mystical experience is more than a hallucination, does the depth-mystical experience dispel any of the mysteries of the fundamental nature of reality? Assume then for the rest of this chapter that there is some substance to these experiences independent of a mystic's prior belief-system. It gives an overwhelming sense of fundamental reality: a sense of the direct awareness of oneness and a fundamental reality—power, ultimacy, immutability, permanence, independence. The experience also gives the experiencer a sense of selflessness—that the everyday ego is not part of the fundamental make-up of reality. We can perhaps add "experiencing the source of an apparent reality" and "a sense of bliss" to the list. That is the starting point for knowledge-claims and metaphysical speculation.

Mystics can remember that the depth-mystical experience is free of any sense of self and filled with another reality, but what is the *nature* of that reality? This leads to the problem of whether we need something conceptual to call anything "knowledge," even if the experience giving rise to the claim is free of all conceptual elements. Only statements can be true, not experiences. But if the transcendent is truly *ineffable*—i.e., inexpressible in words—what is conveyed in the mystical experience? Ineffability cannot be the foundation of any knowledge-claim. An "ineffable insight" is a contradiction in terms: the awareness of the transcendent may be ineffable, but something about any insight must be statable. So what is the depth-mystical insight? What exactly is the knowledge gained? Even if we grant that mystics are in touch with a transcendent reality, the problem is that the experiences still may give no more than a general awareness of its reality. Mystics will know *that* something fundamental exists that makes our ordinary world seem less real, but *what* it is, beyond being "real" and "one," is not given but instead is open to different interpretations outside the depth-mystical state of consciousness.[20] This greatly limits the extent and very claim of "mystical knowledge."

Since claims arising from the depth-mystical experience are open to such wide variations, the experience is less "cognitive" than mystics insist. Even if (contra

Kant) mystics have direct and unmediated access to a reality, the problem of varying interpretations still persists. As with conflicting revelations, the presence of conflicting interpretations that have stood the test of time will remain a barrier to our knowing which, if any, interpretation is correct. We can only test for whether each system is internally coherent and explains all the available experiential data. Theorists from each religious tradition will insist their position is "correct" or "best" or "least inadequate" and offer arguments in its favor, but to those who do not share their doctrinal framework the arguments are never very convincing. This leads to the issue of whether *any* mystical system, each with their highly-laden concepts, actually "corresponds to" or "captures" anything of the reality experienced or whether only mystery remains. Alternatively, one of the possible interpretations may be the best, but the experience itself will not supply the answer.

Mystics may not be able to hold all their beliefs tentatively or open to criticism because of the overpowering effect of their experiences. Mystics may insist that only they know reality's true nature, or that they have knowledge of God, or that the proof of their claims lies within their own hearts. But the problem again is the competing answers from different traditions to such questions as whether what is experienced is nonpersonal or only an aspect of a personal reality. Mystics cannot get around the fact that other mystics apparently with the same experience offer conflicting views and have the same conviction. Mystics cannot say "Sorry, I've had the experience and you haven't" when making claims about the *nature* of what is experienced. The certitude and finality mystics feel from the experience are transferred to their own traditions' beliefs and values, but this does not mean their interpretations are necessarily true or part of the experience—e.g., theistic mystics may believe they experienced an omnipotent reality, but how would they know it is truly omnipotent? The significance of the experience remains an issue after the experience even for the mystics themselves, as does coming up with an interpretation and doctrines that are "adequate" and "consistent" with the experience. There are no "self-confirming," "self-authenticating" or "self-verifying" doctrines of the nature of what was experienced, no matter how powerful the experience giving rise to them. No claims about reality are impervious to error or immune to challenge, even when mystical experiences are involved. Even accepting that the experience is an insight into the world rather than as an illusion requires a decision that the experience itself cannot determine. The experience has a say in such metaphysical matters by adding to the pool of experiences from which mystics make their worldviews, but it does not determine one doctrine or worldview over another. Any experiential claim to certainty about a particular tradition's doctrines is over.

In sum, there is no simple empiricism even in mysticism: the experience may be essential to knowing the transcendent, but it does not dictate a knowledge-claim about the nature of what is experienced—even here knowledge-claims are always more than what can be justified by the experience alone. And even if all "theory" of what is experienced can be totally separated from the experience here (unlike in ordinary experiences), the problem still remains: the "overbeliefs" as William

James called them remain essential to the understanding that the mystics them-selves have of their own experiences in order to lead their ways of life.

Thus, ideas maintain an irremovable role in mysticism. This means that even mystical knowledge will always have a conceptual element that mystics supply outside the experiences. It is also worth noting that many major mystics such as Shankara and Eckhart do not appeal to their own experiences to justify their claims: they appeal to their tradition's *revelations* recorded in scripture—albeit highly interpreted to fit their own ideas—as the source of authority. Shankara said such appeal is necessary since philosophers constantly contradict each other.[21] He also noted the objection that this itself is an instance of reasoning, but he still asserted that the Vedas, being eternal, provide the necessary true knowledge.

In such circumstances, whether mystical experiences can offer *a justification* for any of the claims is open to question since competing alternatives to each claim advanced by one mystic are offered by other mystics having apparently the same experience. In science, scientists can test competing theories with fresh experiential in-put and with the agreed upon criteria for selecting the better theory; thus, eventually a consensus arises. But in mysticism there is no way to test the interpretations and no cross-cultural set of criteria for determining a better interpretation. That is, mystics do not engage the transcendent the way scientists engage the world: there are no new, genuinely novel depth-mystical experiences to challenge or correct previous conceptualizations—there are no tests or fresh in-put as time goes on but simply the same "pure consciousness" event over and over again. Mystics, in short, are not learning more about the transcendent and thus cannot settle disputes.[22]

Thus, the verdict must be that the competing claims mystics have advanced about the nature of what is experienced cannot be resolved by the experience itself. What mystics experience gives them the insight that there is more to reality than meets the naturalistic eye, but they do not know the *nature* of what is experienced any more than do nonmystics. Mystics are in no better position to answer the philosophical questions concerning the nature and status of what is experienced, despite their expanded base of experiences, than nonmystics. In sum, even if we grant that mystics have experienced a transcendental reality, the mystery of the fundamental nature of that reality is not answered. Mystics do not have answers to the basic why-mysteries any more than do philosophers or theologians.

Mindfulness and Language

The experienced sense of "wholly otherness" of the mystical experiences inten-sifies the problem of language discussed in the last chapter and causes mystics to give special attention to language. How does one name the unnameable, express the inexpressible? Is it true, as Laozi said (while writing a book on the Dao), "those who know do not speak, and those who speak do not know"?

It is language that enables us to crystalize our conceptualizations of what we want and what we think should be, thus leading to frustration, anger, and fear when reality turns out not to be as we wish. To mystics, such conceptualizations come to stand between us and what is real. Language-guided perception thus is the opposite of mystical mindfulness: experiencing the flow of an impermanent and connected reality—rather than discrete, independent "entities"—makes the discreteness of any linguistic denotation seem that much harder to reconcile with reality "as it really is." How can we describe a dynamic world in terms that are necessarily static? Naming gives things a status distinct from their surroundings by marking off a referent from what it is not and thus dividing the continuity of reality into a series of distinct "things." That is, naming cuts the flux of reality up into fixed, distinct units—and thereby fixes the mind on "things"—when reality in fact is continuous. Language thus generates a false world of multiple "real" (changeless, independent) entities. What is actually real is not "illusory"—only the conceptual constraints we impose on reality are. The conceptualizations and images we create—in particular, the sense of a self separated off from the rest of reality—become a barrier between us and the fluid, continuous reality we are part of.

But even if language cannot "capture" the connectedness and ever-active flow of reality, the enlightened still use language—even if it only to show how language fails. But how they see concepts has changed: the enlightened now use language without projecting linguistic distinctions onto reality, thereby generating an ontology of distinct "entities." Language remains useful for navigating the world and for leading others toward enlightenment even if there are no permanent referents in the ultimate make-up of reality. This is possible since there are fairly stable configurations in the flux even though they are not permanent: buildings may only be an impermanent assemblage of equally impermanent parts and not be "entities" unto themselves—and thus not "real"—but the term "building" is still useful for directing attention to parts of the present flux of reality as we move through the world. "Buildings" do not exist in a way different than the way a unicorn does not exist—there is some reality there even if the reality is constantly changing and thus there is no permanent referent for the term. All languages make distinctions and categorize things; thus, there is no reason for mystics to try to invent a new language since any language will present the same basic problem of dividing the indivisible. Mystics thus will simply use the language of their own culture; the only change that their writings suggest is in a tendency to use a passive voice more often than an active one to de-emphasize subjects.

The objective of their analysis of language is to see that there is nothing in reality to become attached to and that we are fooled by language into thinking there is something real to value—distinct, permanent "entities" to defend or gain. Still, language is not useless: it works as a tool because of what is in fact real. The Zen analogy that language is merely "a finger pointing at the reflection of the moon in a pool of water" accepts there is a moon and that we can direct attention to it by pointing, if only indirectly. The point is simply not to get attached to the finger, let alone mistake it for the moon—i.e., not to get caught up in the words and miss the true realities. Belittling words as mere "names" or "designations," as Buddhist do, does not change the fact that there is really something there in the world for words to point to. The word "moon" works in the analogy rather than the word "chipmunk" only because there is in reality a moon (albeit an impermanent reality and not an independent entity) being reflected in the pool that we are referring to.

The Depth-Mystical Experience and Language

With the depth-mystical experience, the complete otherness of what is experienced makes mystics emphasize ineffability, silence, the *via negativa*, and paradox. How can we express oneness in a "dualistic" language? How can an Advaitin sentence have the word "Brahman" as its *object* when Brahman is the eternal *subject*? If "existence" applies to things in this world, how can we even say the transcendent "exists"? How do we say it is not "nonexistent" or is "real" without admitting some terms apply? If its mode of existence is wholly different, how does even the worldly term "is" apply? That is, how can even formal concepts (e.g, it exists) rather than substantive ones about its nature apply? So too, if the transcendent is utterly unique and so no terms applicable to anything else apply, doesn't this mean that all substantive characterizations of the transcendent, no matter how abstract ("beingness," "consciousness," "timeless," "transcendent"), are necessarily *wrong* since all our terms are derived for things within the natural realm?

The problem of the ontological otherness of the transcendent is supplemented by the problem of the necessity of another mode of consciousness to experience it. Thus, both the nature of the transcendent and the experience itself are deemed "ineffable." In fact, all experiences are ineffable in the obvious sense that we need to have experienced the taste of a banana to know the taste of a banana—we cannot communicate much of any experience to someone who has not had the experience. Mystical experiences obviously share this problem—especially the depth-mystical experience since it is not like anything else. But the more basic problem is that when mystics speak about a transcendental reality they are in a different *state of consciousness* from that in which they experience it. Thus, whenever we say anything about what is allegedly experienced we are in the wrong state of consciousness to know that what is said is true. Any attempt to "conceive" a transcendental reality makes it into something it is not and thus is a losing battle.

Silence is the obvious response to this problem. That protects both the experience and the reality experienced. But it is hard for anyone to remain silent about something fundamentally important to them. Moreover, our mind's innate tendency to conceptualize takes over after depth-mystical experiences. And the issue arises of *what* the mystics are being silent about. Hence, the paradox of ineffability: in order to claim that the transcendent is beyond all names, we must name it. Merely saying that there is something "unconceptualizable" is not to form a conception of anything, but we in our unenlightened state will form an object for thought about "it." Mystics too speak and write while realizing that whatever they say renders a reality that is not a thing into a thing—hence, they deny whatever they say is actually true. Thus, Shankara said that the word "reality" (*satya*) cannot actually denote Brahman but can only indirectly refer to it, and "self" (*atman*) is qualified by "as it were" (*iti*) to indicate that the word "self" does not apply.[23] All "positive" characterizations of Brahman—reality (*satya*), knowledge (*jñana*) and infinity (*ananta*)—are only meant to remove other attributes: Brahman cannot be the agent of knowing for that requires change and that denies reality and infinity; knowledge merely negates materiality, and the other two negate knowledge.[24] It is all a process of negation. More generally, mystics want to say that "ordinary language" or "human language"—as if there were another kind—does not apply.

Hence, mystics are major advocates of the *via negativa*—the denial of any possible positive description of the transcendent.[25] In the Upanishads, Brahman is described as "not this, not that" (*na iti na iti*),[26] thereby denying all features to the transcendent. So too, the Buddha denied that any concepts about a person "fit the case" (*upeti*) of the state of the enlightened after death.[27] This approach was introduced into the Western theistic traditions through Neoplatonism. Plotinus claimed that all predicates must be denied—even "the One" does not apply to the transcendent since "one" is a number among numbers, and it thus may suggest some duality—and silence is ultimately the only proper response.[28] This approach has never been the predominant trend in the Abrahamic religions, although it has more prominence in Eastern Orthodox Christianity. Even Muslims, who stress the unknowability of God to all but prophets and mystics, do not emphasize this approach. Theists are always eager to attribute positive features to God, and some theologians try to tame the *via negativa* by treating as only a *supplement* to the positive approach. Mystics, however, see the negative way as a *corrective* to any positive depictions of the transcendent since all attributions must of necessity come from the natural realm. It is not affirming merely that there is more to the transcendent but affirming its absolute otherness from all things natural. Positive characterizations may direct our attention away from other objects, but this still makes the transcendent one object among objects. The second step—the negation of all positive characterizations—corrects that and directs our attention away from all objects and points toward the transcendental reality.

The *via negativa* reflects the mystery of the transcendent, and it does direct the listener's mind away from the natural realm. But maintaining a pure *via negativa* is difficult even for mystics and especially for theists who see the transcendent in terms of a moral person. Even Dionysius the Areopagite, the father of the *via negativa* in the Christian tradition, who claimed God is ultimately a "divine darkness" beyond any assertion or negation, wrote a book on the symbolism of divine names. Moreover, for all mystics there is always a basic affirmation beyond the negations—a reality that is experienced. Thus, although the *via negativa* is a movement beyond affirmations, it is never merely a denial—there is a "negation of negation." Certainly the reality of a transcendental reality is not denied by theistic mystics, even if it is Eckhart's "God beyond God." However, one must ask how this differs in the end from the affirmative way prior to the *via negativa* and whether, as Plotinus said, the "sheer dread of holding to nothingness" forces the mystic back to the everyday realm of language.[29] Nevertheless, the introduction of both another state of consciousness and another dimension of reality makes the notion of "ineffability" useful to indicate the direction of the transcendence. But it remains a matter of hyperbole, like the medieval Christian Marguerite Porete saying that God is so much greater than anything that can be said that everything that is said about him is much more like *lying* than speaking the truth.

In sum, the basic danger will remain that the unenlightened will translate anything mystics say into a statement about an object within the world, and so mystics may continue to deny that anything can be said about the transcendent. Nevertheless, if the depth-mystical experience is genuine, the transcendental reality is not literally incomprehensible and indescribable: some of the mystery of the transcendence can be comprehended—and stated—by mystics even if more remains. The chief qualification of mystical depictions is only that their true referent is not an object in the world, not that the meaning of the terms is something non-ordinary.

Paradox

The problems with language also relate to a common practice among mystics: saying something about what was experienced and then immediately denying it—in short, paradox. For example, Mahayana Bodhisattvas exclaim "Countless beings should I lead to *nirvana*, and yet there are none who lead to *nirvana* or who should be led to it." Critics take such remarks to be the height of irrationality since we cannot consistently hold at the same time two beliefs we believe are inconsistent, or they see these paradoxes as a deliberate attempt at obfuscation because mystics know they have nothing to say; thus, paradoxes are grounds for rejecting everything mystics say about alleged transcendental realities. And philosophers in general can ask how we can tell when a paradox points to a profound truth and when it is merely a muddle. As Ronald Hepburn puts it: "When is a contradiction not a *mere* contradiction, but a sublime Paradox, a Mystery? How can we distinguish a viciously muddled confusion of concepts from an excusably stammering attempt to describe what has been glimpsed during some 'raid on the inarticulate', an object too great for our comprehension, but none the less real for that?"[30]

Many "paradoxes" of the transcendent are simply conflicts with everyday ideas, which is only to be expected if the transcendent is "wholly other." Some claims appear paradoxical until they are explained because they conflict with accepted beliefs or have a conclusion that does not seem to follow from accepted premises. For example, the basic panentheistic claim that God is transcendent and yet immanent seems paradoxical to theists who think only in terms of either a transcendental theism or pantheism. But it can be explained in terms of analogies from the natural world (e.g., how a dreamer both is immanent to everything in the dream and yet transcends the dream), and the apparent paradox disappears. Many other paradoxes simply involve statements from two points of view. They are not really paradoxical since two different claims are not affirmed of the same thing from the same point of view. In the case of the Bodhisattva claim, it is the "two levels of truth" for claims within the world: an "everyday truth" involving true statements about beings is juxtaposed with a "higher truth" that there are no beings (since there are no individual substances creating distinct and thus "real" entities). Accepting that language "mirrors" reality also can result in paradoxes: first stating something positive from a worldly point of view and then immediately denying it since the transcendent is not an object within the world. Such paradoxes are simply the positive attribution of a feature and the *via negativa* combined in very a concise form. There is nothing irrational in this process—it is merely the product of wanting to say something about the transcendent but believing that language distorts that reality. If we reject the mirror-theory, the paradox dissolves since there is no need for the second part (the denial of applicability). Thus, in one way or another the "paradox" in a mystical paradox may be eliminatable.

Many people are not bothered by contradictions. In fact, the religious do seem to revel in paradoxes. Christians have no qualms about calling contradictory things "paradoxes"—thus, they can affirm one claim ("Jesus is entirely human") and turn around and immediately affirm the opposite ("Jesus is entirely God") with a smile.[31] Many theologians like to use the alleged paradoxes of particle physics as an excuse to accept any problematic claim in religion as a "paradox." Indeed, many theists believe something of the transcendent precisely because it is impossible—if it made sense, we would not need faith. To quote Tertullian's two most famous remarks, "That the Son of God died is to be believed because it is absurd (*ineptum*)" and "It is certain because it is impossible."[32] At least in science and mysticism, the paradoxes result from encounters with reality and thus are experientially based in a way that theological thought is not. Part of the glorification of paradox in both mysticism and theology may be for soteriological purposes: its shock value may get people out of thinking in ordinary terms, and thus this use of language may help to evoke the experience. The Zen use of *koans* to break the grip that concepts normally have on our mind is a prime example of that.[33] More generally, if the religious treated transcendental realities as mysteries truly transcending this realm (with no revelations countering these mysteries), then the fact that our attempts to understand would produce paradoxes is not surprising.[34]

For mystics, any concepts may make the mystical seem paradoxical when mystics know by experience that it really is not: the sense of paradox only comes from trying to conceptualize what is experienced outside the experience in dualistic consciousness. Consider trying to "express" a three-dimensional cube in the familiar two-dimensional Necker stick drawing. It is a mixture of correct and misleading information—the straight lines and number of vertices reflect the cube, but the angles in the drawing are not all 90 degrees. More importantly, the drawing distorts the cube's basic nature by omitting a third dimension. Being forced to draw in two dimensions introduces this omission and inconsistencies for the angles, but there is nothing we can do about it. The drawing is in fact an accurate representation, but we need the experience of actually seeing a cube to see how the drawing is correct. The drawing cannot convey its own flatness: the missing third dimension cannot be "captured" by the drawing. Thus, studying the drawing is no substitute for the experience of a real cube. Similarly, language does the same to the transcendent: mystical claims do not falsify, but we need the mystical experience to see properly how they apply and are correct, indeed even to understand the claims properly. The problem is not remedied by introducing a new language—a different map projection, as it were—since all languages are "dualistic" and thus cannot reflect the ontological nature of what is nondual. (Also note that we can experience the three-dimensional cube as it is and are not limited by our two-dimensional representations. That is, our apprehension of the real cube is not constrained merely to what is drawable. Similarly, there is no reason to believe the mystics' conceptual systems restrict what they experience—an experience can get through whatever our conceptual systems may be, contrary to what postmodern constructivists believe.)

The drawing analogy does show that we can apprehend a reality even if the result of trying to translate it into language is paradoxical. However, language is one tool that can "draw" the fact that it is a drawing (contra Wittgenstein), and once this is done we can see that reality need not mirror language. Thus, mystics can easily end up speaking in paradoxes: use of "worldly" language leads to self-contradictions when applied to the transcendent, but they have experienced what is real. They know the inconsistencies are not inherent in what is experienced but only in our conceptualization of it in "dualistic" consciousness. That is, mystics can see the drawingness of their "drawings" and see that the "paradoxes" are not really paradoxical, but they also realize that the unenlightened will not see this and so they deny that language can apply. In addition, the cube analogy shows how mystics can still function rationally while using paradox: they can think about the transcendent, believe specific things, understand their claims, and talk consistently about the transcendent even while using language that to those without the requisite experience seems contradictory. They have a good reason based in experience to speak the way they do. To that extent, mystics are like those scientists who have to speak bizarrely as the result of experiments. But unlike in mysticism, scientists try to work past any paradox—indeed, the counterintuitive is an impetus in science to develop new theories. But this does not happen in mysticism since there are no novel experiences to test one claim against another.

In sum, the use of paradox does not mean that mystics are irrational or do not know how to think. Even though reasoning alienates us from the depth-mystical experience itself by introducing another mode of consciousness, reason is not abolished in mysticism: it is a part of the mystical ways of life since after their depth-mystical experiences mystics need to understand the nature of what they have experienced.[35] Moreover, logic applies only to our claims, not to experiences or states of affairs. No experience is "rational" or "irrational." Experiences do not conform to the canons of reason or conflict with them—they just are. Any experience is nonrational; only our attempt to depict or understand what is experienced can be either rational or irrational. Mystical experiences do not differ from other experiences in this regard. Paradox is introduced only when mystics try to understand the nature of what was experienced or even just to indicate what is experienced. It is a product of our use of concepts and of our search for explanations, not of any experience alone.

In the end, mystical paradoxes merely point to the need for an experience to understand the reason for advancing an apparently contradictory statement. Transcendental mysteries resist explication in "worldly" terms and thus contradictions do arise in the mystics' accounts. That the "two-dimensional" linguistic projects of our rational mind cannot capture a transcendent reality is only to be expected. This is true even if the cognitive content of the paradox can be stated in non-paradoxical terms (and then denied because the language suggests that what is experienced is just another item in the universe). Paradox thus is a way to point away from accustomed ways of thinking toward the transcendent.

Mysticism and Mystery

The verdict of this discussion is that, even if we accept that mystics are in contact with some transcendent reality, they, like the unenlightened, must still try to understand its nature in "dualistic" consciousness, and the conflicting claims of the different mystical ways of life indicate that mystics too must use their reasoning to accomplish this. Thus, the bottom line is that mystics, despite their experiences, are still not in a position to comprehend the fundamental nature of what they experience. Thus, the mystery of the transcendent remains even for mystics who are in contact with it. Paradoxically, unlike in science where the mysteries involve the unknown and baffling workings of nature, here is a case of something that is "bright and dazzling" to the mystics and yet is equally deep and mysterious.

For this reason, many religious believers argue that we must rely exclusively upon the transcendent's self-disclosure—i.e., revelations—for whatever knowledge of the transcendent we can have. No other experiences of the transcendent are possible, and all human reasoning is merely our groundless speculation about God's unseen nature. If one or more of the innumerable revelations that have allegedly occurred throughout history and around the world turn out to be true, then a transcendental reality exists and its nature and plan for us is a little less mysterious. But nothing more will be said about revelations here: revelations are matters of faith—to accept them we have to rely upon another person's experience for authority or accept a communication from God unique to ourselves alone, unlike with mystical experiences which we can each cultivate and which do not contain unique messages. With mysticism, others can at least empirically verify the existence of those states of consciousness; whether they can verify various mystics' claims about transcendental realities is another matter.

In addition to the mystery of the nature of any transcendental realities, the basic mysteries of being—the nature of beingness and why anything exists—are not dispelled by having a depth-mystical experience or by developing mindfulness. Beingness may be apprehended in a mindful state or directly apprehended in its simplicity in the contentless depth-mystical experience, but no new information about its nature is provided in either experience. No answer is given to the question of the relation of an underlying "being" and the realm of "becoming," although they share the same being—the problem of "the one" (being) and "the many" (the realm of structures and becoming) remains as profound for mystics after depth-mystical experiences. Emphasis is shifted to consciousness: perhaps the depth-mystical is "pure consciousness," i.e., an awareness of consciousness empty of all content and functions, but it is still open to radically different interpretations—it is hard to claim consciousness underlies "objective" phenomena based only on the depth-mystical experience when that experience is equally open to simply being the ground of the self or simply ordinary awareness void of content. In fact, if anything, the depth-mystical experience shows consciousness to be feature-less—consciousness is what observes and cannot be observed.[36] Either experience may increase a sense of awe and wonder at the that-ness of things but may not—one may serenely or joyfully accept the mundaneness of it all. And the depth-mystical experience may lead to a sense that the natural world is unreal. If anything the mystery of beingness is increased, not diminished, by such experiences.

The mystery of why anything at all exists may also be intensified but not answered by these experiences. The depth-mystical experience may be our most intense experience of reality and even our *summum bonum*, but the experience provides no answers to as to why this particular universe or any universe at all would exist, let alone why a source exists. The peace given in the experience may still our questioning. Or to explain why this realm exists, mystics, like others, must fall back on speculation. In Advaita, creation requires no act of will but is just a natural activity of the transcendent—simply "play" (*lila*). There is no further explanation of why the transcendent exists or why a realm of change is its natural activity. Nor do Advaitins explain why there is orderliness or causation or any structures at all in this realm. Nor has it an answer as to why the illusion of the reality of this realm remains after one is enlightened. All it says about the fate of the universe is that it will slowly disappear as more and more (illusory) selves are enlightened by ending ignorance (*avidya*) and are thereby removed from the (equally illusory) chains of rebirths. Buddhists go in the opposite direction and argue that all speculation about origins and meaning only interferes with ending our suffering and thus such questions should be set aside unanswered. The mystery of being thus does not figure prominently in Buddhism, although Buddhist theorists did end up arguing over such matters as what are the exact components of the experienced world and their nature. Mystics in general also have no explanation for history; mysticism's ahistorical knowledge tends to devalue the significance of history and the natural realm in general in favor of the changelessness of a transcendental reality.

Whatever the answers we devise as to why anything exists or why this realm exists, the important point here is that mystical experiences will not contribute to them. These experiences are void of any content that would provide an answer to such questions as why are we here—at most, all that mystics experience is the bare beingness of reality, with an accompanying sense of joy. The mystery of natural suffering becomes only more impenetrable. As noted above, some mystics have a sense of a loving source of this world—an all-encompassing love that makes all life precious—or a sense that everything is ultimately all right. But, again, how can this be reconciled with the suffering of animals and people for millions of years of evolution? Moreover, even if mystics know something of the nature of the transcendent, mystics typically affirm a depth to the transcendent that is unfathomable and thus not fully knowable. The transcendent is still "deep and profound," to quote Laozi on the Dao, despite being experiencable.[37] The transcendent is not "wholly other" since humans can experience it directly, but the mystery surrounding the transcendent ends up not being penetrated either by our thought or our experience. The depth-mystical experience is no more than a "shaft of light" from a mysterious "darkness." Even participating in such a reality does not help evolved finite beings such as ourselves to know the ultimate depth of reality.

Conflicting Claims and the Limits of Conceptions

Mystics are caught in the dilemma of needing conceptualizations but realizing that any conceptualizations introducing a foreign state of consciousness. There are two problems. First, we literally cannot imagine the transcendent—it is simply beyond our mind. Second, any concepts applied to the transcendent will be misinterpreted by the unenlightened as referring to an object among objects in the universe. Thus, when the unenlightened use mystical concepts, in an important sense they do not know what they are talking about. But mystical utterances are not so much literally paradoxical as simply revealing an inability of the human mind. We have a limited capacity to understand what is experienced in the depth-mystical experience, and even mystics outside their own experiences must try to understand. Our minds need a focal point to comprehend anything—the way we are constructed, we cannot live in a cloud of mystery. Thus, conceptualizations will always be an essential part of mystical ways of life, no matter how central mystical experiences are. Mystical claims to knowledge will reflect these conceptualizations, even if all concepts are absent in the experiences themselves. This means some degree of theory is unavoidable. But again, the understanding can interfere with the experiencing: it directs attention away from the experience and keeps us within the realm of representations. Thus, there will always be tension between the need to comprehend and the mystical thrust to get away from conceptualizations.

That different mystical systems present conflicting claims on the nature of what is experienced should be sufficient to realize our limitations in understanding. Advaita Vedanta, Samkhya, and different Eastern and Western theisms have different answers even as to whether there is one source to natural world. But, due to the power of their experiences, it is difficult for mystics to maintain an empiricist's stance and not ask about what they have experienced or to maintain openness and a sense of mystery when they think they have been in contact with the ultimate reality. So too, it is difficult for mystics to maintain a skeptical stance and renounce all transcendental beliefs. Indeed, the silence of the mystics is the opposite of the silence of the skeptics: it is based on knowing something that cannot be expressed adequately. And the writings of most classical mystics indicates that they were not at all skeptical: they typically believe that their particular system is correct or at least the least inaccurate, and that anyone having a mystical experience will confirm their beliefs. Only the comparative study of mysticism brings out the limitation of human efforts to understand here. As Agehananda Bharati says, the mystic knows very little about the transcendent: as theologians, mystics are as good or bad as they were before they had their mystical experiences.[38]

Critics of mysticism take the doctrinal conflicts as evidence that there are in fact no transcendent realities and that mystical experiences are completely subjective phenomena. But the conflicting interpretations do not rule out that mystics in fact have contact with a transcendent reality, and their experiences will constrain the options in mystical beliefs after their experiences, even though they vary in their interpretations of the nature of what is experienced. But the point is that having that experience does not end the need to speculate in more ordinary states of consciousness about the nature of what was experienced. Nothing new about the transcendent is being discovered today through the replication of these experiences. The only substantial developments in mystical conceptualizations within the world religious traditions since the Middle Ages have been in reaction to science. And since there can be no further original in-put, there is no reason to believe there will be any future changes except to reflect the changing interests within a culture. We are left with very little that we can claim to *know* about the transcendent from mystical experience even if these experiences are genuine.

The Thrust Against Mystery

Mystics may suffer from the same compulsion as the religious in general: the desire to close off mystery. The thirst for transcendence in religion is not necessarily a thirst for mystery. In fact, as discussed in the last chapter, theology can be anti-mystery. As paradoxical as it may seem, in the words of David Burrell, "to know the unknowable God" becomes the "quintessential theological task."[39] Thereby, the sense that we understand is substituted for mystery. The analytical function of the mind is substituted for stillness. Words are substituted for silence. Theological constructs become the center of attention, not the experiences that should keep theological construction tied to the transcendent. The search for explanations replaces the experiential orientation of mysticism. Theologians can easily get caught up in epicycles of their own constructs generating new intellectual "mysteries." And mysticism falls prey to the same problem: over time, doctrines and explanations can become as central to mystical traditions within different religions as to nonmystical religiosity. Affirmative theology can push out the *via negativa* even here. Overall, it is hard to distinguish the mystical and nonmystical sources in any developed tradition's theology when mystics themselves must rely upon their tradition's doctrines to understand their own experiences. Even the mystically-minded are liable to end up with the philosopher's disease.

In addition, the intellectual backbone of a religion may become more and more disconnected from the mystical. This split between "theology" and "spirituality" in the West began at the end of the Middle Ages when theology became a university discipline and mysticism became a matter for monks.[40] The "rational" approach in theology replaced any "experiential" approach. Children's religious education becomes a matter of rational concepts and moral precepts, not inculcating spirituality. God becomes "pure reason," and so nothing mysterious or inscrutable lies in his nature. But without an experiential base, God becomes merely an explanatory posit—an entity to explain the universe or what happens within it—and not a "living presence" of a mystical experience. It becomes the paradigm of the products of thought that mystics condemn. A bloodless, deistic "god of the philosophers" will replace the "god of the prophets"—an entity whose existence can be accepted as a metaphysical explanation without any religious reaction.

The religious impulse may be to sense mystery, but we humans also quickly want the comfort of a familiar image and the concrete symbols to fill out a full religious life and to satisfy our urge to believe we understand. We want to know, but we realize that we cannot know—terms like "transcendence" and "wholly other" only turn our attention away from the world we know and focus it on the fact that we do not know. The tension between the two needs cannot be removed. Mystical experiences are a way to protect against absolutizing beliefs or creating mental idols in an area where silence is more appropriate, but some conceptualization is always needed for any understanding. The *via negativa* preserves the openness and mystery of the transcendent, but we, with our analytical mind, also want to know what is being preserved. Mystical experiences only increase the tension between the openness of reality and the closure of our concepts here.

In the end, mystery cannot but be part of the transcendent (if one exists). However, being what we are it is impossible for humans to maintain such openness. Seeing mystical experience as not providing information but only as an awareness of another dimension to reality—and thus creating only more questions—is not popular even among traditional mystics. That there is an unfathomable mystery at the heart of things is hard for anyone to deal with, even if the religious pay lip-service to the idea. The admission that we do not know is hard for anyone to make. Nevertheless, more humility concerning alleged transcendental realities may be called for, even for those who have had mystical experiences.

Notes

1. E.g., H. Lewis 1961. He uses the term "intuition" to refer to "the apprehension we have of the necessary being of God," although he adds that it should not be treated as any other type of "intuition" (pp. 230-31).

2. Some postmodern theologians today reify mystery into a thing—Mystery with a capital "M"—and, as bizarre as it may sound, they argue that the *absence* of the experience of the transcendent supposedly is our *experience* of the transcendent today.

3. Alister Hardy, *The Spiritual Nature of Man* (Oxford: Clarendon Press, 1983). One problem inherent in any such survey is that the participants may not mean the same thing that the questioners mean by such terms as "religious," "transcendent," "mystical," "feeling," "oneness," and so forth. For example, an experiencer may label a premonition as being momentarily "one with God." Such surveys are also typically framed with a theistic bias; for example, classical Advaitins would not speak of a "presence" of a transcendental reality as if it set off apart from us.

4. Friedrich Schleiermacher, *On Religion* (New York: Harper & Row, 1958). Whether he meant the "feeling of absolute dependence" to be an *emotion* or a *cognitive experience* is not clear—the term "religious experience" in the modern sense had not yet been clearly formed. Another issue is whether this feeling of dependence (or gratitude or obligation) is only a projection from ourselves of knowing the fact that we could not exist.

5. See Exodus 4:10-16; Jeremiah 1:9; II Corinthians 5:20.

6. Recently a gay Episcopalian priest was appointed a bishop in America. He did not speak of revelations, but he did say "Now God seems to be calling me to another journey" and "God is trying something new." Was he receiving some communication from God, or was he just giving a positive religious spin to what he wanted?

7. The term "mysticism" derives from the Latin word "*mysticus*" and ultimately from the Greek "*mustikos*," coming from the same root as "mystery"—*muein*. It meant "closed" or "hidden" and came to mean silence or secret (i.e., what should not be revealed). The term "mysticism" is a modern invention, but the adjective "mystical," as adapted by early Christian theologians, referred to the mystery of the divine and came to refer, not to an inexpressible reality or experience, but to certain liturgical matters such as the invisible God being present in sacraments and to the hidden meaning of scriptural passages. Only in the modern era, when spirituality separated from general theology, did the term come to refer exclusively to certain types of inner religious experiences and states of consciousness. (The concept of "religious experience" is also a modern concept.) Nevertheless, by the Middle Ages "mystical theology" meant the *direct awareness of God* (rather than the knowledge gained through human reason), not the discipline of theology in the modern sense, and the "mystical meaning" of the Bible meant *the hidden message for attaining God directly* beneath its literal meaning and other symbolic interpretations. Thus, "mystical experiences" remained central in the Christian understanding of "the mystical."

8.   This is not to deny that mystical experiences can occur *spontaneously* to a person without any cultivation, and the impact of such experiences may transform the experiencer. But classical mysticism was never about isolated mystical experiences.

9.   E.g., *Dao de jing*, chapters 19, 48, 64, and 81. See Jones 1993, pp. 47-55.

10.   Mystical mindfulness involves an "unknowing"—emptying one's mind of all cognitive matters. This should not be confused with *everyday mindfulness*—e.g., being mindful of others or of our rights or of our place in the society. For a study of the latter type of mindfulness, see Ellen J. Langer, *Mindfulness* (Reading, Mass.: Addison-Wesley, 1989).

11.   One may argue that if the transcendent is completely unknown, then we cannot even begin to speak of it as "unknowable"—we must have some experiential awareness of it even to say that, and so it must be real. But "the transcendent" is also an intellectual construct posited to explain things in the phenomenal world. Thus, we could still speak of it as "unknowable in itself" without any experiential awareness ever being involved.

12.   See Munitz 1965, 1986, and 1990 for a nonmystical defense of this position. See Nieli 1987, p. 69 for the possibility that a mystical experience lay behind Wittgenstein's philosophy.

13.   *Brahma-sutra Bhashya* IV.1.15. Under Advaita, the phenomenal world and the body are the products of ignorance (*avidya*)—they should *vanish* when ignorance is replaced with knowledge of Brahman. The Advaitins' answer is that *karma* that has begun to bear fruit must continue to do so even after enlightenment. But this would mean that *karma* has some *reality* with power even over knowledge of Brahman.

14.   *Dao de jing*, chapters 1, 4, 6, 11, 16, 25, 40, 42, and 51.

15.   See Lloyd P. Gerson, *Plotinus* (London: Routledge, 1994), pp. 192-98.

16.   Edmund Colledge and Bernard McGinn, *Meister Eckhart: The Essential Sermons, Commentaries, Treatises, and Defense* (New York: Paulist Press, 1981), p. 233.

17.   For a defense of this position, see Michael Stoeber, *Theo-Monistic Mysticism: A Hindu-Christian Comparison* (New York: St. Martin's Press, 1994), pp. 95-99.

18.   For discussions of the epistemological issues concerning constructivism, see Steven Katz, ed., *Mysticism and Philosophical Analysis* (New York: Oxford University Press, 1978); Robert K. C. Forman, ed., *The Problem of Pure Consciousness: Mysticism and Philosophy* (New York: Oxford University Press, 1990); Jones 1993, pp. 8-11, 19-46; Jones 2000, pp. 276-79.

19.   For more on the neutrality of science, see Jones 2000, pp. 271-75, 279-88; Newberg, D'Aquili, and Rause 2001. To Newberg, D'Aquili, and Rause, mystics are reporting a genuine neurological event (p. 143), even though they realize that the biology of the event remains neutral on the issue of whether it is merely a neurological event or a genuine insight (pp. 178-79).

20.   This sentence is adapted from Plotinus (*Enneads* V.5.6). He argued that we cannot know the nature of the One because we know only of the One through what comes out of it—its emanations—and hence we do not know its nature in itself. All we can know of the One is that it is the cause of all things (*Enneads* VI.8.11.1-3, III.8.10.32-35).

21.   *Brahma Sutra Bhashya* II.1.10-11.

22.   Nor, as discussed in the last chapter, are religious theorists learning more about how the transcendent. Rather, theologies change as science expands our view of the world, including adjustments to our view of the transcendent. Abrahamic theists' views of the nature of God have evolved over the last three thousand years from a warrior chieftain to a universal being, but mystical experiences no longer offer fresh in-put in the process. One issue is whether any new ideas of God (e.g., as merciful rather than vengeful) are the result of new revelations from the transcendent, or merely the product of our thinking about what

should be the case, without any new experiences.

23. *Taittiriya Upanishad Bhashya* II.1.1; *Brihadaranyaka Upanishad Bhashya* I.4.7. Plotinus said the same of the One (*Enneads* VI.8.13).

24. *Taittiriya Upanishad Bhashya* II.1.1.

25. See Carabine 1995; Turner 1995; Williams 2000. If the transcendent is utterly unlike the components of the world, one might ask why wouldn't negative terms (nonpersonal, and so forth) then apply? The problem is that the transcendent is not a reality to which the idea of "non-personal entity" would apply. Any negative characterizations would still render the transcendent a thing within the world—a thing among the non-personal parts of the universe. In addition, calling it "non-personal" would be siding with Advaitins over theists on the nature of the transcendent and not be merely a formal remark about the logical status of a reality that is "wholly other."

26. *Brihadaranyaka Upanishad* II.3.6, III.9.26, IV.2.4, IV.4.22, IV.5.15. Plotinus echoed this regarding the Good (*Enneads* III.8.11.12-13, V.5.6.22-23). Eckhart said that God is "neither this nor that"—God is a non-god, a non-spirit, a nonperson, a "pure nothing."

27. *Majjhima Nikaya* I.431, II.166; *Samyutta Nikaya* IV.373-402.

28. *Enneads* V.3.12, V.5.6, VI.7.38.4-5. Plotinus said that the term "the Good" cannot apply. Any property is a characteristic of the universe's being, not of the One (the source of being)—even "the One" only indicates a lack of plurality, not one object among objects.

29. *Enneads* IV.7.38.9-10; see also VI.9.3.4-6.

30. Ronald W. Hepburn, *Christianity and Paradox: Critical Studies in Twentieth-Century Theology* (London: Watts, 1958), p. 17.

31. Whether paradoxes resulting from other conflicting *religious doctrines*—rather than paradoxes resulting from trying to express the *experience of a transcendental reality* in worldly terms—can be resolved is another matter. For example, many Christians affirm both our freedom of will (so that we, and not God, are responsible for sin) and that God absolutely controls every event (so whatever happens was ordained by God) without being bothered by the blatant contradiction. They accept that "human reason" cannot resolve that mystery and dismiss the situation as simply "cognitive dissonance." See, e.g., Terry M. Gray, "Biochemistry and Evolution" in Keith B. Miller, ed., *Perspectives on an Evolving Creation* (Grand Rapids, Mich.: William B. Eerdmans Publishing, 2003), pp. 285-87.

32. How can we be said to believe something we take to be absurd or gibberish? If there is no intelligible doctrine, exactly what is it one is supposed to believe? In quantum physics, scientists may come up against phenomena they cannot explain, but they do not contend that they believe some doctrine that they admit is unintelligible. Only in religions based on alleged revelations are believers expected to accept doctrines even if they find them unintelligible—we supposedly will grasp their meaning in the afterlife.

33. Zen *koans* are mental puzzles utilized in a form of analytical meditation designed to force a pupil to see the way concepts control our mental life and thus to attain a sudden breakthrough into our true selfless nature, a state free of the grip of thinking and experiencing through concepts. *Koans* sound like meaningful sentences—"what is the sound of one hand clapping?"—but by reflecting on them the pupil eventually sees that they are like "she is a married bachelor" and eventually sees that all language does not involve reference to "real" (permanent, independent) entities.

34. The vast majority of philosophers think that no contradiction can state a true fact—that, in the words of Quine, someone who maintains that a contradiction might be true just does not know what they are talking about. In dissent, Graham Priest (2002) argues that all attempts at closure at the boundaries of thought and of what is knowable lead to

contradictions—an effect of any conceptual process to cross those boundaries, resulting in paradoxes of self-reference—but that these contradictions state truths, and something contradictory about reality itself renders such contradictory statements true.

35. Mystics do not have any "special logic" or "rationality unique to mysticism" but are rational by "Western" standards. See Jones 1993, pp. 59-78. Of course, what is considered rational for a person to believe will depend on the reasons and beliefs of a particular mystic's culture and era (e.g., it was once rational to believe the earth was flat and did not move). Thus, mystics will differ from modern "common sense" in their arguments since their experiential base is broader than ours. But they can be as rational as nonmystics and can construct a logical argument when the occasion arises. Buddhist logicians such as Dignaga and Dharmakirti even resort to such sophisticated arguments as *reductio ad absurdum* that rely on the law of noncontradiction. Reasoning conflicts with having a depth-mystical experience, but outside that experience some mystics have written works advancing reasons for their understanding of what is experienced and against their opponents' views. Most works by mystics, like most writings, do not contain developed arguments, but mystics can write books of argument if the occasion calls for it, as shown by Shankara's commentaries on basic Hindu texts. Mystics will also use styles of arguing and forms of formal presentation that are particular to their culture or era, but they do not as a rule reject such basic rules of logic as noncontradiction (that something cannot be both x and not x) or the law of the excluded middle (that something is either x or not x). The Buddhist *Questions of King Milinda* and Nagarjuna's works are good examples. Their arguments are implicitly based on these rules of logic. Nothing exhibits an "alternative logic." We may not agree with their premises, but the form of the arguments is logical by "Western" standards. Nor is Nagarjuna "using reason to destroy to destroy reason"—he is trying to show by the use of very logical reasoning (and without appeal to mystical experiences) that any metaphysical system that affirms self-existing, permanent entities involves inconsistencies and thus cannot be held. See Jones 1993, pp. 79-97. Similarly, the Daoist Zhuangzi is not, as is often alleged, being "antirational" or "anti-intellectual" in using reason to show that reasoning cannot establish one point of view either as absolute or as otherwise uniquely grounded in reality.

36. Depth-mystical experiences alone do not constitute grounds for the independence of consciousness from matter or a panpsychism, although it can easily be fitted into such theories. If a "pure consciousness" does exist, this would prove functionalism (which equates consciousness with information-bearing and behavior-controlling functions) wrong once and for all. But it does not prove that consciousness must exist independently of the body. Nor does treating consciousness as a "field" mean it must exist independently of the body or be universal: magnetic fields, upon which the idea of a "consciousness field" is modeled, can be the products of individual magnets and not exist independently of the physical objects creating them. That is, a "consciousness field" may still be produced only by a brain.

37. If a transcendental reality is open to unmediated experience, then there is no Kantian noumenon to this reality, even if there has an unknowable depth. Of course, whether there is a *further* reality grounding what is transcendent to our world cannot be answered by an experience of the source of only this world, even if the overwhelming sense of reality in what is experienced naturally leads mystics to believe they have experienced the ground of this world and to the conviction that they have touched the ultimate reality.

38. Agehananda Bharati, *The Light at the Center* (Santa Barbara: Ross-Erikson, 1976), p. 59. To Bharati, mystical experiences in fact have *zero* cognitive content (p. 48).

39. Burrell 1986, p. 2. John Haught advocates using both the positive and negative ways in religion. Haught 1986.

40. Greeley 2003, p. vii.

— 8 —

Reviving Mystery

A philosophy unaware of mystery would not be a philosophy.
— Jacques Maritain

Humankind cannot bear very much reality.
— T. S. Eliot

The unsettling conclusion to this discussion is that we know far less about the fundamental nature of reality than we like to suppose. At the root of everything, there is mystery. It is central to in our situation in the world. To some, this may seem like merely stating the obvious, but others adamantly deny that there can be any ultimate mysteries to the universe, believing instead that questions of mystery are ultimately meaningless, that the universe is rational to the core and our minds can reflect this structure, or that they know the answers to ultimate questions. But the problem is not merely that there are competing answers to the big questions, but that any final resolution is not likely—we are not in a position to know the answers. We are the one creature that knows we are here and also knows that we do not know why. We cannot just sigh and say "oh, well" and turn to focus on what we know or on those problems that seem amenable to conquest without cutting ourselves off from our total context—we need to realize our dilemma to realize our actual situation in the world. Thus, the importance of mystery today needs to be reaffirmed.

At least five areas remain open: why of anything existing at all, any noumenal realm behind our experienced world, why there is order to nature, the possibility that we will never know all the workings of nature, and the possibility that transcendental realities exist. The origin of life and the origin and nature of consciousness also are current mysteries. Indeed, things have changed only a little since 1880 when Emil DuBois-Reymond listed seven enigmas: the nature of matter and force; the origin of motion and change; the origin of life of the world; the apparently

271

preordained orderly arrangement of life; the origin of sensation and consciousness, and rational thought; the origin of speech; and the question of free will.[1]

Philosophy, science, theology, and religious experiences all fail to close off the mysteries underlying all we claim to know. Indeed, all the attempts to close mysteries in the end only *increase* our awareness of an element of intractable mystery to reality: philosophers by analyzing the issues only show that we do not have a grasp on the big questions; scientists solve more of nature's how-ness puzzles than postmodernists will accept, but they also reveal more and more complexities and levels of organization; theologians try to push the "why" mysteries out of the natural realm, but they can only do this by introducing a whole new transcendental realm to reality and a whole new level of "why" and "how" mysteries; even assuming for the sake of argument that mystical experiences are experiences of a transcendental reality, the mystics' conflicting claims show that they do not have a grasp on the nature of what they experience, and prophets' conflicting revelations do not prove any more helpful.

Thinkers have pounded away at the ultimate questions with reason in the modern era for more than two centuries without not changing our predicament. The end of all explorations has to be an acceptance that the ultimate questions remain alive and unanswered perhaps forever—in the final analysis, there is an impenetrable core to reality beyond our mind's capacity to grasp. Our reason and experience simply cannot resolve them. But we are still left with confronting what we experience. In sum, mystery survives in the modern world, even though we pay less and less attention to it.

The Centrality of Mystery

Of course, we can live avoiding mystery. That is certainly more in keeping with the spirit of our age. We can move through the world perfectly well without dwelling on any underlying mystery. Just as we can use our bodies without reflecting on the mind/body mystery of how we make the body do what we want, so too can we ignore all the other mysteries. We can fill our lives with the details of living and get along perfectly well without asking the big questions or otherwise contemplating the mysteries of reality. (And it is good to remember Søren Kierkegaard's remark that one way God punishes people is to make them philosophers.) Even if we do consider the philosophical problems of reality, we can simply acknowledge that there is mystery to reality with a shrug and focus all our attention only on the known. Or we can treat the unknown within the world merely as temporary puzzles and be confident that someday someone will conquer them, or at least they are solvable in principle by science or philosophy and thus are not ultimate mysteries. Either way, we do not have to take them seriously. In an age of social and cultural uncertainty, the desire to avoid anxiety leads to valuing ways of avoiding any type of mystery (such as turning to religious fundamentalism). In fact, there is no emphasis on mystery today in any of the fields covered. A culture of control prevails in intellectual matters, with the accompanying denial of mystery.

We can ignore or miss what is beyond our comprehension, but it is important to remember that we are aware of the fact that there is more to reality than we now know or perhaps can ever know. Mystery is a large blind spot in our way of looking at reality that is ignored today, but nevertheless it is an irremovable part of our situation in the world. Other animals may well have some knowledge, but we are the one animal that is aware that there are things about reality we do not know. A sense of something more—something beyond the horizon of our understanding—is part of our make-up. But so is our quest to answer "why" and "how." Thus, our search for understanding inevitably leads the philosophically-minded to asking the big questions, not just to a concern about what is necessary for getting along in the world. And when this happens, we are confronted with the unfathomable mystery at the heart of all of reality.

Many people feel that scientists have disenchanted the world—wrenching the mystery out of things, like revealing how magicians do their tricks. But while scientists have solved many puzzles, to date this has always led to many more puzzles—and it has also led many scientists to react with awe at the vastness and complexity of the universe that has been revealed. The interconnection of everything, the interplay of forces, the innovation throughout the cosmic and biological evolution of the universe that leads to intricate complexity, and the improbability of life on our tiny planet are all truly awe-inspiring. Even philosophers who believe that all about nature will be ultimately explained agree. However, many scientists argue that the how-ness of reality will remain deeper than our minds can comprehend. Indeed, the twentieth century has been called the century in which scientists discovered that there are ineradicable mysteries to nature. Moreover, science is limited in a philosophical as well as practical way: even if scientists eliminate all the how-mysteries of the workings of the world, the basic why-mysteries remain untouched. In fact, science adds why-mysteries: why the universe is as orderly as it is or indeed orderly at all, and where do the laws of nature come from?

Most centrally, there is the mystery of mysteries: why does anything exist at all? The very that-ness of anything and why there is a realm of becoming are mysteries. What is the nature of the "stuff" of the universe—"beingness"? In short, why-mysteries surround everything in the world. In addition, basic how-mysteries suffuse the most familiar things around us. But because we are so accustomed to the familiar on a daily basis we are not astounded by their nature. We are desensitized to the mysteries of the ordinary. This includes ourselves. How we can raise an arm—how the process is started by thought—remains a mystery and we have no grounds today to believe that it will not remain so. Why are there new levels of phenomena in the universe? How sense-perceptions and the rest of consciousness fit into the physical universe is not yet understood and may never be. How self-consciousness arose may remain a mystery. Is the mind dependent upon matter? Is it in fact simply a material product? Or is it independent of matter and not entrapped by it at all? For that matter, why are there living, conscious beings around who can pose these questions and ponder mysteries? We are a thoroughly integrated part of a universe evolving over billions of years, but how exactly do we fit in? Are we merely a very complex form of organized matter? Or are we special, even a necessary part of the scheme of things—or is that simply something we like to think? Yet our lives seem so fleeting in the grand scale of things. What happens when we die—is that the end of the matter, or are we reborn in some other realm, or are we reborn again and again in this world?

Add to this all the mysteries of the universe as a whole. Does the universe have an origin? Is there some "final cause" at work in the universe? Does the universe last forever? And what if there is an end of the universe? What happens when the universe "dies"? Are there multiple "universes," or is this one all there is? We also want to understand our place in all of this. What are we doing here? And where did we come from? We want to know our origins and destiny, the alpha and the omega of it all. Do we have a clue about the meaning of it all, if there is one? Is even asking about such meaning merely the result of our evolved desire to search for causes and reasons? Or is there a transcendental source to our world? Does our true environment have some transcendental dimension? Or is it just the natural world? If there is a source, is it an abstract deistic reality that simply automatically endows the universe with certain conditions, laws, and forces and lets it fly? Or is it a personal theistic reality who chooses the conditions of the universe, intervenes in his creation, and cares about humans? Or is there no source at all? Is the natural universe eternal and uncreated, like theists contend God is? Are we simply making up mysteries concerning a transcendental something that in fact does not exist? Did we ourselves create the idea of God, or did God create the brain and mind so that we can conceive and experience him?

All our answers will ultimately rest on something we accept as the final, unexplainable brute fact. Which brute fact do we pick: the universe or the transcendent? The easy loss of a sense of transcendence for so many today shows that this sense is not innate to humans. The default position in the West used to be that a theistic god created and ordered the world, but modernity has changed all that; the religious now need to show that there is a transcendental reality grounding a cosmic meaning to all this, indeed that the very idea is in fact intelligible. Not that science has *disproved* the possibility of a transcendental reality, despite what naturalists may claim, but we have changed our focus and interests. We are no longer sure the universe is created or purposeful. Nevertheless, the conviction that such questions have answers is hard to erase—our mind almost requires supplying answers to quiet its tendencies. Theists and naturalists both think they know the answers, and typically they are quite certain and often arrogant about their positions. A more prudent course is to admit that we do not know enough about such matters to form firm answers—indeed, we may nothing relevant at all. Thus, we should remain open. At a minimum, we are not in a position to know with any certainty the right answer or even how to approach getting the right answer.

Thomas Huxley coined the term "agnosticism" precisely to try to remain neutral on whether there are any unknown and unknowable transcendental realities and to remain free of any doctrinaire "*gnosis*" about whether such realities even exist, let alone their alleged features.[2] Being agnostic is not an act of timidity or denial but is the claim that we do not have convincing reasons to either support or reject a given belief. It has roots in Kant. It need not lead to the rejection of these questions. Arguably, agnosticism is the only intellectually honest position in our situation (if we accept that transcendental realities are at least a logical possibility), but it is hard to maintain: it means not being able to set foot in any camp in a metaphysical dispute. It is not indifference to the questions or refraining from studying them; rather, it is an admission that we have no compelling arguments on the issues, and thus that we should adopt a humble acceptance of the fact that we do not know. We are never satisfied with having no explanation, but we have to accept that beings like us may not be capable of knowing fundamental matters. And even if a transcendental source does exist, we are not capable of reading "the mind of God." (If there are good reasons to question the reasonableness of belief in the existence of a theistic god, this does not rule out a deistic type of source, as discussed in Chapters 5 and 6.) All in all, it is bewildering being the only creature at the intersection of the natural and transcendent worlds—if the latter exists.

However, as a practical matter, trying to remain open about the question of the transcendent or the meaning of life and to leave it as a mystery leads to living a naturalistic way of life in which we live as if there are no transcendental realities or life after death. It also leads to the debate between William Clifford and William James on the right to make belief-commitments. Clifford argued that it is always wrong in every situation for anyone to believe anything upon insufficient evidence. James responded that a leap of faith is acceptable on issues of human existence that are "forced, live, and momentous" when the evidence is inconclusive—we are forced to make a decision, and so we must choose. Are we compelled as a practical matter to take a side on the issue of whether there is transcendental realities or life after death once we are presented with the issue? Can we suspend judgment even in the face of a lack of evidence one way or the other? In short, is agnosticism an option for us? Whether the choice after reasons are exhausted is arbitrary is also a matter of debate.

But placing mystery more centrally in our worldview does not mean being preoccupied only with what we do not know. It is not constantly falling into silence or exclaiming wonder at every turn. Nor could we live in a world of pure mystery: if we are to survive, we cannot simply wander around being amazed by everything or totally absorbed, contemplating the mysteries surrounding what we know. Obviously, we should keep using our inquisitive mind and our investigative abilities to their fullest and always make every effort to learn more. Any conclusion of mystery should be staved off as long as possible. Science may be as Einstein said "primitive and childlike" when "measured against reality," but he adds that "it is the most precious thing we have." Accepting mystery, however, means recognizing the limits of our abilities and accepting that there is much to reality that is beyond our understanding. And it is more than just paying lip service to the idea that there is more to reality than we know: living with mystery means making this "more to reality" a vital part of our frame of reference and not just focusing on the parts of reality we think we have grasped. Genuinely realizing that there is more to reality than our pictures of things can contain will keep us more tentative in our claims to knowledge and more open to the in-put of reality. Mysteries produce a shock to all attempted systems of closure—they reveal something in reality that we cannot get our minds around and yet that we must accept. We resist openness by clinging to what we think we know. But not limiting our focus to what is known or to the next research problem or to what is manageable will give us a fuller sense of the abundance of reality. Facing the mystery of reality reveals the limited nature of what we know by revealing the frame of mystery that surrounds all we do know. It thus reveals the true context of our knowledge, keeps us from limiting our attention only to the known, and reveals the true environment in which we live. All of this can only enrich our lives.

Philosophy's Role

Philosophy can help or hurt this sense of mystery. First, anyone looking for final answers in philosophy will be disappointed—philosophers are much better at pointing out problems and asking questions of purported answers than at finding answers that stand the test of time. Indeed, the failure of the Enlightenment project to find absolute, certain bases leads many to despair of any study of big questions, while to others the fact that we cannot know the answers is exhilarating.

Second, as a rule, philosophers do not like mysteries—they prefer trying to establish Cartesian "clear and distinct" ideas and advancing at least tentative rational solutions to any question. Many would prefer to dispatch all mystery from our view of the world: our rationality demands that there be no ultimate mysteries to reality.[3] Any question that cannot be answered was not a legitimate question in the first place but only meaningless noise. Such questions are mere products of our way of thinking without any application to reality. Even those philosophers who accept unanswerable questions as meaningful want simply to acknowledge that fact, suspend judgment on them, and move on the answerable ones. There is no use dwelling on such questions—it is a waste of time leading only to groundless worry and confusion. We do not think earlier hominids could possibly solve the mysteries of physics—why should we think we now have the ability to answer to all mysteries at just this particular stage of our evolution? Indeed, if we had another hundred thousand years to work on these questions and had a mind with a hundred times more intelligence, we still could not answer basic why-mysteries that we are part of. What more could *any* being within the universe know about why it exists?

We simply are not in a position to attempt to answer any of the great metaphysical mysteries—why anything exists, free will versus determinism, and so forth. The only intellectually honest thing to do is to pass over these questions in silence and get on with our lives.

Nevertheless, this response misses the significance of mysteries for our place in reality: that we are in a position even to have mysteries is among the most important statements of what we are. That we can pose questions that are beyond our ability to answer does not mean they are not legitimate, even if such questions result only in our competing answers backed merely by different metaphysical faiths. This only shows that we have questions that we do not know how even to begin to resolve. Science may progress over the next thousand years or it may soon run up against the final limits of our abilities, but why-mysteries will remain in philosophy. Still, the big questions are hard for a reflective person to ignore, even if there has been no substantive progress on answering them in philosophy. Ignoring the big questions results in only a partial view of all that is real. And simply acknowledging mystery and moving on to more manageable problems leads to a preoccupation with only part of the total picture. In short, these mysteries cannot be banished without truncating our worldview.

But philosophers can do more than just expose hidden assumptions, clarify concepts, remove contradictions and confusions, and find problems with attempts at closure. Philosophy can also play another role: philosophers can keep us aware of the "something more" surrounding us. Indeed, philosophy can be seen as spotting the mysteries surrounding different fields of inquiry. In this way, philosophers have the opportunity to keep us open to mystery and not to close us off. Equally important, studying the big questions does not mean we simply end up back where we began: by reflecting on the big questions, we gain a clarification of the issues and a greater awareness of our situation in a world grounded in mystery. In fact, we may end up with more questions after studying the issues. But knowing that there are unanswerable questions is itself important. Even with no answers, we have a better perspective on the human situation. We can see more clearly the limits of our understanding and its true context. Thus, philosophy can be an antidote to all arrogance on fundamental matters. In short, philosophy can be the cure for the philosopher's disease.

Living Without Answers

Mystery is more than merely the lack of certainty or the existence of options in science, philosophy, and religion—it is the recognition of something more to reality that lies beyond our comprehension and that resolution of some questions is not going to happen. However, there remains the competing trend—our aversion to the unknown and our need to have the sense that we *know*. This tension exists in all fields but more so in philosophy and religion than science. Scientists live more easily with a lack of certainty in their profession—surprising empirical in-put keeps them open. It is with the more encompassing issues in philosophy and religion that we especially want to quiet mysteries. We prefer the comfort of the familiar to the strange and do not like to admit that some things are beyond our grasp. So too, we have evolved to search for causes and explanations in order to function success-fully in the world. Thus, it is hard to be silent about any mystery—our minds will automatically form responses to it. Moreover, a sense of mystery can generate anxiety and disillusionment, and thus the mind will work hard to eliminate the cause, allaying our anxieties with some conceptual shield. Philosophy, science, theology, and mysticism can all be the enemies of mystery. Conceptualizations used to focus our mind can block us off from the fullness of reality—we will focus only on the reality as seen through our thoughts and will be screened from the rest. In this way, what makes sense to us may become the criterion for what we accept as real. Our minds become fixated on the known and not open to what surrounds it. Closure thereby distances us from both the real world and distorts both the world and ourselves.

This is the danger of closure. Daniel Boorstin once remarked that the greatest obstacle to discovery is not ignorance but the illusion of knowledge. However, it is hard to keep an open mind. The poet John Keats spoke of a "negative capability" —of "being in uncertainties, Mysteries, doubts, without any irritable reaching after fact & reason." But resisting our need to substitute concepts for mystery, recognizing the limitations of our attempts at closing off mystery, and accepting that we do not know are all hard to do. Holding our explanations tentatively and seeing their limited nature is not easy but requires real effort. Nor is it easy finally to accept that something will remain a mystery.

Part of living with an abiding sense of mystery is wonder.[4] Most of us lose our sense of wonder by the time we finish school, but a child-like fascination with the world and the question "why?" can only enhance a human life.[5] The emotional response of wonder and the intellectual sense of mystery—wonder versus "wondering why"—are not the same thing, but they are related. Both can include astonishment at the majesty and workings of the world, awe regarding alleged transcendent realities, or astonishment at the sheer that-ness of reality. Such wonder is not just confused bewilderment at all that happens. Wonder itself does not reveal any new truths, but wonder may lead to "wondering why" and to seeing the world in a new light and to thinking in new ways. Wondering why leads to seeking causes. It may also lead to thinking outside the box and thus can lead to new scientific theories. But we cannot live a life of wonder only. It would stop scientific research if we never had any focus. So too, everyone must commit to some beliefs about both the world and the nature of a human in order to lead a life, even if we try to hold those beliefs open to revision. Mystery, after all, is a blank where we want knowledge, not something more constructive. We cannot lead a life based solely on mystery; we need fixed points to live by.

But living without answers on fundamental matters is not impossible. We can easily accept that the human mind is limited and that finite, evolved beings such as ourselves are not capable of knowing the fundamental nature of many things. We can accept that we know less than we would like. We can look at a bigger picture of reality and see our situation anew. This is not celebrating or reveling in ignorance or belittling our knowledge, but it is taking seriously that there is more to reality than we know and probably can know. We can be more open to the world if we recognize this. We should try to expand Nicholas of Cusa's doctrine of "learned ignorance" to the natural realm: the more we increase our knowledge, the more we are aware of the limits of our knowledge and the presence of our ignorance, and that in the final analysis we have no fundamental knowledge.

We live on the shoreline between the known and the unknown. Both must be accepted. Thus, living with mystery is a balance between openness and using our conceptual systems. It involves utilizing our conceptual attempts at closing off the unknown while treating them as just that—our all-too-human attempts at grasping what is ultimately beyond our grasp. Perhaps the professional life of quantum physicists is the exemplar of how to lead a fuller human life in the world: they are constantly encountering the unknown, but they can live with theories despite their problems without closure—indeed, it is only by remaining beyond the control of those theories that they can advance their science by seeing things in a new light. As the physicist Richard Feynman put it: "I can live with doubt and uncertainty and not knowing. I think it much more interesting to live with not knowing than to have answers that might be wrong." He notes that in order to make progress in science, we "must leave the door to the unknown ajar." Science remains very much a way of approaching reality rather than a body of factual claims. In a similar way, a sense of mystery can be very valuable to all our lives without the sense of not knowing overwhelming us. We remain open by asking the big questions and not settling with any proposed answers as closure. The loss of certain answers need not be seen as a loss at all but as a type of liberation. Indeed, life can be more exhilarating if we can live with this openness to the fullness of reality.

How Much Skepticism?

The beginning of philosophical wisdom is questioning what we think we know. But was the oracle at Delphi correct when she said that no one was wiser than Socrates when he claimed he knew nothing concerning profound matters? That is a rather depressing indictment of our condition. Indeed, the emphasis on mystery may lead to an endorsement of a radical skepticism in general. And it cannot be denied that a sustained philosophical skepticism cannot be refuted, even if it is ignored by many. Practical skepticism, on the other hand, does not rule out all knowledge. In particular, in the area of science it is hard to claim that we cannot say anything with assurance about the workings of reality. We do not have to go to the extreme of denying all knowledge—to claim, in the philosopher Frederick Ferré's words, that "our best knowledge is so shot through with mystery that to call it knowledge may some day seem a little arrogant and quaint."[6] Even though most scientific beliefs throughout history have proven to be wrong, the situation is not "everything you know is wrong." We do have some genuine knowledge of the workings lying beneath the surface and inklings of further mysteries. New predictions that prove accurate do strongly suggest that we are hitting some underlying structures and are on the right track and learning more. The expanding range of observation is increasing a core of solid knowledge. Some how-mysteries of nature may remain forever, but we do have some genuine knowledge, not just illusions placed between us and nature. Our knowledge may be forever incomplete, but this does not mean that it is currently totally inaccurate. The only mistake is taking our knowledge to be final and complete. We can accept it tentatively and with the knowledge that there is more we do not understand.

Overall, practical skepticism cannot erode our confidence in all of our claims to scientific knowledge even if we can never attain absolute certainty. However, metaphysical issues are another matter. Philosophical skeptics have a stronger case for our claims concerning these fundamental matters. We must reject one extreme: the confidence of both materialists and people in different religious traditions who are sure that they know the ultimate nature of the natural world or the answer to whether a transcendent reality exists or what happens at death. The practical skeptic's stance toward all our speculation is important for keeping it to a minimum. And empiricists would add that we should keep all our claims as close as possible to actual checkable experiences And philosophical skeptics rightly point out that some topics may remain beyond our abilities forever. However, being what we are we will continue to speculate about all these matters, no matter how groundless our speculations are.

Can we leave all fundamental issues open and hold ourselves neutral? Isn't remaining agnostic about why-mysteries after we have examined the problems the only intellectually honest stance, at least in situations where we can live without having to choose? Skepticism is a philosophical judgment, and thus it can be argued that we can maintain an agnostic stance toward all the fundamental mysteries within its sphere and still lead a life unaffected by that lack of commitment. For example, if one is leading a moral life, one can be neutral on whether there is life after death or not—either belief would not affect how one acts in this regard. Indeed, trying actually to lead a life of pure skepticism about all matters other than what one observes is difficult.[7] For example, we must act as if we have free will and as if the mind is causal, no matter what philosophical arguments we advance concerning materialism. All people have some basic belief-commitments and values in how they lead their lives, and holding basic belief-commitment tentatively and open to revision is not as easy as it sounds. Examining other options may expand our horizon and we may become more open to change, but we still must end up being committed to some option. And being flexible about basic belief-commitments may well lead to a lack of confidence in one's way of life, which may well shake up one's entire life. In short, reconciling commitment and openness may not be easy.

Humility

All of this suggests that we should emphasize how precarious our knowledge-claims are, and we should exhibit a little more modesty in what we assert about the great mysteries of the universe. Despite our accomplishments, humility can be the only reaction when we consider the limitations on our abilities. We may need to speculate in some matters to form a framework just to function in the world, but we should forego any sense of certainty in these speculations and accept our efforts tentatively—blithe confidence in our answers is at least as great an enemy of mystery as any tentative answers in themselves. But what Galileo called "the honesty of 'I don't know'" is hard to maintain. Nevertheless, what Voltaire said of theology applies more broadly to all our deep intellectual constructs: "It is truly extravagant to define God, angels, and minds, and to know precisely why God defined the world, when we do not know why we move our arms at will. Doubt is not a very agreeable state, but certainty is a ridiculous one."

We may not like to admit that there is anything beyond our capacities, but even though our rationality can reveal something of reality, there is little reason to think it can exhaust it. In fact, why should we think that reality to its core must be comprehended to us? It may well not be possible to devise a consistent set of concepts that reflects all of reality. But why would we think that our ability to comprehend and conceptualize could put a limit on what is real? It is not merely a matter of why-mysteries but also how-mysteries. At least some physicists have no problem with the idea that there may well be fundamental laws of the universe that are beyond our cognitive capacities to understand. Compared to the inexhaustibility of reality, let alone any possible transcendent reality, are we much better equipped to figure out the ultimate nature of things than ants are to understand quantum physics? To quote Noam Chomsky: "I'd be as surprised if humans could under-stand all things as I would if a dog could." Moreover, it is not clear what capacities any beings within the universe could possibly have that would enable them to solve all our mysteries. For example, how could any mind become an object of third-person study for us to establish its relation to matter? And this is leaving aside the issue of what we do not know we do not know. Add to this our limited perspective. We are located on a small planet in one corner of a vast universe of billions of galaxies with trillions of stars—and this is possibly only part of an even cosmically vaster multi-world universe. Henri Poincaré pondered the interesting question of what our science would be like if the sky were permanently overcast and thus we could never see through it. The same issue applies to our position on earth even with clear skies. In addition, our sense-experience is accessible to only a fraction of the electromagnetic wave-band. Nothing would lead us to believe that we have a privileged access to the core of reality. To use an analogy from the Daoist Zhuangzi, we are like a frog in a well seeing part of the night sky and thinking that it is all there is to reality. And finally add the possible restrictions to the horizon of our thought that we ourselves have created by our imperfect ability to free ourselves from the strictures of our own older cultural ideas.

All in all, it does not take much reflection to become aware of our severe limitations. Again, this is not to deny that we have gained genuine knowledge of the workings of reality. But remember that this knowledge is hard earned: scientists do not come by their knowledge easily; it is not revealed but rather is the result of very hard work and imagination. We may end up knowing reality as fully as the human mind is capable of comprehending it and our knowledge may still be limited. And there also remains our efforts to comprehend the ultimate metaphysical mysteries of things. There may in fact be correct answers to the mysteries of the ultimate nature of things, but even if we can conceive some of them it is hard to see how we could establish that we are correct. In such circumstances, that things should seem mysterious is only to be expected. Indeed, imagining that we know anything ultimate about reality becomes a case of blatant hubris.

This has not curtailed people in philosophy, science, and religion from pontificating about the fundamental nature of reality. But there is the perennial danger that we will misread our attempts at knowledge and see our constructs as more than they are. Scientists and philosophers who make the philosophical claim that only what scientists can in principle study is real (naturalism) or that science is the only path to knowledge of what is real (scientism) or that scientists can in principle understand all there is to understand are guilty of this. But theists by the nature of their commitments aggravate this problem: by grounding their quest for the meaning of life in revelations from a transcendent reality, they make it that much harder to question their beliefs since the source is removed from any checking by our immediate experience. Their brute facts lie outside the natural realm within a transcendental realm but are no less "brute" or any more "self-explanatory." The religious may well accept that the transcendent reality is mysterious, but they also advance very specific beliefs about it that block the need to confront that mystery. However, it is equally hard for convinced materialists to doubt their belief-commitments, since these are as firmly planted in how they lead their lives as the religious ones are in religious ways of life. It is harder to see why one should hold one's beliefs tentatively when they cannot be questioned by new experiences but only seem confirmed by all experiences. Still, none of this changes the fact that uncertainty in these matters is our lot in life.

This has implications for the religious theory of any theistic or nonthestic tradition. In particular, theists who rely upon the authority of alleged revelations make assertions about the nature of God, even while acknowledging that God is a mystery. Revelation, not science, removes why-mysteries, although both may increase how-mysteries. Science defuses some how-mysteries, leaving intact the why-mysteries; religion is the other way around. In science, the focus is on the questions and the quest for more knowledge; in religion, the focus is on having answers. Science so far had led to things only looking more mysterious, but theists do not accept a radical mystery about the universe: there is an explanation of this world in terms of a transcendental personal creator, and theists speak with confidence of God's nature, acts, and purposes. But the conflicting revelations that have occurred throughout history and around the world of what God supposedly wants us to do must give theists pause. Believers should accept that their answers are based on faith and that faith involves risk; thus, they should be less confident in their positions.

Similarly with speculation apart from revelations in natural theology: even if we could construct versions of the Cosmological Argument, Teleological Argument, Ontological Argument, or the arguments from morality or religious experiences that are formally correct, enough mystery still surrounds the foundational premises of each to doubt that we really know anything of any possible transcendental realities with anything like certainty. That we know so little about the fundamental mysteries of this realm should make theologians hesitate about even venturing into speculating about "the mind of God," but that has not been the case. The problem is illustrated with the Templeton Foundation's "humility theology" project. Even while these theists stress epistemic humility about God, they also say we should not be "complacent" but should seek to expand "our knowledge of the God who would be known" in a form of natural theology.[8] Theologians may acknowledge that such speculations are merely our inadequate efforts, but theologians seldom sound that guarded once they get started. There is always the very great possibility that the religious will simply read their desires, needs, and values into the transcendent—and, of course, always place ourselves at the center of God's plan. Some anthropomorphism may be needed for minds like ours to conceptualize the transcendent, but it is a very real possibility that our views will reflect more about ourselves than any transcendental realities.

In sum, admitting the tentativeness of theological efforts is warranted in religion. The sheer otherness of any transcendental reality would necessarily make any attempts at closure very weak. No one can establish beyond any reasonable doubt even that such a reality exists. Once again the conflict of revelations does not help the religious case.[9] And mystical experiences do not fare any better. Even if religious experiences make contact with a transcendental reality, we should remember what Ninian Smart calls the paradox of religious experience: on the one hand, nothing seems more certain than faith or more compelling than religious experience; on the other hand, nothing seems less certain than any one particular religious belief-system.[10] Less adventurous speculations, more minimalist religious theories, and above all humility are called for in such circumstances.

Indeed, perhaps the chief role of theology today should be to keep the faithful humble before the awesomeness of any transcendental reality that a religious tradition purports exists.[11] Humility certainly goes against the spirit of our age, but even if the practical nature of our minds force us to conclude that there must be some reason for the universe, we have at best only very limited grounds to speculate on that source. Moreover, while intellectually satisfying answers may quiet our minds, more emphasis on how little we know may increase the religious sense today. People will also still need some images to live a full religious life, but these theological constructs can become idols of the mind that stand between us and any transcendental realities. They put their answers in the place of the reality; they think they understand and thus can turn away from the mystery. As Herbert Spencer thought, theology can be very impious.[12]

In fact, all the fields of study discussed here might be helped by more emphasis on the mystery of all that surrounds and permeates our existence. We like to analyze and substitute our concepts for the unknown; we may also like the resulting sense of control. But we must learn to live with some matters as open mysteries. Reality has to be rescued from all our enterprises that attempt to close off mystery. At the fundamental level, the world remains unexplained. Thus, at the end of all our exploring lie doubt and mystery. Separating what we really know from what we mistakenly think we know may never be fully achieved, but the first step is to accept that there are aspects of reality that most likely will remain beyond our comprehension forever. But accepting this does open us to becoming more receptive to the true fullness of reality.

Notes

1. Haeckel 1992, p. 15.
2. See Joshi 2007.
3. Ross 1981, p. 118. Many theists agree: if the creator is rational, his creation must be rational to the core and fathomable by us since our rational mind is made in his image.
4. See Hepburn 1984.
5. Stannard 1996, pp. 198-99. A child's persistent asking of "why?" after an adult thinks the legitimate questions have been answered raises the issues of when questioning should stop and whether we can think outside our normal frame of reference.
6. Frederick Ferré, *Hellfire and Lightning Rods: Liberating Science, Technology, and Religion* (Maryknoll, N.Y.: Orbis Books, 1993), p. 162.
7. See Hankinson 1995, chapter 19.
8. See Templeton and Hermann 1999.
9. Emphasizing mystery can lead to more tolerance of others' beliefs. This tolerance is based on philosophical skepticism. It is not related to a relativism of competing truth-claims but to the recognition that none of us may be correct and therefore that there is no reason to impose our own speculations on others. Whether the competition of conflicting belief-claims about basic matters will lead to progress in our understanding of the fundamental nature of reality is doubtful. There is no neutral way of establishing one alternative between naturalism and transcendentalism, let alone one of the many religious options.
10. Smart 1985, p. 76.
11. See Fuller 2006 on the emotion of wonder.
12. Spencer 1888, pp. 110-13.

Selected Bibliography

Adams, Marilyn McCord. *Horrendous Evils and the Goodness of God.* Ithaca, NY: Cornell University Press, 1999.

Allen, Diogenes. *Spiritual Theology: The Theology of Yesterday for Spiritual Help Today.* Cambridge, Mass.: Cowley Publications, 1997.

Bacik, James J. *Apologetics and the Eclipse of Mystery: Mystagogy According to Karl Rahner.* Notre Dame: University of Notre Dame Press, 1980.

Baggini, Julian. *What's It All About? Philosophy and the Meaning of Life.* Oxford: Oxford University Press, 2005.

Barbour, Julian. *The End of Time: The Next Revolution in Physics.* Oxford: Oxford University Press, 1999.

Barrow, John D. "Frontiers and Limits of Science." In John Marks Templeton, ed., *How Large is God? Voices of Scientists and Theologians,* pp. 203-216. Philadelphia: Templeton Foundation Press, 1997.

_____. *Impossibility: The Limits of Science and the Science of Limits.* New York: Oxford University Press, 1998.

Bloor, David. *Knowledge and Social Imagery,* 2nd ed. Chicago: University of Chicago Press, 1991.

Bolle, Kees, ed. *Secrecy in Religions.* New York: E. J. Brill, 1987.

Bouyer, Louis. *The Christian Mystery: From Pagan Myth to Christian Mysticism.* Trans. by Illtyd Trethowan. Edinburgh: T&T Clark, 1990.

Bradley, F. H. *Appearance and Reality: A Metaphysical Essay.* Oxford: Clarendon Press, 1897.

Brown, Raymond E. *The Semitic Background of the Term "Mystery" in the New Testament.* Philadelphia: Fortress Press, 1968.

Burrell, David B. *Knowing the Unknowable God: Ibn-Sina, Maimonides, Aquinas.* Notre Dame: University of Notre Dame Press, 1986.

Carabine, Deirdre. *The Unknown God: Negative Theology in the Platonic Tradition, Plato to Eriugena.* Grand Rapids: W. B. Eerdmans, 1995.

Chalmers, David J. *The Conscious Mind: In Search of a Fundamental Theory.* New York: Oxford University Press, 1996.

Churchland, Paul M. and Clifford A. Hooker, eds. *Images of Science.* Chicago: University of Chicago Press, 1985.

Collins, Harry and Trevor Pinch. *The Golem: What Everyone Should Know about Science.* Cambridge: Cambridge University Press, 1995.

Cooper, David E. *The Measure of Things: Humanism, Humility, and Mystery.* Oxford: Oxford University Press, 2002.

_____. "The Meaning of Life." In his *Meaning*, pp. 126-42. Montreal: McGill-Queen's University Press, 2003.

Craig, William Lane. "'What Place, Then, for a Creator?': Hawking on God and Creation." *The British Journal for the Philosophy of Science* 40 (1990): 473-491.

Craig, William Lane and Quentin Smith. *Theism, Atheism, and Big Bang Cosmology.* Oxford: Clarendon Press, 1993.

Crick, Francis. *The Astonishing Hypothesis: The Scientific Search for the Soul.* New York: Charles Scribner's Sons, 1994.

Davies, E. Brian. *Science in the Looking Glass: What Do Scientists Really Know?* New York: Oxford University Press, 2003.

Davies, Paul C. W. *The Cosmic Blueprint.* New York: Simon & Schuster, 1988.

_____. *The Mind of God: The Scientific Basis for a Rational World.* New York: Simon & Schuster, 1992.

_____. "Physics and the Mind of God." *First Things* 56 (August/September 1995): 31-35.

_____. *The Fifth Miracle: The Search for the Origin and Meaning of Life.* New York: Simon & Schuster, 1999.

Dawkins, Richard. *Unweaving the Rainbow: Science, Delusion, and the Appetite for Wonder.* New York: Houghton Mufflin, 1998.

Dennett, Daniel C. *Consciousness Explained.* Boston: Little, Brown and Company, 1991.

Dewdney, A. K. *Beyond Reason: Eight Great Problems That Reveal the Limits of Science.* Hoboken: John Wiley & Sons, 2004.

Dionysius the Areopagite. *Pseudo-Dionysius: The Complete Works.* Trans. by Colm Luibheid. New York: Paulist Press, 1987.

Duncan, Ronald and Miranda Weston-Smith, eds. *The Encyclopedia of Ignorance: Everything You Wanted to Know about the Unknown.* New York: Pergamon Press, 1977.

Dupré, Louis. *Religious Mystery and Rational Reflection: Excursions in the Phenomenology and Philosophy of Religion.* Grand Rapids: W. B. Eerdmans, 1998.

Eddington, Arthur. *The Nature of the Physical World.* Cambridge: Cambridge University Press, 1928.

Edwards, Paul. "Why." In Paul Edwards, ed., *The Encyclopedia of Philosophy*, vol. 8, pp. 296-302. New York: Macmillan, 1967.

Eliade, Mircea. *The Sacred and the Profane: The Nature of Religion.* Trans. by Willard R. Trask. New York: Harcourt, Brace & World, 1959.

d'Espagnat, Bernard. *On Physics and Philosophy.* Princeton: Princeton University Press, 2006.

Fagg, Lawrence. *The Becoming of Time: Integrating Physical and Religious Time.* Atlanta: Scholars Press, 1995.

Ferré, Frederick. "In Praise of Anthropomorphism." *International Journal for the Philosophy of Religion* 16 (1984): 203-212.

Feyerabend, Paul. *Science in a Free Society.* London: New Left Books, 1978.

_____. *Farewell to Reason.* London: Verso, 1987.

Feynman, Richard P. *The Meaning of it All: Thoughts of a Citizen-Scientist.* Reading, Mass.: Perseus Books, 1998.

Ford, Dennis. *The Search for Meaning: A Short History.* Berkeley/Los Angeles: University of California Press, 2007.

Foster, Michael B. *Mystery and Philosophy*. London: SCM Press, 1957.

Fraassen, Bas C. van. *The Scientific Image*. Oxford: Clarendon Press, 1980.

_____. *Laws and Symmetry*. Oxford: Clarendon Press, 1989.

Fuller, Robert C. *Wonder: From Emotion to Spirituality*. Chapel Hill: University of North Carolina Press, 2006.

Goodman, Nelson. "The Way the World Is." *The Review of Metaphysics* 14 (September 1960): 48-56.

_____. *Ways of Worldmaking*. Cambridge: Hackett Publishing, 1978.

Grayling, A. C. *Scepticism and the Possibility of Knowledge*. New York: Continuum, 2008.

Greeley, Andrew M. *The Great Mysteries: Experiencing Catholic Faith from the Inside Out*. New York: Sheed & Ward, 2003.

Grünbaum, Adolph. "Theological Misinterpretations of Current Physical Cosmology." *Foundations of Physics* 26 (no. 4 1996): 523-543.

_____. "A New Critique of Theological Interpretations of Physical Cosmology." *British Journal for the Philosophy of Science* 51 (no. 1 2000): 1-44.

Haack, Susan. "The Limits of Science." In her *Defending Science—Within Reason: Between Scientism and Cynicism*, pp. 344-351. Amherst, N.Y.: Prometheus Books, 2003.

Hacking, Ian. *Representing and Intervening*. Cambridge: Cambridge University Press, 1983.

Haeckel, Ernst. *The Riddle of the Universe*. Trans. by Joseph McCabe. Buffalo: Prometheus Books, 1992 (originally published 1899).

Haldane, John. "The Mystery of Emergence." *Proceedings of the Aristotelean Society* 96 (1996): 261-267.

Hankinson, R. J. *The Sceptics*. New York: Routledge, 1995.

Haught, John F. "Mystery." In his *What is God? How to Think About the Divine*, pp. 115-131. New York: Paulist Press, 1986.

_____. *God After Darwin: A Theology of Evolution*. Boulder: Westview Press, 2000.

_____. "Science and Mystery." In his *Christianity and Science: Toward a Theology of Nature*, pp. 19-33. Maryknoll, N.Y.: Orbis Books, 2007.

_____, ed. *Science and Religion in Search of Cosmic Purpose*. Washington: Georgetown University Press, 2000.

Hawking, Stephen W. "Letter to the Editor." *American Scientist* 73 (January/February 1985): 12.

_____. *A Brief History of Time*. New York: Bantam Books, 1988.

Hepburn, Ronald. "Wonder." In his *"Wonder" and Other Essays*, pp. 131-147. Edinburgh: University of Edinburgh Press, 1984.

Hick, John. *Evil and the God of Love*. London: Macmillan, 1966.

_____. *An Interpretation of Religion: Human Responses to the Transcendent*. New Haven: Yale University Press, 1989.

_____. *The Fifth Dimension: An Exploration of the Spiritual Realm*. Oxford: Oneworld Publications, 1999.

Hopkins, Jasper. *Nicholas of Cusa on Learned Ignorance: A Translation and an Appraisal of De docta ignorantia*. Minneapolis: A. J. Banning Press, 1990.

Horgan, John. *The End of Science: Facing the Limits of Knowledge in the Twilight of the Scientific Age*. New York: Addison-Wesley, 1996.

Jeans, James. *The Mysterious Universe*. Cambridge: Cambridge University Press, 1928.

Jones, Richard H. *Science and Mysticism*. Lewisburg: Bucknell University Press, 1986.

_____. *Mysticism Examined: Philosophical Inquiries into Mysticism*. Albany: State University of New York, 1993.

_____. *Reductionism: Analysis and the Fullness of Reality.* Lewisburg: Bucknell University Press, 2000.

_____. *Mysticism and Morality: A New Look at Old Questions.* Lanham, Md.: Lexington Books, 2004.

Joshi, S. T., ed. *The Agnostic Reader.* Amherst, N.Y.: Prometheus Books, 2007.

Kant, Immanuel. *Prolegomena to any Future Metaphysics.* Trans. by Gary Hatfield. Cambridge: Cambridge University Press, 1997.

_____. *Critique of Pure Reason.* Trans. by Paul Guyer and Allen W. Wood. Cambridge: Cambridge University Press, 1998.

Kauffman, Stuart A. *The Origins of Order: Self-Organization and Selection in Evolution.* New York: Oxford University Press, 1993.

Kaufman, Gordon. *In Face of Mystery: A Constructive Theology.* Cambridge: Harvard University Press, 1993.

Kellenberger, John. "God and Mystery." *American Philosophical Quarterly* 11 (April 1974): 93-102.

Kingston, Temple. "Mystery and Philosophy: Michael Foster and Gabriel Marcel." In Cameron Wybrow, ed., *Creation, Nature, and Political Order in the Philosophy of Michael B. Foster (1903-1959),* pp. 297-307. Lewiston, N.Y.: E. Mellen Press, 1992.

Krishnamurti, Jiddhu and David Bohm. *The Limits of Thought: Discussions.* New York: Routledge, 1999.

Klemke, E. D., ed. *The Meaning of Life,* 2nd ed. New York: Oxford University Press, 2000.

Kline, Morris. *Mathematics: The Loss of Certainty.* Oxford: Oxford University Press, 1980.

Kuhn, Thomas. *The Structure of Scientific Revolutions,* 2nd ed. Chicago: University of Chicago Press, 1970.

_____. *The Essential Tension: Selected Studies in Scientific Tradition and Change.* Chicago: University of Chicago Press, 1977.

Lakoff, George and Mark Johnson. *Metaphors We Live By.* Chicago: University of Chicago Press, 1980.

Landesman, Charles. *Skepticism: The Central Issues.* New York: Blackwell, 2002.

Latour, Bruno and Steven Woolgar. *Laboratory Life: the Construction of Social Facts.* Princeton: Princeton University Press, 1992.

Laudan, Larry. *Science and Relativism.* Chicago: University of Chicago Press, 1990.

_____. *Beyond Positivism and Relativism: Theory, Method, and Evidence.* Boulder: Westview Press, 1996.

Laughlin, Robert B. *A Different Universe: Reinventing Physics from the Bottom Down.* New York: Basic Books, 2005.

Lawson, Hilary. *Closure: A Story of Everything.* New York: Routledge, 2001.

Leplin, Jarrett, ed. *Scientific Realism.* Berkeley/Los Angeles: University of California Press, 1984.

Leslie, John. *Universes.* New York: Routledge, 1989.

_____, ed. *Modern Cosmology & Philosophy.* Amherst, N.Y.: Prometheus Books, 1998.

_____. *Infinite Minds: A Philosophical Cosmology.* Oxford: Clarendon Press, 2001.

Lewin, Roger. *Complexity: Life at the Edge of Chaos.* New York: Macmillan, 1992.

Lewis, David. *Plurality of Worlds.* Oxford: Blackwell, 1986.

Lewis, Hywel D. "God and Mystery." In Ian T. Ramsey, ed., *Prospect for Metaphysics: Essays of Metaphysical Exploration,* pp. 206-237. London: Allen & Unwin, 1961.

Louth, Andrew. *Discerning the Mystery: An Essay on the Nature of Theology.* New York: Oxford University Press, 1983.

Lowe, E. J. "Why is There Anything at All?" *Proceedings of the Aristotelian Society* 70 (Supplement, 1996): 111-120.

Lubac, Henri de. *The Mystery of the Supernatural*. Trans. by Rosemary Sheed. New York: Crossroad, 1998.

Macquarrie, John. *Mystery and Truth*. Milwaukee: Marquette University Theology Department, 1973.

Manson, Neil A. *God and Design: The Teleological Argument and Modern Science*. New York: Routledge, 2003.

Marcel, Gabriel. *The Mystery of Being*. 2 vols. Trans. by René Hague. Chicago: Henry Regnery Co., 1950-51.

Margenau, Henry. *The Miracle of Existence*. Woodbridge, Ct.: Ox Bow Press, 1984.

_____ and Roy Abraham Varghese, eds. *Cosmos, Bios, Theos: Scientists Reflect on Science, God, and the Origins of the Universe, Life, and Homo sapiens*. LaSalle, Ill.: Open Court, 1992.

Margolis, Joseph. *The Unraveling of Scientism: American Philosophy at the End of the Twentieth Century*. Ithaca, N.Y.: Cornell University Press, 2003.

Martin, Michael and Ricki Monnier, eds. *The Impossibility of God*. Amherst, N.Y.: Prometheus Books, 2003.

_____. *The Improbability of God*. Amherst, N.Y.: Prometheus Books, 2006.

McGinn, Colin. *Problems in Philosophy*. Oxford: Blackwell, 1991.

_____. *The Mysterious Flame: Conscious Minds in a Material World*. New York: Basic Books, 1999.

McMullin, Ernan, ed. *The Concept of Matter in Modern Philosophy*, rev. ed. Notre Dame: University of Notre Dame Press, 1978.

McPherson, Thomas. "Religion as the Inexpressible." In Anthony Flew and Alasdair MacIntyre, eds., *New Essays in Philosophical Theology*, pp. 131-143. New York: Macmillan, 1955.

Meheus, Joke, ed. *Inconsistency in Science*. Boston: Kluwer, 2002.

Miller, James B., ed. *Cosmic Questions*. New York: New York Academy of Sciences, 2001.

Miller, Jerome A. *In the Throe of Wonder: Intimations of the Sacred in a Post-Modern World*. Albany: State University of New York, 1992.

Morris, Richard. *The Big Questions: Probing the Promise and Limits of Science*. New York: Henry Holt, 2002.

Munitz, Milton K. *The Mystery of Existence: An Essay in Philosophical Cosmology*. New York: Appleton-Century-Crofts, 1965.

_____. *Cosmic Understanding: Philosophy and Science of the Universe*. Princeton: Princeton University Press, 1986.

_____. *The Question of Reality*. Princeton: Princeton University Press, 1990.

_____. *Does Life Have A Meaning?* Buffalo: Prometheus Books, 1993.

Needleman, Jacob and David Appelbaum, eds. *Real Philosophy: An Anthology of the Universal Search for Meaning*. New York: Penguin Books, 1990.

Newberg, Andrew, Eugene D'Aquili, and Vince Rause. *Why God Won't Go Away: Brain Science and the Biology of Belief*. New York: Ballantine Books, 2001.

Nozick, Robert. "Why Is There Something Rather Than Nothing?" In his *Philosophical Explanations*, pp. 115-164. Cambridge: Belknap Press, 1981.

Otto, Rudolf. *The Idea of the Holy: An Inquiry into the Non-Rational Factor in the Idea of the Divine and its Relation to the Rational*, 2nd ed. Trans. by John Harvey. New York: Oxford University Press, 1950.

Parfit, Derek. "The Puzzle of Existence: Why Does the Universe Exist?" In Peter Van Inwagen and Dean W. Zimmerman, eds., *Metaphysics: The Big Questions*, pp. 418-26. Oxford: Blackwell, 1998.

Penfield, Wilder. *The Mystery of the Mind*. Princeton: Princeton University Press, 1975.

Pepper, Stephen C. *World Hypotheses: A Study in Evidence*. Berkeley/Los Angeles: University of California Press, 1942.

Post, John F. *The Faces of Existence: An Essay in Nonreductive Metaphysics*. Ithaca: Cornell University Press, 1987.

Power, William L. "Musings on the Mystery of God." *International Journal for the Philosophy of Religion* 7 (1976): 301-310.

Priest, Graham. *Beyond the Limits of Thought*, 2nd ed. New York: Cambridge University Press, 2002.

Putnam, Hilary. *Reason, Truth and History*. Cambridge: Harvard University Press, 1981.

_____. *Realism and Reason*. Cambridge: Cambridge University Press, 1983.

_____. *The Many Faces of Realism*. LaSalle, Ill.: Open Court, 1987.

_____. *Representation and Reality*. Cambridge: MIT Press, 1989.

Quine, Willard van Orman. *Word and Object*. Cambridge: MIT Press, 1960.

_____. "Two Dogmas of Empiricism." In his *From a Logical Point of View*, 2nd ed., pp. 20-46. Cambridge: Harvard University Press, 1961.

_____. *Ontological Relativity and Other Essays*. New York: Columbia University Press, 1969.

_____. "Two Dogmas in Retrospect." In Roger F. Gibson, Jr., *Quintessence: Basic Readings from the Philosophy of W. V. Quine*, pp. 54-63. Cambridge: Belknap Press, 2004.

Quine, Willard van Orman and J. S. Ullian. *The Web of Belief*, rev. ed. New York: Random House, 1978.

Rahner, Karl. *Foundations of Christian Faith: An Introduction to the Idea of Christianity*. Trans. by William V. Dych. New York: Seabury Press, 1978.

Ramsey, Ian T. "Paradox in Religion." *Proceedings of the Aristotelian Society Supplement* 32 (1959): 195-218.

_____. *Models and Mystery*. London: Oxford University Press, 1964.

_____. "On Understanding Mystery." In Jerry H. Gill, ed., *Philosophy and Religion: Some Contemporary Perspectives*, pp. 295-308. Minneapolis: Burgess Publishing, 1968.

Rees, Martin. *Just Six Numbers: The Deep Forces That Shape the Universe*. New York: Basic Books, 2000.

Rescher, Nicholas. *The Riddle of Existence: An Essay in Idealistic Metaphysics*. Lanham, Md.: University Press of America, 1984.

_____. *The Limits of Science*. Pittsburgh: University of Pittsburgh Press, 1999.

Rorty, Richard. *Philosophy and the Mirror of Nature*. Princeton: Princeton University Press, 1979.

_____. *Consequences of Pragmatism: Essays, 1972-1980*. Minneapolis: University of Minnesota Press, 1982.

_____. *Contingency, Irony, Solidarity*. Cambridge: Cambridge University Press, 1989.

Ross, Steven David. *Philosophical Mysteries*. Albany: State University of New York, 1981.

Rubenstein, Mary-Jane. *Strange Wonder: The Closure of Metaphysics and the Opening of Awe*. New York: Columbia University Press, 2008.

Rundle, Bede. *Why There is Something Rather Than Nothing*. Oxford: Clarendon Press, 2004.

Russell, Bertrand. *Why I Am Not a Christian and Other Essays on Religion and Related Subjects*. New York: Simon & Schuster, 1957.

Saunders, Nicholas. *Divine Action and Modern Science*. Cambridge: Cambridge University Press, 2002.

Schilling, Harold K. "Usages of the Term Mystery." *Iliff Review* 29 (Winter 1972): 11-19.

Schlesinger, George N. "Possible Worlds and the Mystery of Existence." *Ratio* 26 (June 1984): 1-17.

_____. "The Enigma of Existence." *Ratio* n.s. 11 (April 1998): 66-77.

Searle, John. *The Rediscovery of Mind*. Cambridge: MIT Press, 1992.

_____. *The Mystery of Consciousness*. New York: New York Review, 1997.

Sheldrake, Rupert. *The Presence of the Past: Morphic Resonances and the Habits of Nature*. New York: Times Books, 1988.

Smart, Ninian. "Paradox in Religion." *Proceedings of the Aristotelian Society Supplement* 32 (1959): 219-232.

_____. "On Knowing What is Uncertain." In Leroy S. Rouner, ed., *Knowing Religiously*, pp. 76-86. Notre Dame: University of Notre Dame Press, 1985.

Smith, Huston. *Beyond the Post-Modern Mind*. Wheaton, Ill.: Theosophical Publishing House, 1982.

_____. "The Crisis in Philosophy." *Behaviorism* 16 (Spring 1988): 51-56.

_____. "The View from Everywhere: Ontotheology and the Post-Nietzschean Deconstruction of Metaphysics." In Henry Ruf, ed., *Religion, Ontotheology, and Deconstruction*, pp. 43-66. New York: Paragon House, 1989.

_____. *Why Religion Matters: The Fate of the Human Spirit in an Age of Disbelief*. San Francisco: HarperCollins, 2001.

Smolin, Lee. *The Life of the Cosmos*. New York: Oxford University Press, 1997.

Sorensen, Roy. *A Brief History of Paradox: Philosophy and Labyrinths of the Mind*. New York: Oxford University Press, 2003.

Sorrell, Tom. *Scientism: Philosophy and the Infatuation of Science*. London: Routledge, 1994.

Spencer, Herbert. *First Principles*, 4th ed. New York: D. Appleton, 1888.

Sperry, Roger W. *Science and Moral Priority: Merging Mind, Brain, and Human Values*. New York: Praeger, 1985.

Stannard, Russell. Science and Wonders. London: Faber and Faber, 1996.

Stenger, Victor J. *God: The Failed Hypothesis—How Science Shows That God Does Not Exist*. Amherst, N.Y.: Prometheus Books, 2008.

Tegmark, Max. "Parallel Universes." *Scientific American* 289 (May 2003): 40-51.

Templeton, John Marks, ed. *Evidence of Purpose: Scientists Discover the Creator*. New York: Continuum, 1994.

_____. *How Large is God? Voices of Scientists and Theologians*. Philadelphia: Templeton Foundation Press, 1997.

Templeton, John M. and Robert L. Herrmann. *The God Who Would be Known: Revelations of the Divine in Contemporary Science*. Philadelphia: Templeton Foundation Press, 1998.

Trusted, Jennifer. *The Mystery of Matter*. New York: St. Martin's Press, 1999.

Turner, Denys. *The Darkness of God: Negativity in Christian Mysticism*. Cambridge: Cambridge University Press, 1995.

Tymieniecka, Anna-Teresa. *Why Is There Something Rather Than Nothing? Prolegomena to the Phenomenology of Creation*. Assen, Netherlands: Van Corcum, 1966.

Unger, Peter. *Ignorance: A Case for Scepticism*. New York: Oxford University Press, 1975.

Van Inwagen, Peter. "Why is There Anything at All?" *Proceedings of the Aristotelian Society* 70 (Supplement, 1996): 95-110.

_____. "Free Will Remains a Mystery." In Robert Kane, ed., *The Oxford Handbook of Free Will*, pp. 158-177. New York: Oxford University Press, 2002.

Verkamp, Bernard J. *Senses of Mystery: Religious and Non-Religious*. Scranton, Pa.: University of Scranton Press, 1997.

Vernon, Mark. *Science, Religion and the Meaning of Life*. New York: Palgrave/Macmillan, 2007.

Wallenstein, Immanuel M. *The Uncertainties of Knowledge*. Philadelphia: Temple University Press, 2004.

Wegner, Daniel M. *The Illusion of Conscious Will*. Cambridge: MIT Press, 2002.

Weinberg, Steven. *Dreams of a Final Theory*. New York: Pantheon, 1992.

_____. "A Designer Universe?" *The New York Review of Books* 46 (21 October 1999): 46-48.

_____. "Can Science Explain Everything? Anything?" *The New York Review of Books* 48 (31 May 2001): 47-50.

Williams, J. P. *Denying Divinity: Apophasis in the Patristic Christian and Soto Zen Buddhist Traditions*. Oxford: Oxford University Press, 2000.

Wittgenstein, Ludwig. *Tractatus Logico-Philosophicus*. Trans. by David F. Pears and B. F. McGuinness. London: Routledge and Paul, 1961.

_____. "A Lecture on Ethics." *Philosophical Review* 74 (January 1965): 3-12.

_____. *Philosophical Investigations*. Trans. by G. E. M. Anscombe. Oxford: Blackwell, 1967.

Woolgar, Steve. *Science: The Very Idea*. London: Tavistock, 1988.

Young, Julian. *The Death of God and the Meaning of Life*. New York; Routledge, 2003.

Young, Louise B., ed. *The Mystery of Matter*. New York: Oxford University Press, 1965.

Index

About the Author

Richard H. Jones holds a Ph.D. from Columbia University in the history and philosophy of religion and an A.B. from Brown University in religious studies. He also holds a J.D. from the University of California at Berkeley and practices law in New York City where he lives. He is the author of *Science and Mysticism* (Bucknell University Press/Booksurge), *Mysticism Examined* (State University of New York Press), *Reductionism* (Bucknell University Press), *Mysticism and Morality* (Lexington Books), and numerous philosophy of religion and law review articles.